(((LISTENING TO
NINETEENTH-CENTURY
AMERICA

Listening to
Nineteenth-

MARK M. SMITH

The University of North Carolina Press

Chapel Hill and London

Century America

© 2001 The University of North Carolina Press
All rights reserved
Manufactured in the United States of America
Set in Carter Cone Galliard by Tseng Information Systems, Inc.
The paper in this book meets the guidelines for permanence
and durability of the Committee on Production Guidelines
for Book Longevity of the Council on Library Resources.

Library of Congress Cataloging-in-Publication Data
Smith, Mark M. (Mark Michael), 1968–
Listening to nineteenth-century America / Mark M. Smith.
    p.   cm.
Includes bibliographical references (p.   ) and index.
ISBN 0-8078-2657-X (cloth: alk. paper)—
ISBN 0-8078-4982-0 (pbk.: alk. paper)
1. United States—Social life and customs—19th century.
2. United States—Social conditions—19th century. 3. United
States—History—1815–1861. 4. Noise—Social aspects—
United States—History—19th century. 5. Sound—Social
aspects—United States—History—19th century. 6. Silence—
Social aspects—United States—History—19th century. 7.
Elite (Social sciences)—Southern States—History—19th
century. 8. Elite (Social sciences)—Northeastern States—
History—19th century. 9. Sectionalism (United States)—
History—19th century. 10. North and south—History—
19th century. I. Title.
E166.S62   2001
973.5—dc21   2001027541

05   04   03   02   01   5   4   3   2   1

For Sophia, by way of loving welcome

CONTENTS

ILLUSTRATIONS & MAPS

Illustrations

Maps

ACKNOWLEDGMENTS

The University of South Carolina's History Department is home to many listeners who have indulged my musings on the subject of nineteenth-century American aurality with grace, charm, and kindness. The department's fine graduate students have been a constant source of information and support, and I am particularly grateful for the tidbits of evidence thrown my way by Sean Busick, Kathy Hilliard, Aaron Mars, Mike Reynolds, and Rebecca Shrum. For fleeting but helpful conversations I thank Tom Downey, Trenton Hizer, John Hammond Moore, and J. Tracy Power. I was blessed with two superb research assistants in writing this book. Chavvar Penner helped immensely with my research into legal statutes, and my graduate student, Cheryl A. Wells, went entirely beyond the call of duty and often at inexcusably short notice.

This study has also benefited from the advice and leads proffered by several scholars. I remain grateful for the helpful and generous counsel of Ron Atkinson, Joan Cashin, Bobby Donaldson, Walter Edgar, Frank Fahy, Elizabeth Fox-Genovese, Lawrence Glickman, Kasey Grier, John Oldfield, Bill Pease, and Nan Woodruff. Aspects of this book have been presented to audiences at the Nineteenth-Century Studies Association meeting in March 1997; the annual meeting of the American Historical Association in January 1998; the Department of History, University of California, San Diego, in May 1999; the history department at the College of Charleston in November 1999; and a wonderful and enlightening conference titled "Listening to Archives," arranged by Paula Hamilton and Douglas Kahn at the University of Technology, Sydney, Australia, in November 2000. I remain very grateful for the comments offered on those occasions.

Some of the research for this book was funded by a scholarship grant (summer 1998) from the University of South Carolina, College of Liberal Arts, and by the good graces of the former chair of the university's history department, Peter W. Becker, and its extraordinary present head,

Patrick J. Maney. I should also like to thank the editorial staff of the University of North Carolina Press for their customary efficiency, good cheer, and patience in seeing this book through to completion. Charles Grench, Ruth Homrighaus, and Pamela Upton at the press proved utterly accommodating, and I remain in their debt. My sincere thanks also to Lewis Bateman, who got the project off the ground. For her help with the illustrations and maps, I remain in Tina Manley's debt (again). Stephanie Wenzel's copyediting was of the highest order, and this is a better book because of her keen and critical eye.

Several scholars read the entire manuscript, and acknowledging them here is paltry thanks indeed for their insights and suggestions. Few people know more about auditory culture than Hillel Schwartz and Douglas Kahn, and I am supremely grateful for Hillel's extraordinarily close reading of the manuscript and Doug's extremely helpful suggestions. Clyde N. Wilson read an early draft with his southern ear, saved me from a few gaffs, and helped me think on a few matters. Michael Fellman deserves special thanks. He helped shape the book in significant ways, and I trust he will hear the difference he has made. I am also grateful for the encouragement my colleague Paul E. Johnson offered during the course of my research and writing. He kept me straight on several matters pertaining to the antebellum North and was always a source of support. Peter A. Coclanis and Eugene D. Genovese were at once my toughest but most valued critics. Peter asked his customarily penetrating questions, and I hope I have gone some way toward answering them. Gene read the manuscript with an unrivaled toughness, passion, and care. The book is much better for it, and as I hope he knows, I remain deeply touched by and grateful for his valued counsel and example.

Jack's deafness, Tess's protectiveness, and Maggie's inquisitiveness made writing at home a potentially noisy affair. But as always, Catherine kept the peace and gave me moments of quiet. She is my rock without whom this book would not have been completed. As for Sophia, well, I hope the dedication says it all.

(((LISTENING TO
NINETEENTH-CENTURY
AMERICA

Now I will do nothing but listen,
To accrue what I hear into this song, to let sounds contribute
toward it.
—Walt Whitman, "Song of Myself," 1855

(((Introduction: Sounding Pasts

In simple Quaker dress, determined face framed by dark curls,
the thirty-three-year-old South Carolinian stood before a packed Massachusetts state legislature in Boston on February 21, 1838. Nervous and apprehensive, she prepared to persuade her audience why southern slavery
should be abolished and explain women's role in the process. Her jitters were understandable. While sympathetic ears filled the hall, scoff-
ers doubting whether a woman should speak so publicly and politically
abounded. Angelina Grimké was not the first American woman to denounce slavery, but until that day none had spoken to a U.S. legislative
body.[1]

Hush fell. "Mr. Chairman," she began. Her sounds punched the stillness with the force of novelty. She hit stride, regaling listeners with
thoughts on the religious and political enormities of bondage. Her voice
rang with the authenticity of someone who had witnessed slavery firsthand. In her choice of images Grimké conveyed the wretchedness of the
peculiar institution in a way that touched hearts and hardened resolves:

"I stand before you as a southerner, exiled from the land of my birth, by the sound of the lash, and the piteous cry of the slave."[2]

What she said was heard in more than one sense. Enabling and urging her audience to hear not just her words but also the sounds of bondage was a way to tease at her listeners' guts and hearts. For many in the hall who had never actually heard slavery, they could now imagine how it sounded. Of course there were many northerners who could not or did not have the inclination to hear Angelina Grimké's aural representation of slavery on February 21. But actual hearing of what she had heard was not necessary because her speech was reprinted in the antislavery newspaper the *Liberator* a few days later on March 2. Her aural depiction and its authenticity were replayed via print, and readers could now hear, imagine, and reimagine what Grimké had heard and wanted them to hear. In this way her aural construction of slavery began to echo resoundingly in the ears and minds of increasingly powerful advocates of free labor and abolition. Sympathizers already disgusted with bondage found vivid confirmation in Grimké's representations; skeptics who had yet to be persuaded could find the aural projection persuasive, potent, and deeply emotive.

Angelina Grimké was neither the first nor the last to represent slavery aurally. Abolitionist travelers to the Old South and, especially, escaped slaves who recounted their experiences to northerners did the same. Travelers, Grimké, and fugitive slaves were authentic listeners because they had actually heard slavery. While there was doubtless some recognition that they exaggerated the frequency of screams, lashes, and clanking chains, their characterizations of southern sounds gained widespread acceptance among abolitionists of all stripes and, later, among supporters of free soil and free labor. For many northerners the South became a place alien and threatening because of how it sounded.

Grimké did not exhaust the lexicon of aurality used to represent the South. In following years other speakers added their voices and constructed the South as at once resounding with the noises of bondage and the silence of southern political tyranny and economic backwardness. Increasingly, abolitionists, free soilers, and Republicans constructed the South as aurally distinct and depraved. Aural descriptions offered a literal and metaphoric sense of the South as alien, which proved appealing to a wide variety of northern leaders in the years leading to civil war. Doubtless Angelina Grimké and others were unhappy with aspects of northern civilization and dimensions of its soundscape. After all, Grimké herself had been silenced by northern mobs. But she like many others preferred northern sounds to southern ones because the former resonated with free-dom, progress, and humanity; the latter, with only misery, cruelty, and stagnation. Northern advocates of progress increasingly applauded the virtues of their own soundscape, in contrast to the noises of slavery, for in the hum of industry and the buzz of freedom they heard a society that not only was different from the South but reaffirmed their belief in the superiority of industrial, urban, free labor modernity.

William M. Bobo liked to travel. In the early 1850s the genteel South Carolinian jaunted north to New York City, and in 1852 he published his impressions of the place in a brief travelogue, *Glimpses of New-York City, by a South Carolinian (Who Had Nothing Else to Do)*. Although he was not unmindful of the city's odors, sights and sounds guided Bobo in the great northern metropolis. "A stranger" to New York City, he began, "has many things to see and hear, most of which he does not really understand." Part of Bobo's job was to explain. In active, intimate prose he essayed to "give you an idea of the feeling which pervades the very soul of this community, in contradistinction to that which exists in the South," in part by urging his reading audience to hear northern society through his ears. Resounding with "rush and crowd," Gotham was a place where "emphatically large and fashionable hotels" suffered from "too much noise and confusion." The city echoed with the excesses of wage labor and northern capitalism, and its dissonance became more grating the farther in he ventured. Bobo prepared his readers' senses: "Any one who walks the streets of New-York with his eyes and ears open, sees and hears many strange and horrid things." "Poverty, sickness, filth, crime, and wretchedness" echoed in one ear while, in stark aural contrast, "silk rustles" of the sashaying skirts of bourgeois women sounded in the other. Islands of tranquility could be found, of course, but they only accentuated city noise. At Greenwood Cemetery he encountered deliciously "silent tomb[s]," an aural sanctuary "where the dead repose in blissful quietude, from the noisy and perplex-ing confusion which surround the living." "Is this not a sweet place?" he mused. "Here the world, with all its busy scenes, is shut out. . . . We make our beds in peace, and along these peaceful valleys the hum and din of earth's turmoil will never disturb our tranquil repose." Just outside the city, in Yonkers, he found "residences . . . free from the musquitoes, dust and noise of the city." But time was not on the side of such quietude: "New-York will be out here one of these days." The expansive tendencies of northern capitalism would introduce "the noisy and vexatious walks of the living" to Yonkers and places even farther removed. The future sounded bleak to this man of the South.[3]

Back in the city—this time at Five Points—Bobo ventured into one of several "drinking and dance-houses." "There lies a drunken female, screaming and yelling" while men were "cursing and swearing in the most blasphemous manner—a sort of medley which is indescribable." Overwhelmed, he abandoned his narrative: "Let us get out, my senses refuse to behold longer such scenes." Massive immigration, the exploitation of young factory women, and the general misery and wretchedness of wage labor society only "sickens the senses." Northern capitalism, urbanization, and industrialization had introduced more "poverty, prostitution, wretchedness, drunkenness, and all the attending vices, in this city, than [in] the whole South." "This," Bobo remarked, minimizing the extent to which similar sounds could be heard in the urban South, "is a comment upon Northern institutions." Yet northern critics of southern society seemed deaf to the sounds of their own failings. Instead of listening for the screams of southern slaves, abolitionists and free-soil sponsors would do well to turn an ear to their own wretchedness: "When the Abolitionists have cleared their own skirts, let them hold up their hands in holy horror at the slave-holder, and the enormity of his sins." A southerner had difficulty finding peace in such a city, thought Bobo, for southern folk coveted quietude and tranquility. Certainly stillness was found in Yankee taciturnity; though they lived together in close proximity, New Yorkers never talked to one another. A person could "be in the very midst of a half million people, and yet as quiet as if you were ten miles in rear of the Basin Spring, in North Carolina." The comparison seemed to jolt Bobo, perhaps making him hanker for the sounds of his home: "I suppose you are as tired listening to my illustration as I am telling it," and for the time being, he stopped listening and spared his own and readers' ears.[4]

William Bobo's aural representations of New York City and northern modernity were hardly new, and similar examples can be found beginning principally in the 1830s when southern defenders of slavery began to hear the rise of what they perceived as an aggressive and threatening northern society. Often in response to abolitionists' critiques of slavery's evil strains, southern elites and politicians countered with the kind of aural critique of northern society offered by Bobo. Southern representations of the northern soundscape and all that it stood for were expressed in print and so communicated the failings of the North to many southerners who had never actually heard it in operation. Proslavery thinker George Fitzhugh, for example, did not visit, see, hear, or experience the North firsthand until 1855, but he, like Grimké's audiences, had read enough to learn how to listen and what to hear. In all likelihood Fitzhugh had at some point in his life heard sounds of slavery similar to those that assaulted Grimké's ears, but he, like Bobo and other elite white southerners, rarely commented on these aspects of the southern soundscape. Instead Fitzhugh listened for what he and others believed were the keynotes of southern society—tranquility and quietude punctuated with a healthy dose of humming industriousness and the melodies of singing slaves— and contrasted them with what they believed was the destructiveness of northern modernity. In his 1850 *Slavery Justified*, Fitzhugh argued that the social arrangement of slavery and its harmonizing of labor and capital meant that "We have no mobs, no trades unions, no strikes for higher wages" and "but few in our jails, and fewer in our poor houses." The consequence was heard as much as seen: "At the slaveholding South all is peace, quiet, plenty and contentment." Following his visit North, in 1857 Fitzhugh published *Cannibals All!,* a scathing critique of the dangerous tendencies of wage labor. What he had previously read about how northern society sounded was confirmed by the cultural bias of his hearing and selective listening. The competition between labor and capital, Fitzhugh maintained, led to revolution, and its beginnings could be heard in the noises of poverty, wretchedness, and strife that would reach a crescendo in a maddening and destructive cacophony of social dislocation. When capitalists' efforts to tame workers' demands had failed ("we must use violence to keep you quiet," Fitzhugh imagined them saying), "the maddening cry of hunger for employment and bread" would culminate in "the grumbling noise of the heaving volcano that threatens and precedes a social eruption greater than the world has yet witnessed." The rumblings of class conflict and social revolution could be heard in the noises of industrialism, capitalism, and unfettered exploitation.[5]

While exceptional in several respects, Grimké, Bobo, and Fitzhugh were typical in how they understood, imagined, and projected their abstract and actual environments and sectional identities. Most nineteenth-century Americans experienced their worlds through their senses. At times they understood by using—deliberately and unwittingly—all five senses at once (if they had them); at other times one sense took primacy but rarely to the exclusion of the others. It seems almost audacious to point out that in the past, peoples sensed their worlds, their environments, and their places and mediated their experiences sensorially. Obvious though this fact is, however, it warrants stating not least because we are prone to examine the past through the eyes of those who experienced it. While people interpreted their worlds visually, it is also worth iterating that seeing was but one way in which they experienced. Yet for reasons that have to do with the nineteenth-century preoccupation with visuality, the rise of print culture, and the long shadows cast by these developments, it seems fair to say that a good deal of historical work interprets the past principally, if unwittingly and implicitly, through historical actors' eyes. Historians rarely consider in any explicit or systematic way the other four senses, and so a good deal of what we know about most historical experience is really a history of what people saw. In this sense (literally) we understand the past in one-fifth of its texture and scope, and historical analyses of how people sensed—heard, tasted, smelled, and touched— are staggeringly few and far between. Certainly the importance of the eye to nineteenth-century Americans should not be doubted; print, perspective, new technologies of vision, and faith in seeing were important to the construction of ocular modernity. But there is no legitimate reason to read the past solely through contemporaries' vision. This study recovers another way people sensed their worlds—through a faculty emphasized by contemporaries and that they took as seriously as their seeing and other sensory understandings.[6]

This is a study of how people heard the principal economic, cultural, and political—hence social—developments of the United States in the nineteenth century and how their hearing at the everyday level affected their selective hearing and listening to, among other developments, the coming of the Civil War, antebellum class formation, slavery, freedom, modernization, the war itself, and Reconstruction. In this book I do not posit the senses as oppositional; rather, I attempt to recover another, additional, often complementary way people experienced and made sense of their lives, environments, relationships, and identities. Other forms of sensory experience—touching, tasting, and smelling—are important and deserve sustained investigation. But I profile how people heard not least because aurality was important enough to contribute meaningfully and significantly to the construction of what it meant to be northern, southern, slave, or free in nineteenth-century America. When we add sound to sight in our understanding of how historical actors conceptualized and mediated their identities, we begin to understand more fully how important aurality, listening, and hearing were to the process of creating real and abiding notions of slavery and freedom, North and South, especially during the last three decades prior to the Civil War. Without listening to what and how nineteenth-century Americans heard, we will remain only partially aware of the depth, texture, and nature of sectional identity and deny ourselves access to a fuller explanation of how that identity came into being with such terrible resolve. Sectional consciousness was sensed, and hearing and listening as much as looking and seeing were important to its creation.

Sounds and their meanings are shaped by the cultural, economic, and political contexts in which they are produced and heard. Because sound was so embedded in the various fabrics of antebellum U.S. life and consciousness, we must listen as much to the economic and the political as to the cultural if we are to begin to recover the principal meanings that lay in their articulation. Treating aural history simply as a cultural, political, or economic project decontextualizes what must be contextualized, denudes the past of its interrelated texture, and contributes to our deafness by denying us an understanding of how political sounds were shaped by, and in turn influenced, cultural whispers, economic booms, and social screams. Heard worlds, like the seen, were so intimately connected that to reveal their full complexity we should listen to them in their entirety as best we can. Thus while some readers may balk at the binary "North and South" depicted in this study (and the master narrative used to tease out the creation and working of the dialectic), they wince at what and how people heard and the terrible reality of two sections that did in fact go to bloody war.[7]

Although nineteenth-century Anglo-American elites came to define their economic, political, and cultural authority by sometimes doubting their ears and instead emphasizing what James Joyce called "the ineluctable modality of the visual," the heard world still retained considerable currency among ruling classes not least because sounds, particularly as proxies for progress and refinement, were "as much a mark of their nationality as their class," as has been said of the Victorian English bourgeoisie.[8] Antebellum elites north and south similarly appealed to the heard world when defining and reaffirming some of the core values of their class and sectional identity. Accustomed to listening to the health and pulse of their own societies, aurally sensitive antebellum American elites heard threats to their worlds not just from within but also from without in the final decades before civil war. What played on the ear echoed in the mind with disturbing and distorting resonance to the point of warping perceptions, minimizing objective sectional similarities, and ultimately leading to war not just because of what was seen but also because of what was heard, literally and figuratively.

Contemporaries' insertion of aural imagery into the medium of published discourse effectively gave lasting voice to what is sometimes wrongly considered the silent medium of print. Printed aural projections of sectional identity and a variety of other matters were powerful and palpable because the printed words used to convey the various sounds and their meanings rendered aurality permanent and rescued them from the ephemerality of voice. In this way, what would have remained temporary, elusive sound (many arguments were offered in the form of public speeches) gained permanence in the world. Unlike the modern ability to record and thereby reproduce sounds precisely, the antebellum aural metaphor and projection that was communicated through print (and actual hearing) allowed contemporaries to have access to a permanent image of how each section (and other things) supposedly sounded. Hearing the sounds did nothing to contradict these printed images and, in fact, largely confirmed and heightened them. Time and again the imagery of how each section sounded was recorded first in the ear, then in a print version that stripped the sounds of their nuance and replaced them with a clumsy, written representation, thus giving readers access to a captured record of sectional aurality that they in turn could repeat with their voices to other ears. On the whole, aural representations of sectional identity were remarkably clear, candid, and to sectional ears, compelling and required little decoding on the part of the listeners. Aural images of the South and North became crystallized and so helped contribute in meaningful ways to the construction of sectionalism.[9]

Antebellum elites inherited a good deal of what and how they heard from their colonial forebears. Forthcoming work promises to tell us much about colonial "soundways," but in the absence of that and other work, based on exploratory research it is possible to say a few words about how colonial America was similar to and different from the United States of the antebellum period.[10]

As with their antebellum counterparts, colonial Americans shared in national sounds, many of which were part of a transatlantic soundscape. Sounds of cannon in eighteenth-century Williamsburg, Virginia, for example, were linked to public celebrations. Cannon were fired to mark a variety of occasions, including coronations, state funerals, victory celebrations, greetings for various dignitaries, and peace with Britain in 1783. New political realities inaugurated new sounds, and these served to bond Americans. Cannon booms marked George Washington's birthday beginning in 1778, for example, and Williamsburg's Bruton Parish Church rang its bells after the repeal of the Stamp Act in 1766.[11]

None of this is meant to suggest that colonial America was acoustically homogenous. Because different regions were made up of different flora and fauna, they assumed their own peculiar soundscapes. Coastal Florida resounded with "noise we heard from alligators" and the "large number of frogs" that "made a great noise when rain was coming on," according to British sailor Samuel Kelly in 1784. Birds also helped create regionally specific soundscapes, and Kelly compared Virginia and Florida in 1784. "I had often heard of a bird in Virginia that used to cry *whipper Will* in the night," he remarked, "but here a bird of the same description used to cry *Jack Will's widow!* Great numbers of these in fine nights used to repeat this sound all night long."[12]

The soundscapes of colonial America were highly local, not readily distinguishable between north and south, and perhaps best categorized as either urban or rural, the vast majority of which were rural. Since the levels of urbanization did not differ appreciably between the southern, northern, and mid-Atlantic colonies; since industrialization was minuscule wherever it existed; and since slavery and wage labor were present throughout the colonies, the only real distinguishing soundscapes in pre-industrial colonial America were highly local, defined by natural phenomena, and not yet in any real way sectional. Probably the most important and distinguishing factor determining soundscapes for colonists was being within earshot of the sea. These local acoustemologies never disappeared entirely; rather, they were joined and masked in the antebellum period by increasingly sectional soundscapes.[13]

Many aural values embraced by elites, north and south, in the antebellum period were inherited from their colonial experience. In their desire to shatter the silence and smother the aperiodic, erratic howling of the wilderness with the regular, rhythmic sound of white men's industriousness, antebellum elites differed little from their colonial predecessors. The process was still ongoing by the late 1770s. "Extensive wilderness, now scarcely known or explored," reported the *Pennsylvania Gazette* in 1779, "remain[s] yet to be cultivated, and the vast lakes and rivers, whose waters have for ages rolled in silence and obscurity to the ocean, are yet to hear the din of industry." Although elites differed on what industry should sound like in the antebellum period, in the context of a preindustrial, colonial America they agreed that the throb of industriousness of human activity should be inscribed on a wilderness that was both silent and howling.[14]

Colonial elites also concurred in their aesthetic and functional understanding of bells. The ringing of civic bells was reserved for visiting dignitaries and used on important occasions. Bells also became closely associated with freedom. Cities throughout the colonies marked the Declaration of Independence with church bell peels, thus giving God's aural blessing to Americans' secular revolt.[15] Safety was also heard. Williamsburg acquired a fire engine from London in the mid-1750s, and in 1772 the city council appointed "four sober and discreet People, who are to patrol the Streets of this City from ten o'Clock every Night until Daylight the next Morning, to cry the Hours" and also "have care of the Fire Engines." "Bellman" and "watchman" were virtually synonymous. Bells also held profound spiritual and secular civil authority. Their ringing demarcated spiritual and civil congregations within earshot of church bells and alerted God to colonists' piety. These sounds and the objects that made them were possessed and used by elites, and when colonists heard bells, they heard not just their community but also the authorities that governed them.[16]

If colonial elites agreed on what produced sound, they also agreed on who produced noise. Native Americans, African Americans (slave and free), and the laboring classes generally were among the greatest noisemakers in colonial America. Colonial African Americans "disrupted the acoustemology of English speakers in fundamental, frightening ways: they chattered like monkeys, they bellowed like beasts, they mourned in chants . . . they delighted in drumming, they spoke a language that was no language." African Americans, like Native Americans and other nonliterate groups, "defied the surveillance of writing" and made sounds that threatened to fracture the acoustic world of English settlers. Elite seventeenth-century colonists worried about the socially and spiritually disruptive tendencies of dissenters and the possibility that their "rants," because sound itself carried a physical force and gravity, would tear a delicate, evolving social order. While Native Americans could at first be intimidated by the sound of guns, even sounds of modern technology could not subdue what white settlers heard as Indians' "halloing" and their "foule noise." Colonists often considered Indians heathenish and dangerous, an impression confirmed when they heard their "howling" and "screeching" preparations for war. Such groups—African Americans and Native Americans particularly—as antebellum elites later found, could also be worryingly silent.[17]

Noise and sound were delimited temporally in colonial America. What was sound during the day often became noise at night, and civil and religious statutes make clear that a central concern of the Anglo-American colonial elite was to regulate the colonial soundscape, especially in the cities.[18] Eighteenth-century town ordinances from all regions of colonial America indicate clearly who were considered the noisemakers. Civil order was tied closely to religious worship, and inappropriate sound and noise were an affront to both. Consequently the city council of Charleston, South Carolina, tried to minimize noise particularly "on every Sunday, on Christmas day, and the three subsequent days, on New Year's day" as well as at Easter by empowering the guard to arrest persons making or inciting "any riots, noise, or disturbance." Significantly, "whites, as others," could be punished under the statute, which suggests that Charleston's elite believed everyone, but the lower orders especially, capable of disrupting the city's ordered soundscape. Virtually identical statutes applied in northern cities. A Philadelphia ordinance from 1732 was directed against the "frequent and tumultuous meeting of Negro Slaves" to protect "the Inhabitants of this city" from the "great Terror and Disquiet" caused by such gatherings. But the statute also targeted "Children and white servants" who congregated on Sundays and made "disturbances and noise in the City." A 1738 statute was designed to suppress "Tumultuous meetings" by "Negroes, Mullatos, and Indian servts. and slaves."[19] Preeminent among Philadelphia noisemakers in the late 1780s were "the loose and idle characters of the city, whether whites, blacks or mulattoes." Lower orders disturbed the peace, disrupted the solemnity of night, and indulged in "riotous mirth."[20]

Constructions of what constituted noise and sound, of what was gratuitous and what was socially necessary and culturally desirable, were largely a reflection of and contributor to the formation of colonial culture and class relations. Noisemakers, according to reports in the *Pennsylvania Gazette* from 1728 to 1755 from all regions of colonial America, for example, included animals, natural disasters, Indians, criminals, some women, and slaves; things that were sound or produced sound included liberty, correctly pronounced words, good Christians, and genuine coins, whose silver, when tapped, produced a tone unlike that of counterfeits, which emitted "a shriller sound" and so did not constitute sound money.[21]

While antebellum elites inherited a good deal of the values they attached to sounds and noises from their colonial forebears, it nevertheless seems that colonists only rarely spoke of distinctly northern or southern soundscapes. Sounds were local, and sectional soundscapes had to await developments of the antebellum period.

A few words on thesis, organization, and definition are in order. In this book I try to identify and explain how and why antebellum sectional identities emerged, heightened, and coalesced. I explain that process in terms of how contemporaries, elites in particular, heard their worlds—in other words, how their day-to-day listening fed and shaped their listening at the level of sectional consciousness, beginning mainly in the 1830s.

Elite classes within each section varied enormously, and the term "elite" is used here to capture a general but evolving worldview broadly shared by men and women who articulated the principal ideas of their class and section. Whatever their specific differences, northern patricians and the new bourgeoisie shared much not only in their assessments of the northern laboring classes but, critically, in their understanding and depiction of southern slaveholding society. The northern elite—ranging from aristocratic Boston Brahmins to reform-minded abolitionists and capitalists—believed in the virtues of gentility and highbrow culture, and for the most part, the desirability of free labor and virtuous political democracy. On some important matters northern elites disagreed, and their differences were manifest in their formal political party loyalties. But their general preoccupation with gentility and, beginning in the 1830s, the question of slavery muted a good deal of their differences so that in the thirty or so years before 1860, particularly in the 1850s, many northerners of all political persuasions united in a broad agreement that slavery was dangerous to the future of the United States and debilitating to aspects of the American present. Southern elites were similarly variegated and diverse. Southern industrialists, merchants, and the urban middle class sometimes locked horns with the numerically smaller but far more influential planter class. Neither were planters in agreement on all matters (few ruling classes ever are). Again, though, the ways in which merchants, industrialists, and planters were linked through interest and kinship and their broad support of southern slavery ensured that major disagreements began to evaporate in the closing decades of the antebellum period. In the face of the northern critique of slaveholding society, southern elites, the planter class at the helm, coalesced around a simple but powerful credo: southern slavery and all that it stood for were not only desirable but deserving of protection. Whatever nationalist sentiment united northern and southern elites, whatever they shared in common, slavery proved the fundamental issue on which they could not agree.[22]

In Parts I and II of this book I show that sectional awareness was shaped by what elites heard at the everyday level of social, economic, and political interaction. Neither southern nor northern antebellum patricians considered their heard world utterly harmonious—strains of discord were everywhere. North and south, antebellum workers used the transgressive nature of noise and the disturbing power of silence to limit and sometimes end their exploitation. In doing so they initiated strategies and tactics of resistance that threatened to rupture the ideal soundscape and the social, economic, and political security that it represented to elites. For the most part, though, respective elites took considerable comfort in how their societies sounded. In Part II I demonstrate how and why northern elites reveled in the hum of industrialism and the satisfying sounds of free and wage labor. When they listened to class relations in the North, they heard discord and noise but also the tremendous productive capacity of capitalism, and they considered accompanying strains generated by lower orders as sounds necessary for the successful prosecution of their great experiment. Alternatively, as I argue in Part I, southern masters cultivated the hum of slavery and emphasized the quietude and serenity of southern social relations. They, too, saw and heard discord from within the South, but they prided themselves on their ability to levy quietude on their society. Slaveholders did not reject modernity in toto. They cultivated economic productivity and embraced its sounds. Railroads chuffed happily in their ears, and the sound of timed labor anchored them to an idealized past and prosperous future. Even limited urbanization and industrialization were acceptable. But the quietude of plantation life and all the conservatism masters invested in that serenity were sacrosanct. In other words, while northern elites often considered noise a necessary component of modernism, southern slaveholders wanted modernity with quietude, a notion that differed from the image (and, increasingly, the reality) of the ideal soundscape peddled by northern abolitionists and capitalists.[23]

Antebellum elites came to hear one another differently, and in Part III I focus on how and why they constructed an aural sectional consciousness. While ruling classes in both sections agreed on much, they argued vehemently over the preferred form of U.S. social and economic relations. In this debate sounds took on profound meaning. A rapidly modernizing North listened to the South and heard the shrieks of slavery, the awful silence of oppression, and the unmistakable tones of moral, economic, and political premodernity. Masters listened to the North and heard, with increasing volume so it seemed, the disquieting throb of a northern mob made reckless by industrialism, urbanism, wage labor, and passionate democracy. Elites north and south constructed one another aurally; they attempted to change how one another sounded, and they used aural descriptions and representations to try to effect that change. Abolitionists shouted loudly southward, exercising the throat of democracy to critique their perception of silent tyranny and cacophonous slavery. They wanted masters to hear the benefits of northern democracy, the productive capacity of free wage labor, and the undesirability of slavery. They insisted that if the South did not change, then at least the West would sound northern. A variety of ostensibly competing northern voices joined the refrain in the mid-1850s, with Republicans and some northern Democrats applauding the satisfying melody of free labor and castigating the economic and political sounds and silences of slavery. For their part masters shouted back and then applied tools of silencing that they had sharpened in their own society. Indeed, as parts of the masters' world came to sound vaguely northern, they pushed hard to block out northern sounds and deafen their population to the evil strains being flung southward. Sectional ruling classes played on the ear, that "particularly vulnerable organ of perception," to define themselves as legitimate and cast the "other" as reprehensible.[24]

Given that elites invested so much of their identity in how their society sounded, and given that they heard their pasts and futures, it is hardly surprising that they acted with a degree of emotional commitment to the preservation of the aural integrity of their society that might strike some modern observers as excessive and irresponsible. But precisely because they invested such authority in the heard world, their behavior is perfectly understandable. Sounds of social and economic relations, slavery, and freedom were so meaningful that they helped shape the psychoacoustical perception of region. Sounds, noises, and silences took on tactile qualities that proved real, substantive, and palpable for political elites from both sections. Northern and southern ruling classes did not go to war solely because they listened to one another and disliked what they heard (just as the Civil War was not solely a product of what they saw of one another). Aural representations joined visual ones, and once we begin to understand how sectional consciousness was channeled, fed, and articulated through more than one sense, it becomes far more understandable how sectionalism assumed such concrete dimensions and ferocity. A reliance on visuality alone is likely to understate the emotional, visceral quality of the coming of the Civil War, but by adding acoustemological considerations we approach a fuller understanding of how sectional identities assumed such terrible force.

Understanding this reality is difficult if we fail to comprehend the depth of commitment respective elites invested in how their worlds sounded. The heard world was a powerful but clumsy proxy for a host of ideas about self and identity. In this book I attempt to capture both the complexity and the sometimes contradictory nature of heard worlds within each region, to give some sense of how an elite's definitions of an ideal soundscape were shaped by the resistance of laboring peoples and by larger developments in the nineteenth-century world. Whatever challenges each ruling class faced in the antebellum period, whatever threatening new sounds emerged from within their own societies, were never enough to encourage elites to retreat from their preferred model of how their society specifically, and the United States generally, should be heard. Through legislation and the exercise of their considerable economic and political authority, elites in both regions proved skilled at containing and accommodating these internal acoustic threats and what they represented. Increasingly, the real challenge came less from within and more from without, beginning principally in the 1830s.

This is not to deny the relevance of nascent sectional consciousness apparent during and after the American Revolution. In the 1780s northern observers such as J. Hector St. John de Crèvecoeur marveled that the slave owners' "ears by habit are become deaf, their hearts hardened; they neither see, hear, nor feel the woes of their poor slaves."[25] Aural sectionalism built on these few and early representations so that by the late 1820s and early 1830s leaders in both areas began to respond to emerging sectional structural and ideological changes with ever more vivid and potent aural projections of slavery and freedom. In a sense the technological and ideological shifts toward industrialism and free labor in the North encouraged not only a new way of hearing changes within the North but also in the ways of listening to bondage. Abolitionists came to hear slavery in distinctive ways, and advocates of free soil, free labor, and free men in the

1840s and 1850s joined them in many of their aural assessments of southern slave society. Largely in response to this critique but also in reaction to the structural and ideological transformations occurring in northern society, southern leaders defended their society by using its soundscape as proxy for order and safety and critiqued the new world of industrial capitalism by pointing not just to the sights of misery in the North but to the sounds of imminent revolution they fancied they could detect. The aural images deployed reflected emerging tensions, and they heightened sectional consciousness and animosity in profoundly emotional ways.

The degree of commitment that northern and southern ruling classes had in their heard worlds, the ways that they perceived the threats to their societies, and the meaning that they attached to the sounds they heard emanating from without gave meaning to what it meant to be southern or northern. In their choice of aural representations and metaphors for critiquing one another, northern capitalists and southern masters in effect made the heard South and North real places that were clumsy and binary and that subsumed a host of commonalities and complexities. Aural constructions became real. Historian Edward Pessen once remarked that the "terms 'North' and 'South' are . . . figures of speech that distort and oversimplify a complex reality, implying homogeneity in geographical sections that, in fact, were highly variegated." Pessen was right, but he and some others who seek to explain the coming of the Civil War failed to appreciate how binary notions of North and South crystallized. Given the deep emotional commitment contemporaries shared in how their society sounded; the meaning elites invested in their everyday soundscapes; the tenor of attacks that elites in each section launched; the role cutting, profound, and candid aural projections played in those attacks; and the daily reminders each class heard in their society about the security and comfort of their societies, it becomes understandable why, in heard terms, there evolved something that contemporaries heard as a clumsy, homogenous North and a monolithic South. As political parties became increasingly sectional in the 1850s, they drew on aural representations to heighten perceptions of regional differences. Aural sectionalism was not simply a reflection of the emotionalism of antebellum politicians because there was a firm, material basis for hearing one another as different and distinctly dangerous. Neither was aural sectionalism just about material and ideological differences between the two sections. Allan Nevins was not far wrong when he described how during the 1850s, "emotions grew feverish," "every episode became a crisis, every jar a shock," not least because the differences in the acoustic construction of slavery and freedom and the way that those constructions were communicated through aural images served to amplify, in an emotional and abstract manner, what in fact existed in a more tempered reality.[26] Modernizing tendencies shared by both sections were muted in this conversation, and while southern masters courted some of the most economically capitalist features of the nineteenth-century world, they refused to endorse those that threatened the integrity of their society. Hence they rejected the noise of free wage labor and the associated strains of massive urbanization and industrialism even while they coveted the efficient use of machinery and the most modern techniques for managing labor.[27] A key to understanding mounting aural sectionalism is found in the way that sectionalism was communicated, imagined, and projected not just visually but also aurally. Lacking some of the subtlety of the perspective and nuance of vision, aural constructions of sections struck deep chords in each society, challenged and confirmed cherished truths, and in effect generated a degree of tenacity and emotionalism that had a basis in material reality but that rapidly became so abstracted, so irrevocably "other," that aural sectionalism took on a meaning and conviction that served to distort what elites in each section understood of one another.

Sounds of war, noises of loss, shrieks of death, and chortles of success and victory followed this aural sectionalism, as Part IV of the study shows. In a way the Civil War was an aural victory for a modern North not least because it cleansed the southern soundscape of the wretched noise of slavery and paved the way for capitalism's expansion south and west. Yet the noises associated with the Civil War, the boom of battlefields, and the increasing volume of dissatisfied labor on the northern home front only encouraged northern elites to turn eagerly to a quiet, tamed South once the thunder of cannon and tumult of war had ended.

(((PART I

Imagining Maestros

Constructing and Defending
the Southern Soundscape

When antebellum southern slaveholder Alfred Huger wrote of his penchant for "the quiet and retirement of Plantation Life," he referred to a specific set of ideal social and economic relations and forms of behavior embraced by southern slaveholding society. Whatever their ideological bent, nineteenth-century ruling classes heard as much as they saw. Atlantic elites measured their authority and took the pulse of the society over which they presided in part by listening. The Old South's ruling class was not marginal to this tendency, and their heard worlds reveal much about how they understood themselves and the essence of their society.[1]

Men such as Huger spoke of plantation quietude, not of silence. Slaveholders were not foolish enough, even in their most marvelously romantic moments, to believe in an utterly silent slave South (as anyone who has had their ears stung by the roar of cicadas on a southern summer's eve will testify).[2] Their stubborn pragmatism told them that farms and plantations could never be silent. The rural South was punctuated by many, sometimes loud sounds, including braying and snuffling animals, baying hounds, blowing horns, and ringing plantation bells, to name but the most obvious soundmarks of a rural society.[3] True, post-Revolution slaveholders presided over a mature slave society, and they liked to think their mastery absolute. But moments of fantasy aside, realist masters knew that their command was defined in no small part by challenges to their au-

thority from within southern society. Thus they insisted not on silence—
which they knew they could not have and, in fact, did not really want—
but on quietude, tranquility, and sounds they deemed characteristic of
their organic social order. Certain sounds were pleasant and distinctively
rural and, as such, they were considered quiet in an abstract sense. Too
much silence at particular moments rubbed nerves raw simply because
silence was the unheard note that might precede insurrection. As well
as constructing noise and sound, masters also constructed quietude and
silence. Both were rooted in objective conditions, and slaveholders, like
northern elites, communicated (and in the process defined) their core social values with aural images amplified through the skein of slavery's social and economic relations and the physical environment that anchored
them.

Some sounds, such as the hum of industriousness, the register of slave-
based productivity, and the sounds considered preservative of social order, were encouraged. But for the most part, slaveholders imagined themselves custodians of a seamlessly tranquil society where calm reigned and
harmony was heard. This projection of aural order hardly distinguished
slaveholders from other classes concerned with mastery. One need only
think of Victor Erice's 1973 depiction of Franco's Spain, *El Espiritu de
la Colmena,* where all conversations were conducted in precious whispers, to appreciate how far ruling elites in conservative societies will go to
maintain social order (as following chapters show, democratic capitalists
could be equally unforgiving of persons who transgressed their preferred
soundscape).

The slaveholders' embrace of their constructed aural order was not so
much an attempt to regain a putative quieter past but an effort to promote a particular vision of a social order in the face of an emerging northern bourgeois alternative. Training their ears northward, they increasingly
heard boisterous industrialism, unfettered democracy, and wage labor
capitalism. Slaveholders countered with a construction of sound stressing the rhythm of industriousness, not industrialism, and the sober tones
of organic social and economic relations, not the noise of mobocracy or
withering wage labor. They did not reject all modern innovations. Indeed
they actively embraced sounds—such as the sound of time and a host of
other quasi-bourgeois sentiments—that bolstered, disciplined, and improved their social order. Some, especially planters along the Mississippi
River who had to hear steamboats, managed to pastoralize and thereby

incorporate modern intrusions. Instead of hissing steam they heard boat "wheels beat a quick tattoo" and listened to "throbbing engines" that "forced the delusion of beholding a living thing." Modern sounds were incorporated into the southern pastoral idiom through listening in certain ways: "The characteristics of the boats varied so one would know them in the dark unerringly by the sound of the steam-pipe or whistle, as you would recognize unseen the step and voice of friends from those of strangers." Beyond such accommodation, slaveholders refused to countenance alien noises that threatened to disrupt their organic, hierarchic society. In slavery they heard the harmony of capital and labor; in northern capitalism, industrialism, urbanism, and democracy they heard only "revolutions and intolerable distress."[4]

The master class sometimes embraced national sounds. For most southerners as for most Americans, in the years following the Revolution, July Fourth was probably the loudest day of the year about which northerners and southerners throughout the early national and antebellum years commented. Through print and travel, antebellum Americans invented an "imagined community" around an imagined soundscape on Independence Day, when they envisioned all Americans participating in a soundscape that was at once local and national. Juliana Conner, who relocated from Charleston to the rural repose of a North Carolina plantation in 1827, suggested as much. On Independence Day 1827 she commented on the "quiet and stillness of all around" and imagined "the moving throng and busy life which is, *this day* in all our cities." She continued, "Not a fort in our sea port towns but is now reverberating with the echoes of her cannon and on every mast her flags are proudly unfurled—yet here *not* a *sound* is heard." She was a little disappointed but understood why her day was silent: "I always feel a disposition to enjoy this birthday of our Independence as a *holiday* and would have been pleased to witness its celebration in some of the neighboring towns—but the het [*sic*] is so excessive that I suspect it would be more fatigue than pleasure—therefore will remain quietly at home."[5]

A commitment to national sounds and a similarity in the way ruling classes used the heard world to regulate laboring bodies did not preclude a distinctively southern soundscape. Unlike northerners who equated southern quietude with indolence and economic backwardness, slave owners interpreted their plantation quietude as betokening honesty, pastoral contentedness, simplicity, studied efficiency, and above all, social order. Slaveholders wanted the quiet life, facilitated by their nonindustrial, comparatively quieter, rural society. The preservation of this organic society was, masters believed, contingent on keeping the plantation soundscape tranquil. In their efforts to protect their pastoral soundscape from external and internal enemies, slaveholders experienced profound and troubling tensions.[6]

The plantation was very quiet, with that stillness which broods over broad, clean acres that furnish no refuge for so much as a bird that sings.

—Kate Chopin, *Bayou Folk*, 1894

"WHERE'S YOUR TOWNS?" The question was so characteristic, and was uttered with such a meaning look and gesture, we could not refrain from turning aside to have a quiet laugh. And yet at least one half of the Northern people, used all their lives to the bustle of cities and towns, and the noisy clatter of mechanical trades, if similarly situated with our earnest New-York acquaintance, would propound just such a question as he did—never once reflecting that cotton, sugar, rice, wheat, corn, tobacco, and all other agricultural products, grow only in the country, and very quietly too at that.

—D. R. Hundley, *Social Relations in Our Southern States,* 1860

(((1. Soniferous Gardens

In part, constructions of a highly stylized and ideal antebellum southern soundscape took shape in postbellum nostalgia. Listening back, postbellum southerners heard the South of old. Writing in 1901, James Battle Avirett attempted to retemporalize his readers through an appeal to sound. "What noise is that we hear over in that direction?" he asked, reimagining his Old South plantation. "That's the song of the boys on their light carts hauling the marl." The "supper bell" and the work bell mingled to create a romantic tapestry of southern serenity. "What is that suspended high up in the air, just between those two houses? That is the old plantation bell which, in the hands of Uncle Jim, regulates the movements of the servants, calling them to and from labor and telling [*sic*] out the hours for the various duties." Cue the sound of plantation industriousness: "The hum of the spinning wheel, the noise of the loom, with the stirring whiz of the weaver's shuttle, all accompanied, many times, by the melody of plantation songs." Such sounds, "heard from January to December," were inextricable from the plantation soundscapes because, slaveholders fancied, slaves acted at the behest of a master/maestro who conducted his plantation like a well-governed orchestra.[1]

Postbellum southern women's ears were similarly trained. Writing in the 1920s in the best tradition of the Lost Cause, Susan Bradford Eppes began to compile her memories on "a life-time spent among the negroes, both as slave and free." Among the raft of reminiscent works by former plantation owners and mistresses, perhaps none captures both the imagined and actual soundscape of an Old South plantation as well as Eppes's. For her, memories of plantation quietude rang loudly. "In the palmy days of the Old South," she declared, "country life was the one thing to be most desired; only those who were forced by the trend of events lived in the cities." Plantations, after all, were quieter. The big house was "the very sanctuary of domestic felicity and peace," and the workforce was quietly obedient. With frightening aplomb Eppes described her antebellum Florida plantation in the following terms: "Many a time have we been good-naturedly laughed at by our Northern visitors for the many servants needed to carry on the work; but to us it seemed really needful and we loved to see them moving around so quietly and so cheerfully taking from us every menial task." Herein lay the essence of the ideal plantation soundscape: a quiet world staffed by obedient, dutiful labor. Industrious workers added the satisfying timbre of sober productivity: "All winter the wagons were busy hauling wood, the axes of the wood cutters rang clear upon the air." Plantation sounds were harmonious, not unlike ideal slave-master relations. "A summer's day in Florida" was one where the "bees hummed drowsily as they hovered in and out of the honeysuckle vine." Southern heat added to the serenity: "The canaries are preening their feathers and hopping lazily from perch to perch, but it is too hot to sing: on the door mat the yellow cat is sleeping and within the house all is quiet, for the post-prandial nap is an institution in the Summer Land." Even those not privy to the afternoon shut-eye were deliciously tranquil: "From the distant laundry comes the subdued murmur of voices." Eppes was not alone in her assessment of the heard southern past, for even former slaves remembered plantation life as quiet in several respects. Writing from antebellum Philadelphia, Charles Ball remembered of his South Carolina rice plantation at night, "all was quiet, and the stillness of undisturbed tranquility prevailed over our little community."[2]

Antebellum southern elites appreciated the soundscape in ways similar to those of their postbellum brethren because they, too, heard their pasts and ideal South. Sound was tied to place and memory was the tether.

William D. Valentine, for example, heard the history of his South through birds. "I hear the birds of Spring," he wrote in 1852, "now reminding me of childhood and early friends. . . . Bugle notes of the past whisper in my soul." So, too, with Georgia's Robert Habersham, who was reminded of yore in 1832 by the sounds of the plantation: "How pleasant is the sound of rain[.] Its soft monotonous pattering on the roof,—and windows is soothing and puts one in a meditative humour. . . . To take a book upto the carriage house,—and hear the rain,— . . . was the delight of my boyhood days."[3]

Although there was a good deal of wishful thinking embedded in these projections, and while they never served to obliterate slaveholders' everyday anxieties, such idealizations nonetheless had their basis in reality. Because the antebellum South, even by 1860, was still a predominantly rural society, it retained sounds of ancient pedigree inherited from the colonial period and slave societies of old. Sometimes southerners constructed this soundscape in absentia while abroad. When in southern Europe they commented favorably on the soundscape of countryside and towns and, in the process, heard their home as well as their European pasts. James Johnston Pettigrew heard "the old plantation" while trekking through Spain in the 1850s, harvesting reassuring registers as he went. It was a place where "birds sung" and "a Southern breeze rippled gently." He also marveled at the "quiet of the solitude" and hinted at comparisons between this chivalric, silent country where the "watchman cries the hour in your hearing" and his quiet North Carolina plantation home. One writer for the *Southern Literary Messenger* in 1857 commented on the "soothing tranquil[ity] and beauty" as he traveled through the Campagna surrounding Rome. Greeting him was "alluring solitude, unbroken by any objects that are not in perfect harmony with its sweet and mournful desolation." This was a place unscathed by the ravages of modernity, a place where "silence has here its perpetual home, a deep and ancient silence that the song of the lark and the occasional report of a sportsman's fowling-piece seem scarcely able to disturb." Rome was even better, "the most delightful place of residence on earth." Here the "repose of the city and its isolation from the great, throbbing, active world of Europe and America, render it especially attractive to the quiet, meditative person who has no great projects of ambition to work out, and an easy competency in his affairs." True, a "liberal government, stimulating the energies of the people and giving freedom of thought and opinion would no doubt work an important change in the aspects of the city; it would make the Campagna wave with golden harvests, and cause the banks of the Tiber to resound with the hum of industry." But was this really wanted? Here was a parable for the brewing sectional American conflict: "But the clash of engines would jar upon the eloquent silence, and the hand of improvement." This mapping of the southern soundscape onto places foreign (and, indeed, the idealization of those places as models for the South) extended to Naples and in interesting ways. "No hum of industry, no shock of ponderous machinery comes upon the slumberous air." In Naples "all is tranquility." Therein lay the danger, as southern masters well knew. For while the Italian city had no stealthy, conniving slaves, it had its fair share of swindlers who threatened to wrest precious money from unwary travelers. "All is tranquility—but not peace, for the quiet of Neopolitan life is a treacherous quiet, which, gathering with the shades of evening, is characterized by the frequent play of the stiletto."[4]

Southerners did not need comparisons with Europe to confirm the virtues of their soundscape. They knew the benefits and courted the tranquility of slave-based country life. North Carolina lawyer William D. Valentine suggested as much in 1838. "I love a country seat where I can study in quiet," he wrote, à la the English aristocrat, concluding, "where I can enjoy the mellow, mild air of fine autumn days." And if some southern counties did not fit the mold, then others would. Harriet Randolph hoped her family would find "many tranquil and useful days" and "a peaceful haven" after they moved to Florida from Virginia in 1829.[5]

Southern sounds were romantic. From New Smyrna, Florida, Connecticut native Stella Hull wrote in 1804 that her "favorite spot is a little arbour [which is] the resort of a great number of beautiful Birds who entertain us with their wild enchanting melody—This rural spot is pleasantly situated on a bank of the little river Hillsborough." Countryside by seaside was also romantically peaceful: "I listen by moonlight to the hoarse and distant murmurs of the vast Atlantic." Moreover, there was something about country life in the early national South that this northerner found therapeutic. Her country home she considered a "quiet retreat" and their living offered a comfortable "sober quiet." Georgia minister William W. Yarbrough agreed. Of his antebellum childhood he recalled, "A little well-chosen solitude . . . soothes the asperities, quiets the nerves, closes the ear against much that it ought not to hear."[6]

William J. Grayson, a South Carolina proslavery supporter, pastoral writer, and antisecessionist, often pointed to the nature and significance of the southern soundscape. Born in 1788 in Beaufort—a small village of perhaps 200 souls at the time—he idealized village life generally, north and south. Writing in 1862 Grayson remembered Beaufort as "quiet, healthy, religious." It was not silent, for it had its own geographic and temporal soundmarks, such as the drum fish, which "makes a singular noise, in Spring of the year like the tap of a drum." He recognized that these pastoral sounds were both local and national, for they "are found as far North as New York." He heard northern villages—beginning in 1801 he taught school in one—similarly "quiet" and pastoral. There similarities ended. Compared with Charleston the burgeoning metropolis of Philadelphia roared by 1817. "We arrived late in the evening," recalled Grayson, and on "our way in a carriage to the hotel we encountered a tumultuous uproar such as I had never heard til then. There was an alarm of fire" and nothing "could exceed the din and clatter." Although this was at the beginning of the North's march to the modern, and although the "North had not yet reached the riotous period of enormous wealth," Grayson heard the early registers. This was the period when "steam began to be seen and heard."[7]

If southern elites ever flirted with forgetting the nature and virtues of their soundscape, jaunts to the North and subsequent returns even to the urban Upper South reminded them of what they had. "The atmosphere of age and repose about New Castle [Delaware]," commented Cecilia for the *Southern Literary Messenger* in 1856, "was most grateful, after recently witnessing the smart, new air, and the bustling life of northern towns." Commerce in large urban environments—northern cities particularly—could be deafening to southern ears, even for those long removed from Dixie's soil. As Edgar Allan Poe described in "The Man of the Crowd," the shouts of venders on city streets were "full of a noisy and inordinate vivacity which jarred discordantly upon the ear."[8]

Many southerners who visited the North came away with the conviction that northern bustle was not for them. Following a trip north in 1858, Thomas R. R. Cobb remained anxious and "yearning for the welcome & quiet & rest of home" in Athens, Georgia.[9] For a young southern woman to venture north, as did Elizabeth Ruffin in 1827 (she was twenty at the time), was an aural as much as a visual experience. Traveling from her rural Evergreen plantation in Prince George County, Virginia, to a number of northern cities was an acoustic education. Her first night in a Philadelphia hotel, for example, prompted the following remarks: "Strange customs here indeed; tho' we are boarding here the very capital Hotel in the city . . . we never see or hear a living creature except one or two servants and they appear not but at meals, or particularly rung for: everything still as midnight." This was a curious instance of the North being quieter than the South—quieter servants, quieter hotels generally—and initially Ruffin found it "very agreeable to me." Within five days, though, this particular brand of social silence proved grating because it bespoke isolationism: "Everyone keeps within his own limits and manifests no disposition to encourage sociability. Oh! I would not live here for a trifle." Social silence became a northern hallmark for this young mistress: "I've often heard these northerners borrowed their frigidity of temperament from the climate." To a southern ear, such taciturnity bordered on the frigid and rude. But if servants, hotels, and northern guests were too quiet, the same could not be said for Philadelphia the city. The day after her arrival, Ruffin was disturbed by fire alarms, the second of which prompted the following diary entry: "Am closing the day amid a scene of confusion: another fire! Bells ringing, people hallowing, engines rattling." Elizabeth moaned, "Really, I am getting right tired of Philadelphia. . . . My eyes have seen till they are saturated and can do their part no longer; my ears satisfied with the sound of bells, rattling of drays and other music a little more melodious and now am ready, willing and anxious to be off." New York City was little better. Night's sacred silence was disturbed, and sleep was made fitful by the "noisy, rattling drays . . . which seemed to me to continue most distressingly loud all night" and left her with the impression that the city was in "constant confusion."[10]

Ruffin considered even the North's most celebrated health spas less than serene. Saratoga Springs, for example, attracted too many people to ensure peace and quiet; she complained bitterly of "a fiddling and droning" from a fellow hotel guest. "My ears," she grumbled, "are almost deafened with the sound." Ruffin found northern society noisy and noisome. The northern mob, whose moral degeneracy she saw and heard, came in for particular criticism. On her way to Albany she witnessed a crowd awaiting a hanging; the participants she heard "yelping and howling like bears and dogs" and attached a moral dimension to such noise because it apparently "griev'd me to witness so much depravity and insensibility in human nature."[11]

It was highly unlikely that slaveholders would forget the virtues of quietude, for at base they understood that sounds were markers of gender, race, class, and virtue. Gentility demanded serenity and serenity was, by definition, quiet. Antebellum and early postbellum courtesy cards, for example, were frequently inscribed with mawkish homilies applauding the quiet life and the virtue of silence. "A Tranquil Heart," touted quiet, while others highlighted the goodness of "silent deeds." [12] The "high breeding" of gentlemen, it is not surprising, was defined by "the quietness" of their tone and remarks. Harriet Martineau found Andrew Jackson and Daniel Webster "quiet" and stately. Henry Clay's "quietness" commanded her "respect and admiration." Their noisy lower-class counterparts always existed as aural opposites: "rowdies or gentlemen," as James Johnston Pettigrew styled the binary.[13]

Because women were important custodians of southern quietude, their supposed serenity was part of the larger expression of southern conservatism. The "quiet of her nature," George Howe told young women at an address in South Carolina in 1850, was a reflection of her tendency to "lean upon others," and her "meek and quiet spirit" was also, "in the sight of the Lord," welcome, for it reassured men that all was right with their world and, indeed, gave them the mental space to think about how to preserve it. The loud woman, conversely, "unsexes herself." When women "meet together in organized bodies and pass resolutions about the 'rights of women,' and claim for her a voice and a vote in the appointment of civil rulers," concluded Howe, "she is stepping forth from her rightful sphere and becomes disgusting and unlovely, just in proportion as she assumes to be a man." Southern women were women by virtue of their quietude, silence, and submission.[14]

Part of this insistence stemmed from men's perceived need for day-to-day serenity. Women were responsible for disciplining not just themselves but also some of the most emotional members of southern society: children. Travelers commented endlessly on the noise of children and woman's responsibilities in quieting them. Passenger boats unlucky enough to be caught in rough waters acquired a distinctive noisescape that, according to one man on a ship on a choppy Potomac River in 1827, consisted of "heaving, grunting . . . besides six children bawling most loudly and refused to be quieted till attended to by the mothers." Children forced to endure wagon rides, as Harriet Randolph found when she relocated from Virginia to Florida in 1829, "solace themselves with quarrelling and fighting all the way, so that between them & the negro children (who are equally amiable) there is a continual *yell* from the back of the wagon." [15] Even in calm conditions children were, to slaveholders' ears, prone to make noise because their faculty of reason was as yet undeveloped, their control of the passions unchecked, and their propensity to ejaculate noise likely. That children lacked self-control was obvious and even understandable, but it did not stop southern men from complaining. Southern women were responsible for quieting such outbursts, protecting, as it were, their ears and those of the master class. Indeed southern women, who took their responsibilities seriously, were highly critical of mothers who failed to silence children, particularly in social and sacred settings. Of North Carolina's countrywomen in 1827 Juliana Conner remarked, "During service, the ladies often leave the church to quiet their children—had the discourse interested me I should have been much annoyed by the crying of the infants whose mothers vainly attempted to hush to silence."[16]

Elite southern women were expected to maintain a dignified quietude in the presence of men. The admonition that children should be seen and not heard applied in almost equal measure to southern women. Evidence suggests some were. While traveling with her husband through Tennessee in 1827, Juliana Conner "rested quietly in my room" while her husband "engaged with gentlemen on business." Southern women criticized men, but few did so publicly and aurally. Private diaries were often the preferred, and silent, mediums in which elite southern women expressed their thoughts. Conner, for example, was less than taken with a sermon preached in Mecklenburg County, North Carolina in 1827. Her criticism was confined to silent, private print, though: "[Mr. Williamson's] style was verbose—his language an attempt at the sublime, in which he totally failed, and his reasoning superficial and unconvincing." Good southern woman that she was, Conner acknowledged that the other parishioners "all said it was an excellent sermon and I was silent."[17]

However idealized and romanticized, the southern soundscape was not acoustically flat in either perception or reality. It had its variations and rhythms according to place and time. In unsettled areas, for example, proud and disturbing stillness reigned. Throughout the antebellum period much of Florida retained the sound of the wilderness. It was both quiet and full of strange and wondrous sounds. Travelers, even those familiar with Florida's various soundscapes, commented often on the region's registers. On his way to St. Augustine from Tallahassee in 1833, for example, judge and later territorial governor Robert Raymond Reid remarked in his diary, "It is raining, the old deserted huts, the gloomy pine forest, secluded fountain, the thick and tangled swamp, the ancient figtrees, and the screams of wild birds, excite feelings of awe and melancholy."[18] Yet human attempts to tame the wilds, especially when they involved silencing the people who lived there, could be noisy and unsettling. Bartholomew Lynch, a private in the Florida war of 1837–39, found the noises of battle in Florida's wilds incongruous. During the Battle of Lake Monroe he commented, "The Constant firing of 1000 Rifles and muskets the reverberation across the bosom of the Calm and placid beautiful lake Monroe" disrupted "the stillness" of untamed Florida. Quiet returned once the countryside had been subjugated and its inhabitants removed from earshot. But violent wilderness had to be mastered before appropriate quiet could be imposed on the southern frontier. The aural disruption of the wilderness with the sound of war was acceptable because it paved the way for the reassuring keynotes of southern slave society.[19]

Antebellum southern soundscapes were also seasonal and temporal. Crisp December air carried sound some distance on plantations, and James Battle Avirett "could always tell when the neighbors were butchering by the noise of the salt pestles which could be heard for miles on a clear, cold morning." Similarly the sound of the axe chopping wood, while hardly alien in a region dependent on wood for fuel, was most common in winter not simply because these were the coldest months but because the "season selected for clearing" was winter. Rainy days were often reserved for carding and spinning, and at these times the "snap" of the reel being cut resounded. Such was the seasonal nature of the plantation soundscapes that, while varied among southern regions and crops, were nonetheless united by the sounds of slavery.[20]

Southern urban noise counterpointed rural quiet. The southern literati rejected the noise of urbanism categorically and embraced the mournful silence of nature. William Gilmore Simms had no ears for the "crowd and noise of city life," preferring instead the melancholic but spiritually pure quietude of nature. What he wrote in the *Southern Literary Gazette* in 1829 would have sent shivers up the spine of northern boosters:

I think there is more sublimity in barrenness, in ruin, in decay and desolation, than any other collection of objects in nature. . . . The gloom of silence tends perhaps as much as any thing else to invest such objects with that kind of sublimity we speak of. We feel a sense of awe for which there is no accounting, while listening to and hearing only the sullen and continuous murmurings of ocean, or the swelling mutterings of the wind among the tree tops of the forest. The hum of men and cities is puerile and childish to this great and solemn voicing of nature. We feel the contrast immediately, and by our own silence and awe, we seem to acknowledge ourselves in the presence of God. I could never enter into any kind of conversation while traveling through the pine barren of our Western country.

"Stillness . . . interrupted only by the cooing of the dove, and the murmurs of the breeze as it played through the tops of the long-leafed pines" even for travelers to the region constituted the sounds of a pastoral South.[21]

Although highly idealized, these renditions of the southern soundscape contained a kernel of truth and objective conditions that fed into masters' subjective constructions of their world. Southern slavery and the predominantly rural society that undergirded it was literally quieter than the emerging industrial capitalist society to the north. Although I discuss the structural basis for increasingly divergent soundscapes between the North and the South later, here it is worth saying a few words about the material basis of cognitive southern and northern aural worlds.

Consider, for example, the soundscape of the antebellum urban South. Objectively there was a good deal of noise in Old South cities, and to the people who lived there, they probably sounded like cities everywhere. Critically, however, far fewer southerners than northerners lived in such noisy environments. According to widely accepted figures, in 1820 just over 10 percent of New England's population lived in an urban environment; the urban population of the South Atlantic states was about half that. In the Middle Atlantic states 11.3 percent of the population was urban. Differences increased over time. By 1860 New England's urban population had more than tripled, while the urban population in the Middle Atlantic states stood at 35.4 percent. The urban population of the South Atlantic states, however, remained relatively small; from its 1820 level it had doubled to 11.5 percent, still two-thirds less than New England's. The East South Central states were even less urbanized: 0.8 percent of the region's population lived in towns and cities in 1820, and the figure was only 5 percent forty years later. Manufacturing reflected similar trends. The per capita value of manufacturing output in New England in 1850 was $100.71; in 1860, $149.47. In the South the value was a pitiful $10.88 in 1850 and $17.09 a decade later. To be sure, the majority of the North was still nonurban and agricultural by 1860, and many parts of it sounded in some respects like the rural South. But those similarities decreased over time, and pockets of rural quietude in the North became scarcer as the antebellum period wore on. Qualitatively (and critically) the different sounds produced by slavery and free labor made the soundscapes of the two regions, in the ears of contemporaries at least, increasingly different, a divergence accentuated by the much more rapid and widespread urbanization and industrialization of northern states.[22] Southern masters thought that the timbre of their own society remained much the same (some obvious modern innovations notwithstanding), while the North's rapid urbanization and industrialization and the free labor used to power modernization had upped the volume and changed the tenor of northern society. The enormous productive forces of industrial capitalism unleashed in the antebellum North silenced colonial slavery, replacing it with louder and qualitatively different keynotes. By 1860 not only were there many more ears and mouths in the North—not only was there much more sound to hear—but the timbre of the sound that was produced and incorporated in the North was, in some critically important respects, qualitatively different from that of southern sounds.

Qualitative evidence supports the quantitative. Benjamin F. McPherson, for example, testified inadvertently to the clarity of his Texas soundscape in the 1840s. In his rural school, "Every student read out or spelled out, and of an evening . . . you could hear them distinctly a mile."[23] Similarly, foreign travelers, even those hard of hearing, hinted at the quieter southern soundscape. Harriet Martineau found the South's rural soundscape tranquil and quieter after the din of northern cities. "The American," she wrote (meaning the northerner), "can conceive of nothing more dismal than a pine-barren on a rainy day; but the profound tranquility made it beautiful to me, whose rainy days have been almost spent in cities, amid the rumbling of hackney-coaches, the clink of pattens, the gurgle of spouts, and the flitting of the umbrellas."[24]

Slaveholders' ideal of quietude was built on the inclusion, exclusion, and occasional enfranchisement of new sounds (made at particular times and in particular places). Combined, the collage of decibels and timbres made up the reality and imagined ideal of the southern soundscape. Planters, like agriculturalists everywhere, were familiar with the role of sound in running affairs. As John S. Claybrooke advised his brother entrusted to run his Tennessee farm in his absence in 1833, "You must get a bell to put on our steers." Large herders on the southern frontier did not necessarily use bells to locate animals; cheaper methods were available. Of a herdsman in the "Indian Nation of the Choctaws," in 1820, Adam Hodgson "was amused by his method of summoning his herd from the neighbouring woods. This he effected by means of a whip. . . . Swinging it round his head, at first slowly, but with accelerated velocity, till the lash had acquired a momentum, which sustained it in a horizontal position, he cracked it suddenly with such dexterity, that I mistook it for the explosion of a gun. . . . In about half an hour we saw the cattle coming slowly out of the surrounding forest."[25]

Unlike on northern farms, the sound and quietude of the enslaved were central to the idealized southern soundscape. What slaveholders wanted—and what they got for the most part—was something akin to a plantation orchestra: an acoustic environment that they conducted with precision. Sounds made after a certain time (often delimited by the sound of the bell) were noises, and slaves' production of particular sounds at particular times indicated southern quietude.

All orchestras rely on timing for their coordination, and the southern plantation was no exception. In place of the metronome was the southern bell. The sound of time, at once the soundmark of the modern and the echo of the medieval, was perfectly consonant with southern quietude. "No sound ungrateful to the ear breaks upon the charm of quietness," lulled one writer in 1838 for Charleston's *Southern Rose,* "except the musical chime of the distant bells as they signal the passing hours."[26]

Regulating work through the sound of time was essential to slaveholders. On one hand, aural time was a traditional part of the southern soundscape and was not an obvious northern import. On the other hand, management by aural time was essentially modern: northern factories used it to regulate free laborers. According to a *Farmers' Register* correspondent, efficient agricultural societies in Europe used aural time, too. German agricultural workers, for example, began work "early in the morning, and with short intervals of rest continue till eleven o'clock, when the various village bells suddenly strike up a merry peal, which is a signal to the laborers to come home to their dinners." Herein lay the beauty of regulation by aural time for planters: connected to a romanticized past aural time was also a guarantor of an efficient, well-regulated, modern slave future. The remainder of the correspondent's observations are worth quoting not least because they expressed the ideal embraced by antebellum planters: "It is pleasing to remark the rapid progress which the several parties are making; . . . in the middle of this simple, rural, busy scene, it is delightful indeed, to hear from the belfry of their much revered churches a peal of cheerful notes, which peacefully sound 'lullaby' to them all. In a very few seconds the square fields and little oblong plots are deserted."[27] The transposition to the slave South was quick and easy.

Because planters aimed to improve plantation agriculture without jeopardizing slavery, their past, or their relationship with nature, aural time, which was not a danger to any of these, proved useful.[28] Time and sound were important for the careful modernization of plantation agriculture. Published by J. W. Randolph in Richmond in 1861, the *Plantation and Farm Instruction, Regulation, Record, Inventory and Account Book,* in this instance used by Beldale planter Philip St. George Cocke, stated explicitly the links between plantation management, agricultural regularity, clock time, and sound. The account book recommended that at the beginning of a day's work the plantation manager "blow a horn for the assembling of the hands" and "require all hands to repair to a certain and fixed place in ten minutes after the blowing of the horn." At the end of the workday, "a horn will be sounded at 9 o'clock after which every negro will be required to be at his quarters." Lest slaves managed to manipulate the regularity of aural plantation time, the account book further recommended that "the manager will frequently, but at irregular and unexpected hours of the night, visit the quarters and see that all are present." Should the manager be unfamiliar with the science behind the practice, the account book explained it:

Sound passes through the air uniformly at the rate of 1142 feet in a second, or through a mile in $4^2/3$ seconds.

Therefore any distance may be readily found in feet by multiplying the time in seconds, which the sound takes to arrive at the ear, (from the time a flash of lightening or of a gun, for instance, is seen,) by 1142.[29]

Aural time maintained plantation order and regulated the behavior of slaves. While taking account of the variation in working hours throughout the year, one planter's "Notions on the Management of Negroes" centered on aims and methods. "Recollecting that a 'stitch in time saves nine,'" he established Sunday as the day for a review of his slaves. "I appoint a certain hour for attending to this matter," he explained. On Sundays at nine o'clock in the morning, "every negro distinctly understands that at this hour he will be reviewed." The mechanism for ensuring this understanding and coordination was aural: "An hour or so previous to the review, I make it the business of the driver to sound the horn, for the negroes to prepare themselves and houses for inspection." Sometimes, as on William W. Brown's plantation, "the ringing of a bell" told slaves to rise at four in the morning; thirty minutes later a horn was blown, "the signal to commence work." James Bolton's master "diden' have no bell. He has 'em blow bugles fer to wake up his hands, an' to call 'em from the fiel's." This acoustic discipline was good for slaves. An unidentified Georgia clergyman maintained that slavery had blessed Africans with peaceful quietude. Should abolitionists doubt it, he advised, "Go back, then, into the wilds of Africa, and see two of her savage tribes engaged in the din of deadly battle."[30]

The spatial dimensions of plantations facilitated aural communication. According to Winn Willis of Texas, "Master Bob's house faced the 'qua'ters' where he could hear us holler when he blowed his big horn for us to rise." The reason for the holler? "When the bugle blowed you'd better go to hollering so the overseer could hear you. If he had to call you, it was too bad." Bells were also important for regulating labor in the plantation household. Indeed, planters already knew what southern industrialists told them in this regard. In 1857 William Gregg, one of the South's few great industrialists, advised planter James Henry Hammond that when building a house, "a well regulated set of bells" governed by "bell wires" was "indispensable to a well regulated family residence."[31]

While masters coordinated plantation labor through sound, they also imbued slave labor with a particular tempo. Slave singing in the field was a sound of industriousness with a built-in rhythm designed to increase productivity. Whatever the slaves sang, that they sang at a desired tempo was the critical matter for masters bent on making money. Although rural laborers could be heard singing in the antebellum North, songs of unfree labor were heard only in the South. Even in southern factories manned by the enslaved, singing provided tempo. Tempo increased productivity and bolstered the slaveholders' aural idiom of the serenity of slavery and the orderliness of southern society, and it reaffirmed to masters the realness of the slave as the "happy singing subject."[32] To be sure, the plantation soundscape was a product of slave and free, black and white, with slave singing in particular percolating through white households. But as long as this singing was deemed pleasant and, perhaps, illustrative of slaveholders' imagined plantation Elysium, it was encouraged. Singing slaves were, to masters' ears, happy slaves, and slave songs—sung at proper times and in appropriate places—helped reaffirm the conviction among masters that not only was all right with their world but their world, in fact, was all right. This is why slaveholders waxed eloquent over "the Negro quarters, from whence on a still night came the faint tum-tum of the banjo, accompanied often by a rich, sonorous voice of melody."[33]

Still, singing was regulated. As Charles Dickens observed of a Virginia tobacco factory: "Many of the workmen appeared to be strong men, and it is hardly necessary to add that they were all labouring quietly, then. After two o'clock in the day, they are allowed to sing, a certain number at a time." Greater quietude and even temporary silence were preferred in other southern factories. Scottish traveler Alexander McKay noted of slave workers in another Richmond factory, "The utmost silence was observed amongst them." Plantations were little different not least because the aural imperatives of slave-based productivity applied there with particular force. "At 9 o'clock horn blows," mandated Hugh Davis on his Alabama plantation, signaling that "all communications between house servants and those at the cabins [was] strictly forbidden after that hour." On other plantations a different regimentation of sound prevailed. "When at work I have no objection to their whistling or singing some lively tune," explained a planter in 1851, "but no *drawling* tunes are allowed in the field, for their motion is almost certain to keep time with the music."[34] Regulated sound tied to work-time efficiency in field and factory constituted an important part of the southern soundscape in the ears and minds of slaveholders.

Because slave songs and religious expressions were so tightly regulated, reminiscences by whites on slavery sometimes invoked the consonance between slave songs and spiritual shouts and plantation idyll. A mixture of patronizing curiosity and whimsical nostalgia suffused Susan Bradford Eppes's description of how slaves sounded on her Florida plantation. She remembered fondly the cadence and tones of slave religion, "How they shouted, and how they sang—what queer rendering of Sacred wit was given." Good slave preachers regulated the aural behavior of the congregation. "Dave, a solemn-eyed boy of nine," for example, was an effective preacher because at his command "all noise ceased and the whole crowd ranged themselves on imaginary benches to listen to a sermon. Dave harangued them with great eloquence, but great paucity of words, and then the shouting began." Christmas was particularly attractive because then "old plantation songs resounded through the Quarters." As long as the sounds slaves made fell within a context defined and regulated by masters, they were deemed notes becoming master-slave relations.[35]

Planters fancied that, like "the Swiss, Tyrolese, and Carpathanians," requiring slaves to sing while in field "lighten[ed] their labor." Many remembered the singing fondly after the Civil War, retroactively bestowing slave plantation songs with a meaning integral to the identity of plantation life.[36] Provided they controlled the timing and volume of slave songs (when they were aware of them), the masters found not only beauty in such sounds but teased reaffirmation of their plantation ideal from the earthy strains. "In the evening," mused John Tyler's wife, Julia (an elite northern woman with profoundly southern sentiments), in 1845, "we sit upon the piazza and listen to the *corn song* of the work people as they come winding home from the distant fields." Such songs were so reassuring of orderly idyll that they were tolerated even when they drowned out the president's voice: "The reapers have come to their labors in the field about five hundred yards from us and their loud merry songs almost drown the President's voice as he talks with me." Depending on its frequency, loud singing at 3 feet is roughly 75 dB, and the sound of an automobile engine at 25 feet can reach 80 dB. If these songs could drown conversation at 500 yards, their intensity probably measured around 80 dB or above.[37] Even ear-splitting volume was permitted and incorporated into the pastoral ideal of southern quietude provided the context was deemed fitting by the elite southern ear.

Part and parcel of slaveholders' vision of plantation tranquility came from the sounds made by slaves themselves and by interaction between slave and master. The plantation calendar, particularly at Christmas, harvesting, and shucking times, was marked by the sounds of black and white celebration and interaction. Others commented on the beauty of slave hymns. One traveler who was not unsympathetic to slavery listened to hymns sung at a slave funeral. It was "the most solemn and yet the sweetest music that had ever fallen upon my ear." Its profundity was amplified by a southern soundscape in which sounds traveled far: "The stillness of the night and strength of their voices enabled me to distinguish the air at the distance of half a mile." Combined, these sounds and moments of quietude were considered good for slaves. As William J. Grayson put it in *The Hireling and the Slave:* "Calm in his peaceful home the slave prepares / His garden spot and plies his rustic cares."[38]

Some masters allowed slaves' noise at particular times. On James Battle Avirett's plantation, corn shucking was accompanied by the blowing of whistles, "bantering . . . boasting," and singing. Saturdays and Sundays were days when slaves could fill the air with their sounds, their time to shape the soundscape of the plantation. Beginning on Saturday evenings, often the "air was filled with happy shouts." "I must not omit to mention," advised a planter from Mississippi in *De Bow's Review* in 1851, "that I have a good fiddler, and keep him well supplied with catgut, and I make it his duty to play for the negroes every Saturday night until 12 o'clock. They are exceedingly punctual in their attendance at the ball, while Charley's fiddle is always accompanied with Ihurod on the triangle, and Sam to 'pat.'"[39] Planters made clear that oral and aural expression on the part of the slaves and aural reception on the part of the planter was a decision to be made by the latter. Sound was sliced into discrete packets and parceled out as masters saw and heard fit.

Acoustic order came with costs, and planters understood that to regulate plantations by sound in order to concoct an acceptable plantation soundscape demanded sacrifice on their part. An English traveler noted the compromise slaveholders had to strike to preserve a relatively quiet plantation soundscape. To keep the noise of slaves tolerable, Charles Lyell observed, slaves "have separate houses provided for them." It was "an arrangement not always convenient for the masters, as there is no one to answer a bell after a certain hour." Slaveholders themselves had to respond to the sound of the slave-rung bell, but provided that it enabled them to eat on time and coordinate life, even members of the master class did not object to being summoned to supper by bells. The servants' ringing of bells especially regulated southern mistresses. A few grumbled but apparently did not resist. "We were aroused at an early hour by the breakfast bell," commented Juliana Conner in 1827, "for their hours are almost primitive, dine at noon and sup at sunset, and like good sober folks retire at 9." Mistresses and masters readily allowed themselves to be summoned for meals "at the ringing of the bell."[40]

Because the soundmarks of the antebellum plantation South—of southern society generally—reflected a particular set of social and economic relations extant in a particular historical context, the sounds deemed essential to preserving order were not simply natural. True, planters wallowed in "the lowing of the kine and the plaintive bleating of the sheep." But the slaveholders' preferred sonic environment was more than merely pastoral. Planters' need for some degree of self-sufficiency meant that the sounds of clanging metal and thumping wood required deliberate incorporation that devolved on slaveholders' distinction between sounds they considered desirable and appropriate and those they did not. Listen to the heard recollections of James Battle Avirett:

The bellows in the blacksmith shop began to puff and blow as Robert and Washington ranged themselves for the day's work, and the hammers and the saws in the carpenter shop told that George, Virgil and Jim were at work. Thus it was that by the time the breakfast bell at the great house had rung this hive of industry was buzzing, each and all at their own work. No unnecessary noise, no confusion, but all in the quiet order with which each had gone to his own work, showing what the discipline of a superior mind over servants could and did accomplish.[41]

Southern masters courted the sound of industriousness while deploring the noise of industrialism and the social dissonance it represented. These men wanted progress defined by their terms, and it had a particular register and connoted disciplined activity. "The enterprise of our people," mused proslavery thinker Henry Hughes, "will chain the gigantic powers of nature, and compel them to turn unceasingly the busy wheels of industry. . . . In every quiet valley, a cottage will nestle," and "our citizens [will find] . . . happiness in concord and industry."[42]

Typical of this industriousness was the sound of the blacksmith. This keynote, actually quite loud in terms of decibels, was freighted with pastoralism. It also proved enduring and was protected by law. An Alabama court, for instance, refused to grant an injunction against the building of a blacksmith's shop in Tuskegee in 1846. In answer to the plaintiff's argument that "the noise, smoke, &c." would be "a source of perpetual annoyance to himself and his family," the court asked and answered, "Is a blacksmith shop, in a small village, of this character? In our opinion, it is not." In the judge's ears this ancient activity and its attendant smoke and noise did not threaten materially the plaintiff or his property. Blacksmithing was in concord with a "small village, of this character."[43]

Diarists echoed the sentiment. "Men of industrious natures," thought North Carolina lawyer William D. Valentine in 1837, "will never forget business habits—they regret every moment of compulsive leisure." But how to remain industrious? "The very birds of the forest teach a good lesson in behalf of industry," Valentine explained. "How eagerly and cheerfully they chirp and fly from one branch to another. . . . Is this not industry?" Valentine heard industriousness in southern nature.[44]

The hum of activity in all its guises, notes, and registers was integral to the slaveholders' soundscape, and the silencing of the hum was ominous. Recession in the South was heard as much as seen. The "dullness and monotony of the sickly season," noted the *New Orleans Bulletin* in July 1838, was predictable and heard: "The hum of business is silenced. The squadron of drays that formerly dashed along so furiously, has dwindled down to a few teams. The steam boat wharves are nearly empty." Harvests in the fall would reanimate the city with the hum of business. When storms destroyed human progress and threatened to return conquered land to the silent or howling wilderness, southerners kept an ear strained for signs of recovery. After a dreadful storm in Florida in 1833 one diarist commented, "The noise of the hammer is heard in the land repairing the wrecks of the past night, and the cocks are crowing. We hope . . . that we may not be visited by another storm."[45] The silence of economic depression, especially when national in scope, was hardly regional in meaning. When agricultural, industrial, or commercial production was interrupted, America generally fell into disquieting silence.

In their calls for modernization, southern boosters sorely tested the soundscape of the agrarian master class. Nature itself begged for the throb of the modern, so it seemed. "We listen daily to the murmur of streams that invite us to appropriate them to manufacturing purposes," maintained a writer for *De Bow's Review* in December 1851. Similarly South Carolina's improvement in manufactures, argued one booster in 1853, was "rapid and perceptible" and audible: "Everywhere the busy hum of industry resounds, and the demand for new laborers is increasing."[46]

South Carolina factory owner William Gregg complained most loudly and in the process tested the bounds of the master class's willingness to tolerate alterations in the soundscape and the rending of organic relations that the new sounds foretold. Gregg believed that the South was doomed to remain a dependent vassal of the North should it not make the best of its abundant resources, and he pushed hard for southern industrialization. For Gregg, industriousness and industrialization were virtually synonymous. He explained to unhearing South Carolinians in 1850 that they must "learn the difference between indolence and industry." The state's resources must not "lie idle" but, rather, resonate with industriousness. If not, suggested Gregg, listen to the noise of importing Yankee goods, listen to "the crowds that are continually thronging the northern cities." "If we listen much longer to the . . . *croakers* against mechanical enterprise," he warned, South Carolina and the South generally would fall behind, the number of poor would increase, and who knows what such a class might do? Charleston and the men who ran it came in for particular criticism. The city's ban on steam power was particularly galling: "and this power is withheld lest the smoke of an engine should disturb the delicate nerves of an agriculturalist; or the noise of the mechanic's hammer should break in upon the slumber of a real estate holder, or importing merchant, while he is indulging in fanciful dreams, or building on paper, *the Queen City of the South.*"[47]

Instead, recommended Gregg, stop importing northern manufactures and let the South fill the vacuum. He was confident it would. Within two years of banning the importation of northern locomotives, the South would make its own: "The puffing of stationary engines, the noise of the trip-hammer would be heard, and those comparatively stagnant towns would feel the cheering influence of a busy throng of mechanical engine-builders." Moreover, "One man ardently engaged in such a work will be worth far more to the South than a thousand noisy politicians."[48]

J. D. B. De Bow agreed and in 1850 made the point that southern improvement must be genuine and initiated from within, "not in the rhetoric of Congress, but in the busy hum of mechanism, and in the thrifty operations of the hammer and the anvil." William T. Sutherlin, the largest manufacturer of tobacco in Danville, Virginia, in the 1850s, also urged the South's modernization, castigating the region's traditional agrarian classes. "Which seems the more elevated position," Sutherlin asked, "to hammer the red iron on the ringing thundering locomotive, or sit lazily in some professional office and talk politics, or study idly at the street corner and discuss with learned criticism the dresses of the ladies as they pass?"[49]

Most southern modernizers, though, feared alienating the master class. Only rarely did they advocate the wholesale adoption of free labor. Instead they presented their arguments for the South's modernization in part by using a lexicon of aurality and visuality that the planter class would understand and endorse. As Danville's manufacturing of tobacco in the late 1840s gave rise to a skilled black and white working class and all the problems of social instability and intemperate behavior associated with it, manufacturers and merchants attempted to stanch the rise of a potentially dangerous pocket of wage earners in their section of the South. These southern Whigs, who applauded economic modernization but feared unfettered, passionate democracy, turned to temperance. Refusing to issue liquor licenses for the town, magistrates in the 1850s created a citywide prohibition. The capitalist elite was pleased: "If the worst apprehensions of those who favor the license system were well founded, the question would then be, between a flourishing trade and degraded morals, on the one hand and a crippled traffic with peace quiet and sound morals on the other." Southern capitalists opted for the reassuring keynotes of the latter. Because it encouraged quiet, sober behavior, temperance reaffirmed the southern elite's social order. And when southern industrialists assured planters that using white wage laborers in factories in mill villages where temperance was strictly enforced would empower the South economically and anchor poor whites to southern institutions, slaveholding elites proved accommodating, as William Gregg discovered in the 1840s.[50]

Provided they did not threaten the soundmarks of slavery, registers of industry (and even limited industrialism) could be tolerated. The South's modest efforts to industrialize were generally greeted as generating positive sounds. Alabamians who urged the modernization of their state commented favorably on the sound of a cotton factory near Tuscaloosa in 1838: "The dashing energies of water power saluted our ears, accompanied with the whirling sound of diversified machinery." Indeed, antebellum men and women welcomed even ear-splitting booms because they occurred in particular contexts as hallmarks of progress. Juliana Conner heard as much when she commented on the "blasting rock" paving "an excellent road" through the North Carolina mountains in 1827. The saw- and gristmills and various other kinds of machinery that began to alter the southern soundscape in and after the 1830s offered other evidence of acceptable progress even as some of the louder innovations, like steam-driven cotton gins, remained few and far between in the Old South.[51]

For all their championing of northern-sounding sounds, even strident southern boosters wanted to industrialize or urbanize on their own terms. Arguments for the region's urbanization and industrialization infrequently touted the benefits of wage labor in that process because even boosters winced at the implications of a propertyless wage labor proletariat in the region. Thus a writer for the *Southern Literary Gazette* in 1852 applauded the "sound of the steam" whistle while still considering slavery the guardian against "revolutions . . . civil commotions" and "kindred strifes." Even *De Bow's Review,* which encouraged the South's modernization, was by 1860 beginning to fret over the northernization of the South's soundscape. This was particularly the case with the region's perceived rapid urbanization between 1840 and 1860 (perceived because, in relative terms, southern rates of urbanization were nowhere near those of the North). In 1860 the *Review* published a piece titled "Country Life," which described with exquisite detail and consummate nostalgia the changing soundscape of the South. "Within the last forty years," argued the anonymous writer, "country life has quietly and almost imperceptibly undergone great changes." This was not an aural revolution with a sharp bang. Nevertheless it had altered the "whole tenor and complexion of country life." The countryside had become "more and more dependent on the towns." The change could be heard. The countryside's "private social festive board is rarely spread; the barbecue, with its music and its dance, is obsolete and almost forgotten; the report of the fowling-piece disturbs not the slumber of the woods or fields; the huntsman's horn is not heard, the cry of the hounds, and the clattering hoofs of the pursuing steeds enliven but rarely the dreary monotony of country life." The familiar, comforting aural cues of southern society and all they meant were being threatened by the growth of cities. The lure of towns had created an aural vacuum in the country sufficiently powerful to suck the sounds of industriousness from the rural: "The anvil rings no longer under the sturdy strokes of the stalwart smith . . . and the sounds of the shuttle and the spinning-wheel are forgotten." This was the silence of economic recession on a massive scale. The emergence of southern towns "rob[bed] the country at a distance of its wealth and civilization." The promises of city life were draining the countryside of men and boys; women and girls who remained behind "mope and pine at home." The solution was simple: "We must have many small towns. . . . We must make country life tolerable, nay, fashionable, by bringing the country nearer the town." But this did not mean fashioning southern towns in the northern image, even though southern towns sometimes sounded like their northern counterparts: "Disunion alone will not cure the evil—non-intercourse must be superadded to disunion." Blend the virtues of town and country, advised the writer, and in so doing, reinvest the southern countryside with its familiar, distinctly southern timbre.[52]

Should people have trouble remembering the essence of the southern register, William J. Grayson was on hand to remind them. Instead of the groan of the northern multitude whose hunger pangs echoed with increasing volume, Grayson advised listening to the plantation, "the cheerful song that rings in every field," and "the long, loud laugh, that freemen seldom share." In the slave South,

> No city discords break the silence here,
> No sound unmeet offend the listener's ear;
> But rural melodies of flocks and birds,
> The lowing, far and faint, of distant herds,[53]

greeted those lucky enough to inhabit plantation paradise.

As long as they did not indicate northern free wage labor, other modern sounds were acceptable to the South's master class. Railroads, for example, sounded like progress. Upcountry Georgia farmers did complain about the noises of the railroad and their tendency to spook livestock and horses. But railroads, because they were integral to the South's market economy, made positive sounds. Steam locomotives that "puffed and snorted" were nevertheless "sublime" because of their importance to the southern economy.[54]

The transport and market revolutions resulted in similar aural marks north and south. Riverboats and railroads used bells and whistles both to guard public safety and to announce arrival and departure times.[55] Because fast travel, even if it was loud, was preferable to quiet, slow travel, southerners accepted the clanging of steamboats as necessary to their immediate travel requirements.[56] The market and commercial revolutions could also inaugurate quietude, thanks principally to print. That one of the hallmarks of the modern was the silencing of commerce is revealed clearly in a lecture delivered to the Petersburg Library Association in 1854 by John Tyler. The precommercial man of medieval Europe, believed Tyler, was one whose "hall resounded either with the noise of his bacchanalian revels or the songs of the troubadour." Even "as late as the beginning of the 17th century," he continued,

> if the merchant was found stationary in London or elsewhere, his shop, we are told, was but little better than a booth or a cellar, generally without a door or window. No sign emblazoned his name or employment, and no editor's columns made known his list of merchandize. His station was at the entrance into his shop, where with all the zeal that the desire to sell could inspire, he recounted to each passer-by, the articles he had for sale. When thoroughly exhausted, he was succeeded by his clerk, who pursued the same course until exhausted in his turn. So that, as we are told, and can readily believe, "London was a babel of strange sounds by which the wayfarer was dinned at every stop."

Although market bells still played a critical role in southern towns, the modern age had quieted commerce, and "instead of rendering himself hoarse by proclaiming the contents of his store to every passerby, he addresses the world through the columns of the newspaper press." But commerce had been quieted, not silenced. Tyler applauded the international progress of merchant capitalism, the arrival of whose goods were announced by "the whistle of her steam engines," "heard all over the sea." Southern poets such as Henry Timrod found the region's "streets echoing with trade" and indicative of orderly commerce.[57]

Commerce was further quieted with the coming of the low-frequency hum of the telegraph, which replaced the louder post horn. "Nor is the telegraph clamorous," ventured a writer for *De Bow's Review* in 1854: "We may hear it breathe, listen to the pulsations of its mechanisms; but we pause in vain to catch any other sound when it addresses its language to the four corners of the earth." "Eloquent, yet voiceless," the telegraph "cleaves space in silence." The master class willingly embraced technology that quieted their soundscape and furthered their commerce, which serves as a healthy reminder that not all technological advances are loud.[58]

Slaveholders heard as much as saw their South and were keenly aware of the threats to their way of life, increasingly so as the antebellum period wore on. Many of these threats were heard originating outside the region. Increasingly, though, planters heard noises that emanated from within southern society, registers that threatened to fracture the delicate acoustic fabric of what it meant to be a southern master.

So long had he been used to the continuous hum and
noise of a large city—so long had he been accustomed to
being jostled about at every turn—that to him *unrest*
seemed to be the only species of *rest* of which he new
anything.

—D. R. Hundley, *Social Relations in Our Southern States*, 1860

(((2. Creeping Discord

The southern elite's definition of a viable soundscape and all
that it stood for incorporated a modest industrialization informed by
notions of industriousness. The same cannot be said of southern urbanization and the threat of internal class conflict. Though in many regards
civilized, exciting, and urbane, and while necessary to the South's intellectual life and integral to its export economy, the region's cities were
nonetheless hubs of disorder and vice. Hearing towns and the class conflict they seemed to heighten, southern authorities attempted to regulate
the urban soundscape. In addition to urbanization, the southern frontier, the position of poor and nonslaveholding whites in southern society,
and the emergence of radical strains of evangelicalism all threatened to
introduce the passions of class conflict to the quiet South. Slaveholders
had heard many of these sounds before in the colonial and early national
periods, but they now heard them in a new context, beginning mainly in
the late 1820s and 1830s. With their society under increasing attack from
vocal abolitionists intent on pointing out the evils of slavery and the desir-

ability of a system of social and economic relations premised on freedom, the slaveholders heard the day-to-day developments in their own society with sharpened ears lest what was beginning to take place in the North start to take root in the South.

Challenges to the slaveholders' vision of quietude and the social decorum implicit in it came from various quarters. Rowdy poor whites, particularly when fighting, threatened to disrupt the southern peace. The master class counterpointed the duels of the elite, ritualized combats conducted with grave quietude, against the boisterous fights of the rabble. Masters disliked the orality and loudness of poor whites, but in a slaveholders' society dependent on the support of nonslaveholders for its continued longevity, this noise, while unpalatable, was something they sometimes suffered in relative silence.[1]

If slaveholders disliked the South's permanent poor whites, a class that survived by laboring for others, how did they understand and accommodate their sounds of political participation? Here we must be careful not to conflate planters' perceptions of different southern classes. Planters respected and courted the political support of yeomen and hard-working nonslaveholders. Many planters, after all, sprang from such humble origins. Planters often acquiesced to some of their political and economic demands and employed familial metaphors in an effort to cement the yeomanry's support of slavery in the abstract. Large slaveholders nonetheless quietly held a degree of fearful contempt for poor whites, whom they considered marginal to the disciplining tendencies of southern slaveholding society and disruptive of its tendencies. Planters wanted a "representative" and not excessively "democratic government" and harbored grave fears that further democratization would erode, as one South Carolina newspaper editor put it in 1860, "the barrier to radical and vulgar aggression."[2]

From propertyless tenants, laborers, and poor whites, planters sometimes heard the groans of the *Lumpenproletariat*. Masters' repeated claims that black slavery preserved white freedom contrasted with a reality in which in some southern regions in the 1850s landless poor whites constituted between one-fifth and one-half of the population. Here was a dissonance that seemed to undermine the equation of skin color with independence. While poor whites never mounted a sustained political challenge against the master class, their occasional protests and frequent interactions with free blacks and slaves proved troubling to elites and caused them to worry about what might happen, particularly if northern sentiments reached southern ears.[3] When poor whites protested perceived or real unfairness on the part of slaveholders in a context in which slavery was coming under increasing attack by northerners as undemocratic, the master class heard landless and dependent southern whites as the "mob"; in illicit trade and contact between slaves and poor whites, planters perceived the seeds of social disruption and petty capitalism, even abolitionism.

While many planters concealed this contempt, their private assessments reveal their true estimation. James Henry Hammond of South Carolina called poor whites the "mob." The noise of poor whites, though, served an important function: it kept planters on their toes because within their own society they could hear the strains of social discord. But because these sounds were relatively faint and because masters exercised some control and gained support from the yeomen and small slaveholders through kinship ties, various appeals to white men's republicanism, and the use of familial metaphors that yeomen seemed to find compelling enough to maintain their support of a broadly defined "southern" way of life, planters remained confident in their authority.[4] Although mob riots in the Old South were not uncommon, southern riots were less often a conflict between capital and labor. More likely, southern mob noise was heard in support of slavery and against abolitionism.[5]

To be sure, slaveholders believed their lower orders capable of disciplined quietude, but only if their betters trained them in aural niceties. Although sometimes considered especially prone to making noise, antebellum soldiers were also assumed to exercise precise and admirable aural discipline, not least because the military functioned through heard cues that new recruits and volunteers learned quickly. Soon after enlisting as a private in the U.S. Army during the Florida War of 1837–39, Private Bartholomew Lynch acknowledged the importance of aural cues to army life and discipline. "I likewise understood the calls," he noted, "whether on the bugle or drum—that is revalee at day light, Fatigue half an hour after, Breakfast 7 oclock, Doctors call 8 oclock," culminating in "tattoo at 9 oClock."[6]

Southern elites found particular satisfaction in the aural discipline of the military. Not only did soldiers' behavior inspire confidence that lower orders, provided they were properly instructed, were capable of disciplining themselves, but slaveholders recognized that soldiers were needed to quiet the southern frontier and thereby usher in the serenity of slaveholding civilization. The southern frontier, after all, set men's passions free.

Here southern elites heard "unmitigated rowdyism" and the collapse of "the nicer harmonies of the moral world." Robert Raymond Reid could hardly wait to leave Tallahassee's "noisy, senseless crowd" and return to the relative refinement and discipline of St. Augustine, "the metropolis of the East." Occasional urban noise was sometimes preferable to frontier noisy rabble. Removed from the sobering influence of southern slave-holding society, frontier dwellers inevitably began to sound like animals, and their lack of internal discipline manifested itself in unpredictable yells, grunts, and hollers.[7]

Quieting the southern frontier presented acoustic contradictions for southern elites, particularly when they had to deal with Native Americans. On one hand, patricians wanted to quiet the rowdy, yelling Indian frontier by civilizing it so that it hummed with sober activity and echoed with dignified quietude. On the other hand, Indians were considered a quiet people in harmony with the soothing registers of the natural world—but worryingly so. Soldiers in the various Florida campaigns, for example, noted the military efficacy of Native Americans' silence and noise during engagements. During the 1837–39 Florida War, for instance, Bartholomew Lynch confessed that on "the evening of 7th of Feb.y" 1837 his company stationed at Fort Barnwell, Volusia, had no "idea that indians were in the vicinity" until they "were disagreeably surprised by the hideous yelling . . . of over 600 savage Seminole indians led on by the Celebrated Oscola." Indian stealth got them close, where their yells then paralyzed unsuspecting victims.[8] Native Americans, in other words, could either "keep us in an uproar all the time" or render everything disturbingly quiet. Indians' sensitivity to their aural environment cost white lives. One white traveler near Picolata, Florida, was found shot after "having the Indians hear him fix his flint" in preparation for a scrap in 1840.[9]

So worrying was Indians' ability to control the heard environment (as well as detect the acoustic faux pas of whites) that their reduction to noisy savages offered paradoxical relief to frontier southern whites. Although the introduction of alcohol into Native American communities transformed them into the noisy rabble of the South's lower class and thereby threatened the peace of the frontier, at least drunk Indians could be heard. If the Moravians who ventured to the Cherokee nation in the early nineteenth century expected to find peace and quiet, they were mistaken, at least after other white men had arrived and plied quiet Indians with noisy drink. In August 1805 one missionary remarked, "Thus far they have not harmed us physically but we have lived in constant fear and many a night we had to listen to almost constant noise." The Floridian "state of quietude" was often "disturbed . . . during the months of January and February 1836 by apprehensions of Indian outrages." Benjamin F. McPherson's ancestors ("Quiet, law-abiding people") ventured to Texas in the 1830s from Alabama only to encounter "drunk, singing and dancing" Indians whose main preoccupation seems to have been "screaming the war whoop." McPherson much preferred the natural soundscape of the Texas woods, where "the wind would sound 'woo, woo,' so mournfully that my mind turned heavenward."[10]

Even when Native Americans were sober, southern whites heard them as less than desirable. Juliana Conner encountered a Cherokee woman in western Tennessee in 1827 who "spoke only in her native tongue. . . . Their language seems to be a muttering of sounds which they utter with great volubility in a kind of singing croaking voice—nothing can possibly be more disagreeable—you would think that they were in a violent passion."[11] It was plain that these were a people, like lower orders generally, given to passion, a weakness manifested in their noise.

Thus the control of the Indian and the quieting of the southern frontier—the cleansing, as it were, of the frontier's soundscape in preparation for the sounds of southern slavery—meant war. Contradictions plagued the decision. War was mostly a noisy affair as it entailed the loud assembly of an overwhelming military force in an effort to encourage the quiet submission of the enemy. The "six additional companies of U.S. troops" that were sent to Florida in 1835 were "considered sufficient to procure the quiet removal of the Indians." Although removal was anything but quiet, southern elites scarred their frontier with the temporary noise of war in anticipation of reassuring calm.[12]

Following the rancorous and bloody Seminole wars, white Floridians were determined to preserve the hard-won peace and quiet of their world by arguing against the army's forced removal of the "remaining Seminole and other tribes of Indians now living in South Florida." In 1853 the leading citizens of Hernando County contended that not using force would ultimately succeed in "quieting the apprehension of Indian disturbances," and they argued "that we felt gratified that the Indians have not heretofore been invaded, and that our peace and quiet have not been disturbed." The search for the quiet life sometimes meant peaceful coexistence with a much reduced enemy whose silence could be worrying but whose forcible removal bordered on bedlam.[13]

Once removed, Native Americans were taught the dignity and desirability of social quietude. "There is a quiet demeanor in some of these educated Cherokees," noted army officer Ethan Allen Hitchcock during his journey in the Southwest in 1841. The most civilized could hold meetings where "all was perfect order and silence." Gifts worked just as well. "I am now happy to tell you," wrote Stella Hull to relatives in Connecticut from New Smyrna, Florida, in 1804, "that its peaceble [sic] times in Petunxes Wigwam—the murmurs of the Indians have been silenced by the late distribution of their annual presents."[14]

Antebellum southerners fancied that many Indians had been quieted by the sound of the progress of white men. Traveling the North Carolina Appalachians in 1827, for example, Juliana Conner reflected that "doubtless there are many now who from experience could point to us the horrors of an Indian war and the sufferings and hardships attendant upon their settling in this region of mountains—yet now it is as calm and peacefull as if its echoes had never resounded to the War hoop—or the cries of suffering whites."[15]

For Native Americans, removal to areas new and unfamiliar and the introduction of the gun changed how they heard their world. Relocation altered their understanding of their sense of place. Seasonal time, in Indian as much as white culture, was partly heard. Chief Luther Standing Bear of the Sioux recalled that he was born in December 1868, "when the bark of the trees cracked." Extraction from this acoustic environment meant estrangement from place and identity—something slaveholders themselves understood and feared.[16]

If southern elites could quiet their frontier principally through physical removal of the aurally ambiguous Indian, the control of people within the Old South that could not be removed proved more challenging. Southern towns and the class dynamics they encouraged and the excesses of southern evangelical Christianity presented thorny challenges to the master class's efforts to preserve the integrity of southern slaveholding society.

Southern towns had positive aural marks, of course—sounds indicative of order, discipline, and southern progress. Echoes of the Middle Ages rang loudly in the urban South. Market hours in both fourteenth-century Europe and the antebellum South were delimited by the sound of the bell to prevent engrossing and forestalling by the unscrupulous before markets opened.[17] St. Augustine, for example, allowed market traders access to the city's market bell during the day but not before the ringing of the morning bell and not after 9:00 P.M. An 1853 ordinance explained that the city marshal left the bell rope accessible during the day so that "after the Ringing of said Bell in the morning for Market . . . persons Coming from the country or others having articles of provisions for Sale, out of the Market hour, may by ringing said Bell, (which they are hereby authorized to do) give notice thereof to the public." St. Augustine's market must have been loud at times, but for all the volume, such sounds were necessary for coordinating city life and preserving urban order.[18]

Where southern elites faced their most serious challenge to the preservation of the southern soundscape was in the activities of urban lower orders. Ideally the progressive urban southern soundscape sounded industrious, free from plebeian noise. One such place was Baltimore, at least according to southern Federalist James Stuart in 1814. "This place," he thought, "is well situated for trade, although the heavy ships cannot come above Fells point, still there is a most beautiful basin above, for smaller vessels." "What is called Fells point and Baltimore," he continued, "are completely united, and all the bustle of the Shiping being at this place, free's [*sic*] the City from the riots of Sailors, and the noise of drays."[19]

Ideals proved elusive. For the most part, towns were inherently noisy places. It was in southern cities that the master class heard rioting and mob activity that sounded not unlike the cacophony emanating from northern conurbations. Robert Raymond Reid said of Tallahassee in 1833, "How far preferable is St. Augustine to Tallahassee!" Explanation followed exclamation: "The latter place is full of filth—*of all genders.* I never knew such dirty houses and indifferent people. . . . My time passes in Tallahassee very unpleasantly—passed at the public house where noise and dirt prevail to a disgusting degree." William J. Grayson agreed. "The negroes from the country," he recalled of Sundays in South Carolina, "assembled in town and broils were common among them. It was not much better with the lower class of white." By the mid-1850s the turmoil and strife associated with northern cities could be heard clearly in several of the border South's largest towns. Urban areas were like "a jewsharp, two thirds brass and the other tongue. The brass is glittering, the tongue noisy." While southerners liked the idea of progress as represented in towns and cities, the noisy tongues that accompanied urbanization sometimes caused the South's master class to question the wisdom of urban development, especially along northern lines.[20]

Thomas R. R. Cobb's first Sabbath in New Orleans, for example, was aurally sour. His experience with the city was sufficient for Cobb to desouthernize New Orleans and lump it with the noise-ridden city in the abstract: "Oh! How thankful the children of Athens [Georgia] should be that they are not subject to the temptations which surround the children of this & other large cities." Rev. William McCormick, a Presbyterian minister in Micanopy, Florida, apparently had a hard night on September 8, 1860. "My poor head reels & every thing seems moving," he wrote, continuing, "My heart is too full of grief because of the abounding wickedness here. Last night although excessively fatigued I could not sleep because of the noise & disturbance in the streets." Even small southern towns had their share of the wickedness that caused sleep deprivation.[21]

Custodians of the urban South tried to keep noise to a minimum. "It shall be the duty of all drivers of public vehicles when employed in conveying passengers," mandated a Mobile, Alabama, statute in 1859, "to conduct themselves in a quiet, peaceable, and orderly manner." Public houses, common sources of plebeian noise, were similarly regulated. Publicans were advised "to prevent all persons who may be there after ten o'clock at night, from disturbing by cries, noise, songs, or otherwise the peace and tranquillity of any of the neighbors." Similarly, theater patrons engaged in "loud and boisterous talking, whistling, swearing, or hallooing" were ejected.[22]

Officials in antebellum southern cities crafted ordinances designed to quiet their laboring populations. Patrols were charged with preserving at night "the peace and good order of the city," which was threatened by "whites and blacks." Day and night, police preserved "peace, quiet, and good order," and authorities did not hesitate to arrest blacks or whites for breaches of the peace. "A white man" was committed for two days to a Pensacola, Florida, jail in 1844 "for being noisy in the streets." He was arrested by the "constable" whose "duty is to look after the peace and quiet of the city . . . to ring the city bell on all proper occasions . . . and to take up all slaves found in the streets without a pass after the bell has been rung."[23] One of St. Augustine's earliest antebellum ordinances was directed expressly toward quieting the city's "People of Colour." In 1822 city authorities declared "that whenever any noise or disturbance shall take place at any meeting of people of Colour within the limits of the city," anyone "feeling aggrieved & disturbed" could call on the city marshal or "other peace officer" whose duty "it shall be, on being satisfied of the noise and disturbance" to arrest and arraign offenders. The offender would receive "not more than forty Lashes and [a fine] not exceeding Ten

Dollars." Although leaving the definition of noise up to the arresting officer, the statute prohibited people "of Colour" from holding any meeting after midnight, suggesting that although the city's blacks were likely to be considered noisy at any time, the sounds they made after midnight were considered illegal.[24]

The heightened rigor and precision of St. Augustine's later ordinances suggests that the 1822 statute failed to keep the city as quiet and as orderly as officials hoped. Free blacks and slaves were apparently the cause. In 1845 the city council added muscle to the 1822 statute. Again it was directed toward "any negro or Mulatto, bond or free," and required that such persons carry permission slips if absent from "his or her proper residence . . . after the hour of nine o'clock P.M. from the first day of September to the first day of May, and after the hour of ten o'clock, P.M. from the first day of May to the first day of September." This time the legislation spoke explicitly to the problem of slaves and free blacks generating noise through recreation. Playing "at Cards, dice or any other game of chance for Money or other wise" caused the city's black population to "assemble themselves together," which led them to "conduct themselves in an improper" and "noisy riotous or tumultuous manner." Should such noise be "to the annoyance of the Citizens," "any such negro or Mulatto" was subject to an unspecified "fine and imprisonment or whipping." But parts of this statute also applied to anyone, whites included, who threatened the "public peace" by keeping "a disorderly house" and selling "intoxicating liquors." Urban lower orders were admonished to keep quiet, even during their ostensible leisure time.[25]

Slaveholders were not foolish enough to howl at the moon. Their own urban soundscapes, they understood, were acoustically dense and muddled because trade, commerce, and economic and social coordination demanded and generated public sounds. But masters were as careful as possible to prevent the sounds of commerce, religious worship, and social coordination from being appropriated by slaves and the laboring classes generally. Woe betide those who used legitimate sounds for devious purposes. No one, stipulated an 1859 Mobile, Alabama, ordinance, "shall ring any bell within the city of Mobile, larger than an ordinary hand bell, except it be church bells, tavern, shipping, or steamboat bells, and bells giving an alarm of fire, insurrection or riot, and the market bell, under penalty of fine of ten dollars; and if the offender be a slave, he shall receive not exceeding twenty lashes, unless the person having control of him will pay the fine." At work here was an attempt to control the use of particular sounds, and anyone heard imitating "any of the signs, signals, devices adopted and used by the city watch" was subject to punishment. Such aural larceny carried a hefty fine of $50.[26]

Here we must distinguish between noises that posed long-term threats to the abstracted social order of the Old South and sounds that posed immediate dangers. Mobile's officials, for example, fretted about the short-term impact of the appropriation and mimicking of civil sounds. The city's entire warning system depended on sound. "Alarm sentinels," mandated an 1859 statute, "shall be at all times on the bell tower, to watch and observe the city. He shall cry the hour of the night, at each half hour through the night, till daylight." Spatial orientation was aural: "Whenever there is any alarm of fire, he shall strike the bell to indicate the alarm for all parts of the city south of Dauphin and east of Franklin, with one tap. All south of Dauphin and west of Franklin with two taps. All north of Dauphin and west of Franklin street with three taps. All north of Dauphin and east of Franklin street with four taps." "For riots or civil commotion," the statute continued, "rebellion or insurrection, double the number." Should sentinels remain confused, city officials provided them with a diagram indicating the number of taps for respective compass points.[27]

The preservation of an orderly aural urban soundscape had much to do with ensuring elite white safety. In Montgomery, Alabama, in 1850, "Every evening, at nine o'clock, a great bell, or curfew, tolls in the market-place . . . after which no coloured man is permitted abroad without a pass."[28] An 1845 Georgia statute wisely prohibited slaves from using "drums, horns, or other loud instruments, which may call together or give sign or notice to one another of their wicked designs and intentions." Drum beating, after all, could act as a coordinating device among insurrectionary slaves and also could serve to confuse unsuspecting whites who associated drums and bells with fires.[29] Masters also attempted to regulate the sounds made by urban slaves in part because of their belief that slaves' unregulated expression sounded too much like democracy. Frederick Law Olmsted noted the equation after witnessing a particularly loud slave service in New Orleans in the 1850s. He heard the "shouts, and groans, terrific shrieks" and offered a comparison that struck fear into the hearts of the master class: "The tumult often resembled that of an excited political meeting."[30]

In addition to legislation, urban authorities and slaveholders had other options available to them to quiet public noise, or at least to protect themselves from it. One possibility was their use of sound-absorbing material.

Some noises were absorbed almost by accident. That southern streets were often unpaved probably helped muffle some urban sounds. At other times the choice of building materials seemed deliberate, for in addition to the pleasant aesthetic of brick homes, wallpaper, and carpets, such materials, while obviously insulating homes against cold, also helped protect internal spaces (and those who inhabited them) from external noise. Moreover, the use of drapes and textiles in parlors and heavy "door and window curtains [that] muffled sound, softened light, and enhanced domestic quiet" were also used in expensive antebellum homes and hotels in the 1850s and probably earlier. Elite and, increasingly, middle-class households were more insulated against public noise than houses inhabited by slaves and workers. The cumulative effect of the southern elite's choice of drapes, carpets, and textile furnishings suggests that they insulated their homes acoustically as well as for aesthetic reasons.[31]

The wealthiest slaveholders had additional options. As long as planters could escape to the tranquility of the country, they put up with the bustle of the city as a necessary evil. As John Basil Lamar wrote to his sister from his Georgia plantation in 1835, "I took your letter from the office in Macon yesterday, & should have answer'd it forthwith, but preferring the calm of Swift Creek to the boisterous crowd of the Central hotel, for that purpose, I deffer'd it until now." James Henry Hammond of South Carolina was offered a summer residence for sale in 1833 with quietude in mind. Situated near Barnwell Court House, the house was peddled as "sufficiently near the village for all business purposes, and far enough to be retired from the noise and bustle of public times." Planters, not unlike elites of ancient times, bought quietude.[32]

Southern elites defined their soundscapes in large part through the sounds of piety, which they related intimately to their sense of community. Christianity's bells helped define communities and established the parish as an acoustic and spiritual space. Church bells unified communities socially, drew man and God together, and repelled evil spirits. Bells in colonial and antebellum America generally did all of these things and more. The sound of the church bell served as an acoustic calendar, announcing holy dates and feasts; it was also the courier of clock time. Although the bell drew people together and united man with God, it also regulated bodies. The church bell was also the proxy for civilization, which explains why missionaries used it to delimit acoustically "the civilization of the parish from the wilderness beyond its earshot."[33] Most of all, though, the sound of God's bells helped delimit southerners' sense of place, history, and social values. Consequently, challenges to the elite's conception of what true godliness sounded like were also threats to southern slaveholding society.

Travelers littered their journals and dairies with references to bells pealing and tolling in southern hamlets and cities. Indeed church bells, which served both secular and religious functions, were among the first and most important items placed in churches along the antebellum frontier. Good religion was rooted physically in space because the sound of the bell delimited the religious sphere of worship. Bells called God-fearing congregants to church, and once inside, they were often bade to listen and sit in silence. This was social order of the highest degree, an order worthy of God's presence.[34]

While bells disciplined the bodies of virtually every antebellum American, some had more aural leeway than others. Workers who ignored first bells were deemed troublesome and resisting; parishioners who came after the first ringing of God's bells were tardy. But should a southern man of influence such as the Reverend William McCormick go "to church when the second bell rung & found a large congregation out anticipating," then this was a sign of authority itself. Only the most powerful could ignore first bells.[35]

Southern elites who prescribed this acoustic theology viewed other God-fearing southerners (as well as foreigners) as precisely that: others. Mainstream, especially genteel southern Protestants often heard what they considered the quirks of other religions. When Virginia's Elizabeth Ruffin first attended a Roman Catholic service in Philadelphia in 1827, she remarked, "Strange does it seem to me and not only strange but nonsensical . . . such a repeating of prayers, crossing themselves, ringing of bells under the tail of the Priest's robe." Such was the case with other religions. Ruffin described the singing of "Shaking-Quakers" as "strong, loud, discordant, without melody."[36]

There was an ambiguity in how southern elites heard evangelicalism. Evangelical religion simultaneously challenged and reinforced the discipline of the self in the Old South. On one hand, aspects of evangelicalism rejected "reason and restrained sensibilities" in favor of "the democracy of emotion," which narrowed the social and spiritual distance between audience and preacher and, in the process, evoked "tearful, passionate outbursts." Such clamor was dangerously like that of slaves and Indians who could not control themselves. Antebellum South Carolina planters contrasted the "calm and seriousness" of Episcopalianism with the "wild enthusiasm," the "contortions of body," the "ejaculations of the congregation," and "frantic shouts" typical of Methodist and Baptist churches. Such religious noise reaffirmed some planters' suspicions that even yeomen were emotionally suspect, since they lacked the discipline of quiet gentility. Evangelicals in the 1830s, recalled William Grayson, were consumed with "the contagion of excited emotion or passion, such as sometimes maddens a mob." But because evangelical Christianity also stressed the values of inner strength and self-control, genteel elites also heard evangelicalism positively.[37]

The sounds and noises of the Second Great Awakening reaffirmed a tradition that placed an importance on the aural in religious expression and communication in the belief that by "making an impressive noise, man hoped in his turn to catch the ear of God."[38] The aural dimension of religious practice, especially though not exclusively in the Christian world, had never been lost. Bells communicated God's time and his spatial influence, and antebellum revivalism invigorated the aural tradition of religious worship. In this respect revivalists represented both a continuity—insofar as they amplified the aural importance of religion—and a change, since their mobile circuits and meetings tended to prohibit them from using large, heavy bells. The Second Great Awakening in the United States was partly a resurgence in sonic theology and represented a renewed desire among many Americans to hear God and for him to hear them.[39]

Some evangelical preachers resurrected the emphasis on sacred sounds by appealing to the spirit through the ears, not the intellect, of what they often called "hearers." Circuit riders, such as Henry Bryson in 1826, were delighted when "the house was crowded with hearers," for it suggested that conversions would follow.[40] Less hostage than genteel Protestants to scriptural interpretations based on the printed word, evangelicals of all stripes were especially attuned to aurality because they believed sinners could be converted if they heard correctly and listened attentively. "Popular preachers," those "disdainful of technical theology," "indifferent to cultural refinement, and preoccupied with the stimulation of emotional excitement," were common among Methodist and Baptist clergy in the early antebellum South. They suggested that people were passive hearers and that to get their attention and save their souls, one had to turn them into active listeners. Hence shouts by preachers were entirely appropriate and even necessary to convert the spiritually deaf. Methodist circuit rider John Early surely discriminated between mere hearers and attentive listeners during his travels in early national Virginia and North Carolina. On May 27, 1807, for example, he hinted at the aural passivity of his audience: "I spoke to some hearers that night." In September Early encountered still more calloused ears: "I spoke at the accepted time and day of salvation but hard hearers and not much religion myself. Two joined the society." Good listeners were not necessarily those who sat attentively but those who joined Early in his audible praise of an aural and oral God. Early believed *"Blessed are the people that know the joyful sound"* and measured his conversion success in terms of what he heard and its volume. This is why he sometimes referred to a "crowd of listening hearers" who actively heard the word of God. In June 1807 he reported in his dairy, "I preached and had a happy time. Some got powerfully converted and cried for mercy aloud." Early took enormous satisfaction when he heard his communicants proclaim devotion to the Methodist church. He was exhilarated when his "Christians shouted aloud for joy. Sinners trembled and rolled in the dirt." "Presbyterians," by contrast, "stood and looked," no doubt bemused. True conversion success was measured both in the aural abstract and concrete cacophony. "I desire," confessed Early in 1808, "to make the earth ring with loud Hosannahs to King Jesus." "Though I spoke very loud," he wrote after a service in 1811, "yet the people drowned my voice with their shouts and cries."[41]

For men like Early, a successful camp meeting was heard as much as seen. Using "awful," by which he meant awe-inspiring, he declared in 1807: "Oh what great things my God can do. A great institution, that of camp meetings. An awful sight to hear preachers preaching and prayers and singing, crying and shouting in every direction." What genteel southerners of all denominations heard as a lack of self discipline, Early heard as testimony to the efficacy of God's word, delighting in "the cries that were heard night and day."[42]

Early, as mere man, could not force his congregation to hear, convert, and repent. Only God could open their ears, but he sometimes spoke through Early to do so. In September 1807 Early "spoke on the poverty of Christ and the riches of the Christian and the Lord (blessed) me in speaking and the people in hearing." Early's mortality did not stop him from trying to create a soundscape appropriate to religious conversion. In July 1808 he commented that while preaching he "hollered very loud (as I commonly did) and stamped with my foot." It worked: "The people appeared to hear as for eternity. . . . Some [sinners] poured their complaints aloud into the bossom of a complaisant God. He heard the cry of two and gave them oil of joy." Early was satisfied when sinners "shouted aloud" because "I long to see and hear sinners crying in every direction for mercy." Sporadic shouting and ranting, provided they occurred in a context ordained by God, were sounds pleasant to the ear for this man. He was especially gratified when he encouraged a "shouting time" on his circuit. When he inspired listeners so that they "shouted and jumped and jerked and rolled and for three or four hours my eyes or ears hardly ever saw or heard such a time among Christians," Early was ecstatic.[43] In short, evangelical circuit preachers believed that passion in expression was a sure way to worship God and an even more certain way for them to ensure that he heard their entreaties. Loudness at revivals indicated that "the work of the Lord seems to be prospering." True godliness was marked by "shouting and praising God aloud" and being unable to "refrain from shouting aloud." Vented spiritual passions were good; the devil, by contrast, "whispered."[44]

Class and race complicated genteel southerners' evaluations of evangelical Christianity. William D. Valentine of North Carolina thought about such matters at some length in 1851. A Methodist, Valentine believed that the "true religion is of the heart," which was "persistent, loving and calm." "Many persons find fault with praying and shouting aloud as no religion in it," Valentine observed, declaring, "and I confess I am of the number." That lower orders generally, black and white, embraced a noisy religion hardly surprised him. Such people, after all, were "unable to restrain themselves," but at least their shouting and carrying on was sincere and "of the heart." Valentine thought southern blacks were particularly susceptible to evangelical passion. In November 1851, for example, he visited "Bethel Church (Methodist). . . . Went into the house and saw it full of negroes." This was a new experience for Valentine, and the remainder of his observations, visual and aural, are worth hearing: "Took a seat in a corner. . . . The negroes coming in more and more, the house was literally crowded. Mr Tom Wright Hayse preached a sermon to the negroes. It was suitable and well adapted to them. His conclusion roused their religious feelings. He touched the cords of the heart of hope, which spontaneously vibrated. . . . Such fierce, rough struggle of gesticulation and shouting I never before encountered on so large a scale." Valentine believed that the minister, who at the beginning of the service "was calm and guarded for the purpose of imposing decorum and propriety . . . and restraining them from unnecessary boisterous shouting and noise," upped the volume of the gathering. Valentine and men like him expected slaves and lower classes generally to behave in a noisy, passionate fashion in church. While not necessarily condoning such behavior, elites nevertheless thought they understood and tolerated it, as would any good patriarch. In Valentine's opinion, "The humble, illiterate, ignorant class should be allowed to enjoy religion in their own way lest check be put to true religion." When Valentine wrote "class," he meant it: "The difference between the negroes and illiterate ignorant whites in this respect is simply in the degrees of refinement and their respective natures." Although he thought that "the negroe has a genius for music" and "sends through the heart a religious music," Valentine associated noisy evangelism with the South's lower orders. But as long as the "preaching and expounding was evangelical and earnest," the people "serious and earnest," then "noisy excitement" was permitted. Occasional shouts were a small price to pay for such a potent opiate if, in fact, that is what it was.[45]

Slaveholders remained ambivalent about the impact of evangelical Christianity on their slaves. With good reason, some were concerned that proselytizing slaves in the evangelical vein bestowed on them an undue agency and subjectivity. Especially worrisome in the context of the 1830s when the northern agitation against slavery was beginning to take on ominous tones was the fear that slaves might put their new religion to subversive uses and listen more closely to the revolutionary implications of religious teachings while turning a deaf ear to its disciplinary tendencies. The unfettered vocal and aural expression that evangelical Christianity encouraged confirmed masters' fears by making their own admonitions for disciplined quietude seemingly subordinate to a deity who enjoyed immense authority among the enslaved. Their God seemed to encourage rather than condemn loud emotionalism. This explains why many masters sought to circumscribe black religious gatherings. The Presbyterian Reverend Charles Colcock Jones spoke for many slaveholders when he counseled that the "public worship of God should be conducted with reverence and stillness on the part of the congregation," which meant no "exclamations, or responses, or noises, or outcries of any kind during the progress of divine worship; nor boisterous singing immediately at its close."[46]

In addition to saving their souls, there were other reasons to proselytize slaves, even if doing so generated nervousness on the part of masters and mainstream Protestants. Though proselytization did invite resistance and a sense of egalitarianism, the formalization of slaves' religious instruction allowed masters to manage black noise and spiritual expression to some extent. This they did by "forcing black religious practices into public," overseeing African American services, and in effect, institutionalizing the teaching and expression of black religion.[47]

Sober evangelical worship was, in fact, to be encouraged. In 1832 William Meade, the third Protestant Episcopal bishop of Virginia, advised a colleague that evangelicalism was not necessarily bad. Writing in May 1832 Meade counseled, "I could wish indeed that some of our brethren who are much afraid of the tendency of these frequent meetings . . . had been present with us. . . . They would indeed have seen overflowing assemblies four times in each & heard animated sermons & exhortations, & the responses of the Liturgy from hundreds of voices." But, he maintained, "they would have witnessed nothing bordering on enthusiasm . . . no anxious seats no calling." Sober evangelical reason reinforced faith, he suggested, and was worthwhile.[48]

On balance, excessively passionate evangelicalism worried southern elites less for spiritual and more for secular reasons. Although genteel clergy agreed with the notion that "were man silent, God would not want praise," while their God-fearing nature sometimes encouraged them to agree with the admonition of Paul the Apostle that "faith cometh by hearing" (Rom. 10:17), and although they believed that "man's soul is audible, not visible," unrestrained revivalistic fervor worried genteel clergy and slaveholders not least because they believed that what they heard on such occasions was a proxy for liberal social relations.[49] Gentlemen theologians often scorned southern revivals and translated what they heard at camp meetings into aural indictments of the lower orders' tendency toward passionate excess. Genteel clergymen of all denominations fancied themselves as self-conscious guardians of a rational orthodox religious tradition that stressed the desirability of social hierarchy and disdained the passionate excesses of the multitude. They disliked the democratic rhetoric and fervor of unschooled circuit preachers and endeavored to restore reasoned theological discourse by relocating their denominations, Baptists and Methodists included, to the South's cities. There these learned men essayed to promote calm reason from the pulpit instead of the "noise and blunder" the evangelical laity had come to expect. To realize this reform of spirit and social discipline, the urban clergy appealed to the ear. Without educated ministers, observed a Georgia Baptist in 1832, "the number and respectability of our hearers must and will continue to decrease." Distrustful of the rhetorical flourishes of popular preachers and critical that such men "never tax their hearers with the drudgery of thinking," during the 1830s in particular the gentlemen theologians began to challenge growing revivalism and the threat to order it represented by adding rationality and discipline to emotional worship and thereby "maintained the tone of an upper-class defending itself against levelling tendencies." Controversies naturally followed, and "Primitive Baptists" and others attacked the "chaste and classic" sermons as elitist and intended "to please . . . refined ears." Orthodox clergymen nevertheless made considerable headway in taming the excesses of evangelical passion from the 1830s on.[50]

Juliana Conner's remarks after a camp meeting in North Carolina in 1827 offer insight into the genteel mentality. "A feeling of devotion I will candidly confess, was not my motive for attending to the Preacher, the novelty of the scene distracted my mind too much to admit of it—yet I listened to learn the merits of that style which had such powerful effects on the passions," she observed. What she heard was the noise of the vulgar and chronically plebeian: "It certainly exceeded anything I ever heard and if there was any regular gradation by which *nonsense* could be measured this (I think) would have reached the highest degree." The sound of evangelism was the noise of the ill-disciplined to Conner and the noise she and other genteel southerners associated with the masses. Of the preacher she sneered,

> There was neither connection, good sense, or even pertinent remarks, his sole aim, end and object was to make the people groan, and the intonation, rise and fall of his voice and almost crying manner produced what words alone had failed to effect—they were called to the altar to be prayed for—and such a confusion of tongues that Babel only could have equalled it—I had various fears for their lungs if not for their "souls"—and I was glad to escape from the crowd . . . my curiosity sufficiently gratified and my pity . . . strongly excited at witnessing such fanaticism.

"My sense of hearing," she lamented, "was nearly destroyed" by the "ranting bombastic speaker," and she and her husband left the "supposed place of worship" and "returned home with a feeling of satisfaction at once more being restored to quiet." Feverish evangelicalism, in other words, injected the noise of the masses into the elite's quiet South. While Conner admitted that "these meetings doubtless have a beneficial influence of minds of a certain cast," that class was not made up of "well informed, intelligent people."[51] These were not theological arguments about the appropriateness of particular religious sounds made by particular denominations. Rather, they were social in nature and exposed the elite's suspicions that the sounds of evangelical fervor were but a few tones and decibels removed from the shout of a mob that seemed to be gaining ground in the society to the north.

There is probably no clearer statement of southern conservatives' fear of the passionate masses and their threat to sober democracy than the Reverend James Boyce's 1848 speech concerning South Carolina's "bill to confer upon the People the power to chose electors" in the state legislature. Boyce, a Baptist theologian, resisted even this modest extension of democracy in terms distinctly aural. He reasoned, "If . . . you make this a popular election, you inflict upon us, a people, the turmoil, the passion, the convulsion, with all the demoralizing effects arising therefrom. . . . And for what purpose? . . . Will the People, after all the turmoil, the noise, the distraction, the ill-feeling, excited by such contests, [be] any better off? . . . Shall we, therefore, force upon the People an excited, angry, tumultuous national election every fourth year, for the benefit of a few aspiring demagogues? I hope not." This was as much a defense of the southern political status quo as it was a scathing critique of northern democratic excess. Boyce went on: "We may introduce into the Government of the State a more Democratic spirit than we imagine. This is the age of progress; the public mind in the United States is verging to extremes in questions of government and social order. We see it in the ultra Democracy of the Northern States. Their Agrarianism, their Anti-Rentism, and, above all else, in their Abolitionism. In the free States we see them constantly destroying the conservative elements in the Constitution, giving up the ideas of Republicanism." Enact the legislation, he maintained, but be warned: "The voice of the multitude will be heard louder and louder for change. They will tear from your Constitution its conservative elements; they will, before many years, be thundering at the doors of the Senate Chamber for equality in representation. . . . We will no longer be a Republic; we will be a Democracy—a tumultuous, numerical Democracy!" Should they doubt it, Boyce asked his readers and reading hearers to listen more closely: "Already we hear the rumbling of distant thunder. . . . The ground is moving under us. . . . Shall we, at this moment of all others, choose to descend to a low, tumultuous, and deranging scramble for the spoils and patronage of the Federal Government[?]" No, for "this is the South Carolina Democracy!" To "allow the people a voice in the choice of the President of the United States, is to debauch them, to inflame them, to infect them with ultra Democracy, with the extreme spirit of progress, to nationalize them, to imbue them with Federal feelings." Harmony was kept through political silence, and the true custodians of the southern soundscape and all that it stood for were the elite: "Let the poor, ignorant, imbecile people, sleep in peace."[52] Quietude was good for them, even if they did not know it. Silence, though, was rarely golden.

Those old grey-haired darkies and their noiseless,
automatic service, the result of finished training—one
does miss that sort of thing when away from home.
—Mary Boykin Chesnut, *Diary from Dixie,* November 28, 1863

If they want to kill us, then can do it when they please,
they are noiseless as panthers.
—Mary Boykin Chesnut, *Diary from Dixie,* October 7, 1861

(((3. Dreadful, Silent Moments

Slaveholders experienced great successes and terrible failures
in sculpting their preferred soundscape on their plantations. In a sense
they succeeded too well in imposing quietude on their aural estates. Slaves
did not so much buck slaveholders' demands for quietude as accommodate them by deploying silence as a powerful tool of resistance. To be sure,
the enslaved gained much sustenance and solace and managed to convey practical information through their shaping of the audible world with
song and prayer. Historians have "analyze[d] the sounds," of slavery, as
Frederick Douglass termed it, and shown that songs and chants were an
important form of cultural resistance to bondage, an indispensable way
for the oppressed to carve out inner and outer space through their manipulation of the audible world. Slaves' songs could slow the tempo of
work to minimize exploitation while inviting northern sympathizers to
hear the true suffering of the enslaved.[1]

Less recognized is the effect such sounds had on antebellum slaveholders and the lengths to which they went to regulate them. While mas-

ters were fairly adept at shaping slave soundscapes, the sounds made by bondpeople influenced masters' understanding of their world and sensitized them to the sounds emanating from the modernizing North. Slaves' noises were too similar to how masters imagined northern working classes sounded, and this helps explain why slaveholders attempted to regulate slave soundscapes with such precision. At base this was a political question devolving on slaveholders' identity and the future of their world. What masters heard on their plantations reaffirmed their core beliefs about the volatility of a free working class and the desirability of slavery. While masters took much solace in what they believed to be their quieter laboring classes, they were increasingly aware during and after the 1830s of a potential northern future when they heard their bondpeople.

Masters went to great lengths to discipline plantation soundscapes by insisting on quietude and trying to delimit slaves' sounds. The essence of plantation serenity was not silence but carefully regulated quietude. Slave songs were part and parcel of that imagined quietude. Sung in and at the right time, songs were active confirmation of slaveholders' soniferous gardens and metronomes of slave productivity. Sung at the right time and in the wrong place, slave songs sounded like excesses of passion. Herein lay the dreadful irony. For the most part antebellum planters got what they wanted from slaves: songs sung and sounds made at appropriate times in appropriate places. Many things went unheard, and many pieces of vital information were conveyed by slave songs. Masters probably recognized much of this, but on the whole they interpreted the aesthetic beauty of slave songs to demonstrate their bondpeoples' supposed happiness and the superiority of their own social order. The irony lay, however, less in what masters heard and more in what they failed to hear: slave silence. In slaves' silence, masters had the most to fear.[2]

Historians of southern slavery rarely recognize the role that silences— the inaudible and the unheard—played in slaves' resistance to bondage. In their silence, slaves resisted effectively and shaped masters' worldviews and the articulation of those views at the level of national political discourse. Slaveholders' fear of silence—engendered by slaves themselves— made masters what they were: skittish, nervous, worried, and aurally sensitive beneath a tough, confident exterior.

In the slaveholders' world of probabilities, slaves were more likely than most to make noise. Evidence suggests that some white southerners racialized the acoustic and constructed blacks as innately noisy, especially sensitive to the acoustemological environment, and prone to vent passions. Josiah C. Nott, that exemplar of biological determinism, suggested, not unlike some recent geneticists, that black people were predisposed to hear particularly well: "In animals where the senses and sensual faculties predominate, the nerves coming off from the brain are large, and we find the nerves of the Negro larger than those of the Caucasian."[3] Samuel A. Cartwright of New Orleans went further and made the following points on the slimmest evidence: "The negro's hearing is better. . . . His imitative powers are very great." Black peoples' nerves connecting body and brain, he maintained, "minister to the senses" in a purely intuitive sense: "Thus, music is a mere sensual pleasure with the negro. There is nothing in his music addressing the understanding; it has melody, but no harmony; his songs are mere sounds, without sense or meaning—pleasing the ear, without conveying a single idea to the mind; his ear is gratified by sound, as his stomach is by food." Such were the curses of this "excess of organic nervous matter." Nevertheless, maintained Cartwright, the wise guidance of the master class enabled slaves to lead "a happy, quiet and contented life" and protected their sensitive ears by giving them an aurally tranquil environment.[4] Such fantasies belied consternation. The suggestion that blacks were acutely attuned to the heard world had important implications for the way southern whites behaved and sounded themselves. The belief, in effect, made an often confident master class apprehensive, forever regulating their own sound in the presence of blacks and always listening to telltale rustles from slaves who knew how to hear and listen better than they.

Particularly exuberant slaves embarrassed planters, especially when they shouted at church. Anxious to repress his slave's noise, one master promised him a new pair of boots if he kept quiet. Try as he might, the slave could not contain the spirit, and the noise could not be suppressed. Doubtless this confirmed the master's view of ill-disciplined lower orders of which black slaves were exemplary. Robert Habersham's diary entry following his return to his Georgia plantation in 1832 suggests that the mentality was not uncommon. He spoke of his "triumphant arrival amid the shouts of Ethiop hosts as black as my boat."[5]

Planters' logic concerning the oral and aural proclivities of laboring peoples generally and black slaves particularly were confirmed when they listened to the workings of the slaves' economy, that sliver of wage labor that slaves insinuated into their mode of production. While a good deal of the slaves' economy was tolerated—not least because it seems to have saved slaveholders some money and because the enslaved themselves pushed hard for the inviolability of their own property—governing authorities worried much when they began to hear the logic of class conflict under wage labor in the trading activities of the enslaved. When competition over property and bad temper over debts erupted into a brawl in Anderson District, South Carolina, in 1846, the presiding magistrate and white witnesses heard the slaves' economy not as a harmonious exchange of goods and services between sellers and buyers but, rather, as a product of materialist greed and conflict within the slave community with possibly negative implications for relations between blacks and poor whites. Aggrieved slaves behaved in a "very disorderly manner in drunkenness and rioting fighting & Quarrelling," and earwitnesses heard tumult, "noise Cursing and Quarelling" in the ruptured skein of these small market transactions gone awry. These sounds, which could be heard within the bosom of the slave South itself, and the conflict that accompanied them added credibility to slaveholders' logic and bolstered their determination that burgeoning northern wage labor capitalism should remain muted and not be allowed to penetrate the South beyond what was an already disturbing tendency.[6]

Plantation mistresses also wondered whether African American religious shouts "may have been some latent instinct of the wild that caused such ebulitions." A more disturbing alternative was the idea that suppressed sound in the form of "muttering" was, some believed, "a habit many of the race have." Aware that particular slaves were apparently prone to "quarrelsomeness . . . & fighting and misbehaving in every way" and disposed to "getting drunk & raising a row on the lot," planters designated such disrupters of the plantation soundscape worthy of surveillance. But in their racialization of black noise, southern masters differed little from other Western elites. English-born architect Benjamin Henry Latrobe could hear only the "incredible noise," squalling women, and "an abominably loud noise" when he listened to slave singers and musicians in New Orleans in 1818. Before (and even after) the rise of abolitionism in the 1830s, many northerners agreed.[7] Some felt black people especially inclined toward oral expression. As New Yorker Henry Cathell commented while in Quincy, Florida, on Christmas Day 1851, "The Negros assembled in town to day, drinking, singing, dancing to their hearts content. . . . [They] cant stand still when the fidle begins. . . . None but the Ethiopian can do it."[8]

Slaveholders believed that some slaves were more likely to make noise than others. Mentally unstable bondpeople, such as Kate, a habitual runaway in Louisiana, were considered prone to vocal excess. In 1834 a New Orleans sheriff elaborated on Kate's case and testified that he had seen Kate "act as a crazy person—she holloed danced all night and could hardly answer when spoken to." Kate ignored his admonitions for quiet and so was deemed noisy and unsound. In a similar case, Melly was considered an idiot in 1841 because "she whistled after answering any question put to her by a white person." Conversely, when slaves feigned ignorance of what whites were saying by pretending to be hard of hearing, masters considered them either willfully ignorant or deaf.[9]

Because masters believed slaves were unable to regulate passions, however, they often turned a deaf ear and allowed weak-willed bondpeople to make noise. Masters sometimes let tumultuous moments pass unpunished because of their largesse. This was especially the case during annual holidays, Christmas particularly, when as Chief Justice Thomas Ruffin of North Carolina argued in 1849, masters were not responsible for their slaves' noises. These periodic sounds were customary and had long been a feature of the southern soundscape. Moreover, Ruffin contended that slaves' "simple music and roaming dances" kept them happy. Masters "may let them make the most of their idle hours, and may well make allowances for the noisy outpouring of glad hearts, which providence bestows as a blessing on corporeal vigor united to a vacant mind." There were still other twists to the stereotype. Ever enthusiastic to construct the image and preserve the reality of the serene plantation soundscape, some white southerners were apt to cast bondpeople as quiet by nature and not prone to riot. "Nearly all of them," believed Ruffin, "are the least turbulent of all men."[10]

When reality did not coincide with ideal, slaveholders brought to bear powerful tools for imposing quietude on the bodies and activities of plantation slaves. Racial stereotypes helped shape forms of punishment. Aural deprivation and stimulation were powerful weapons in the planter's arsenal. Slaves, they believed, "dread solitariness, and to be deprived from the weekly dances and chit-chat" was punishment enough. Conversely, dances with fiddling were a reward.[11] Slave sounds were both permitted and prohibited at designated times. The production of sound served as an incentive; its prevention, a punishment.

Some slaves were partly responsible for maintaining the quietude of plantation life. "Uncle Jim," for example, trained boys, slave and free, on the Avirett plantation to fish by advising them to "be very quiet." Slaves who disturbed the fishing efforts were advised, "why doan' yuh keep less noise?" As masters well knew, some slaves, especially those who hunted, possessed a "well-trained ear." Slave custodians of plantation quietude could even render whites tranquil. Nurses, for example, were allowed to frighten "us into silence when we were unduly noisy," as J. G. Clinkscales recalled of his plantation childhood. There were also contractual obligations of a sort. Southern law prescribed that whites keep their end of the bargain: quiet, peaceful slaves should not be punished by patrols. So said a Florida Territory act governing patrols in 1845: "If any white man shall beat or abuse any slave quietly and peaceable [*sic*] being in his or her master's plantation," he was fined. Beating bondpeople for being quiet would, in itself, produce noise and din inappropriate to a pastoral slave soundscape.[12]

Slaveholders were ultimately responsible for maintaining plantation serenity. Masters expected slaves to keep mum at their request. Slaves were not to discuss certain subjects, especially politics. Slaves heard democracy in the South, but they were not permitted to interrogate the meaning of its sounds. Louis Hughes recalled hearing "every white man we met . . . yelling, 'Hurrah for Polk and Dallas'" but noted, "We were afraid to ask them the reason for their yelling, as that would have been regarded as impertinence." Slaveholders also understood the potential ramifications of allowing slaves to see and hear an impending sale of kin. "Howls and screams and groans" alerted slaves that offspring were being rounded up for sale. Masters knew the dangers. The sounds of separation could wrench guts and induce desperate attempts to save family members, which is why one master insisted that "those black b——s" were put "out of sight and hearing." Sometimes, though, "mothers heard the wail of their children, and came running through the fields." Masters wisely and readily kept their mouths shut and accepted their own enforced silence as the price for perpetuating human bondage and their own authority.[13]

Should masters' injunctions for silence be disobeyed, slaves could find themselves subject to the whip. James W. C. Pennington's master exclaimed, "I am master of your tongue as well as of your time" after Pennington's willful refusal "to hush." The extent of the masters' power was clear, and it was not unknown for even freedpeople to get "a few lashes one day for whisperin'." "In the early part of June," snorted Charles Ball, "our shad, that each one had been used to receive, was withheld from us, and we no longer received any thing but the peck of corn and pint of vinegar." "This circumstance," he lamented, "in a community less severely disciplined than ours, might have procured murmurs." Instead his slave community greeted the breaking of the implicit social contract with silence. Such was the power of flogging.[14]

Deliberate silence could mean the difference between a beating and its cessation. The experience of J. W. Loguen, or "Jarms," a slave who eventually escaped bondage, is instructive. In a drunken stupor his master, Manasseth, beat him. Regaining consciousness, Loguen began to moan within his earshot, polluting Manasseth's aural space and peace. His mother was there to teach him the importance of aural strategy. "Hush!" whispered his mother; "don't groan!—your groans will make him mad, and he will come and kill you! . . . For my sake—for the sake of your poor mother—don't make a noise—don't bring him on again." His mother tried to help, keeping eyes on master, ears on son. While the slaveholder "was reading his Bible lesson, her ear watched intently in the direction of Jarms, to catch the least sound that might proceed from him." Lest unintentional moans disrupt Manasseth, his mother, "desirous to imprison his aching breath, had prudently closed the door upon him." But pain has a way of making noise, and Jarms could not help himself. "Manasseth had but begun his form of prayer, amid the most perfect stillness—when Cherry [Jarms's mother] fancied a sound floating on the brooding silence and indistinguishable therefrom, which awakened her auricular nerves to painful intenseness. Again, and again, and again, the sound came at intervals, with increasing distinctness, until it was certain it entered the ears of the praying man." Disturbing a praying master with noise had a predictable price: "'I'll make you grunt for something, you black devil.'"[15]

Other masters went much further. Charlie Johnson offered chilling testimony to slaveholders' desire to control the bodies of slaves: "En de on reliable ones, dey'd bell'um—Tempie en Snip wore a bell—en make um work in de fiel's so they couldn't run away en de overseer could keep up wid where dey was at. Dey'd de a iron band roun' de waist, en another iron band roun' de neck, en a thing stick up de back, way up, wid de bell on hit what dey couldn't git hit off—warn't no way ter reach hit."[16]

Another master attempted to control the behavior of his slaves when they were beyond his field of vision by designing a contraption that clamped onto a slave's body. About "five feet over their heads he hung a brass bell on the top where they can't reach it." The master warned, "Don't let me hear that bell leavin' this place, or God have mercy on your black hides." "To prevent my running any more," recalled John Brown of his Georgia master, "Stevens fixed bells and horns on my head. This is not by any means an uncommon punishment. I have seen many slaves wearing

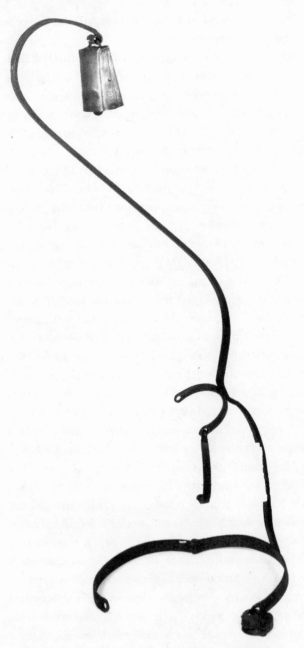

Belling slave bodies (Historical Society of Pennsylvania, Slave Harness, accession no. 1000.1214)

them." Weighing "not . . . less than twelve to fourteen pounds," the bells restricted and betrayed bodily movement.[17]

Alert to the aural betrayal of their physical acts, slaves muted bell clappers with mud. But slaveholders preempted this, too. One day, "three or four miles from home" collecting turtle eggs in a swamp, slave Charles Ball "heard the sound of bells, similar to those which wagoners place on the shoulders of their horses." Such sounds put Ball on the defensive: "At first, the noise of bells of this kind, in a place where they were so unexpected, alarmed me, as I could not imagine who or what it was that was causing these bells to ring. . . . I thought the bells were moving in the woods, and coming toward me." He crouched down "and at length I heard footsteps distinctly in the leaves." From the undergrowth emerged a slave and "elevated three feet above his hair" was an arch "beneath the top of which were suspended the bells, three in number." Here was the fabled device used by masters to control fractious slaves—the contraption that betrayed betrayal as masters heard it. This one was worse than most, for the three, presumably enclosed, bells could not be muted with mud. The poor wretch found that the bells denied him survival. "He had not been able to take any of the turtles in the laying season," explained Ball, "because the noise of his bells frightened them, and they always escaped to the water before he could catch them." Moreover, he "had been afraid to travel much in the middle of the day, lest the sound of his bells should be heard by some one." Try though he did to save him, Ball later found the man's body strung from a tree. Death by bell was the result, but nothing tolled to mark it.[18]

Mistresses could also be callous. Following a severe beating, Charles Ball writhed in pain, and his mistress "bade me not to groan so loud, nor make so much noise." Punishment itself produced mewling, and as such, it had to be reconstituted as necessary noise by aurally sensitive masters. It was a prospect utterly intolerable to free wage labor republicans. Take, for example, Ball's description of the punishment of some captured South Carolina runaways: "The wretches were unanimously sentenced to be stripped naked, and bound down securely upon their backs, on the naked earth, in sight of each other; to have their mouths closely covered with bandages, to prevent them from making a noise to frighten away the birds, and in this manner to be left to be devoured alive by the carrion crows and buzzards, which swarm in every part of South Carolina." Such perversity carried its own cost: the disruption of the plantation soundscape. The victims' "groans and death-struggles were heard in the darkest recesses of the woods, amidst the flapping of the wings of vultures, the fluttering of carrion crows, and the dismal croaking of ravens." Moreover, "in the midst of this nocturnal din, the noise caused by the tearing of the flesh from bones . . . floated audibly upon the evening breeze." Sometimes the price to pay for long-term quietude was short-term aural horror. But masters paid it, of necessity.[19]

Some purported forms of aural control of slave bodies probably were not real. Although southern historic house museums sometimes claim the existence of "whistle walks" in antebellum plantation homes—a passage from the kitchen to the dining room where house hands carrying food "were once required to whistle, thereby making it impossible for them to sample the food"—little evidence exists to support such claims. Other forms of aural control were real, however. When dogs tracked runaways, their barks were an aural compass for catchers. Charlie Johnson recalled, "When de niggers run away, dey'd chase em wid dogs, en maybe dey'd be up a tree lack a possum, en de dogs bawling at de foot when de overseer come up." Masters' dogs betrayed, through barking and howling, illicit slave meetings and the hiding places of runaways.[20]

Slaveholders' efforts to impose safety and order on their plantations succeeded at a price because slaves understood the power of their own silence and learned to manage aspects of plantation soundscapes. Slaves heard better and listened more closely than masters, not so much because of their African legacies, which placed a premium on the aural world, or because of their putative genetic disposition to hear better, but because survival and escape were contingent on an acute appreciation of the southern plantation soundscape. Although their masters were a close second, the people most sensitive to the aural world in the Old South were slaves. For them the ability to control sound and silence could mean freedom.

Former slaves remembered the potency of plantation soundscapes. Some of these memories comported with those of plantation masters and mistresses. Ex-slave George Fleming of South Carolina, for instance, described the sound of plantation industry precisely: "de looms—boom! boom! sho could travel. . . . Dey peddled dem looms, zip! zap! . . . Hear dem looms booming all day long 'round de weaving shop." The same could be said of the seasonal sound of cotton gins. Like their masters, slaves also heard the sound of the market revolution. Berry Smith, for example, recalled jumping the first time he heard a train whistle. Former slaves also heard the plantation pastoral. In the "stillness of night" Charles Ball listened to rain. There "is a deep melancholy sound of the heavy drop as it meets the bosom of the wave in a dense forest at night, that revives in the memory the recollection of the days of other years, and fills the heart with sadness."[21]

Most memories of the sounds of slavery, though, were more sinister. Charley Williams recalled the sound of time and work on his plantation. "Bells and horns! Bells for this and horns for that! All we knowed was go and come by the bells and horns." Slaves remembered the horrors of slavery aurally. Lou Williams recalled that slave sales were marked by the "yellin' and a cryin'" of distraught family members and that she could hear whipped slaves "holler and holler 'til dey couldn't holler no mo' den dey jes' sorta grunt every lick 'til dey die." Witness, too, the reconstruction of slavery in the memory of Jacob Stroyer, so different from masters' fond nostalgia: "In closing this brief sketch of my experiences in the war, I would ask my readers to go back of the war a little with me. I want to show them a few of the dark pictures of the slave system. Hark! I hear the clanking of the ploughman's chains in the fields; I hear the tramping of the feet of the hoe-hands. I hear the coarse and harsh voice of the negro driver and the shrill voice of the white overseer swearing of the slaves. . . . I hear the fathers and mothers pour out their souls in prayer." Similarly, "noise of horns blown," barks of dogs, and shouting men were the sounds of the slave hunt that struck terror in the ears of runaways years after they had escaped.[22]

Resistance to the realities of plantation life took various forms. Some occurred in the mind with slaves constructing alternative definitions of what they considered sound and noise. For bondpeople, plantations were rarely quiet places of shaped serenity. Rather, "everything was in a bustle —always there was slashing and whipping. . . . It was awful to hear the cracking of that whip . . . so loud and sharp was the noise." "I have often laid and heard the crack of the whip, and the screams of the slave," remarked William W. Brown. He remembered in particular his mother's sounds, her shrieks carried to his ears with painful clarity by the plantation's soundscape: "I heard her voice, and knew it, and jumped out of my bunk, and went to the door. Though the field was some distance off from the house, I could hear every crack of the whip, and every groan and cry of my poor mother." Although the "sound of the whip ceased," echoes resounded for years. Charles Ball similarly remarked on the hypocrisy of southern masters' efforts to construe as genteel what was plainly noisy.

When "several persons are quarreling, brawling, making a great noise, or even fighting," noted Ball, "they say, *'the gentlemen are frolicking!'*" When slaves brawled, though, different aural aesthetic standards applied.[23]

Slaves interpreted the aural activities of their owners and judged them accordingly. When slave owners attended southern religious revivals, the aural purity of the events was debased by their hollow shouts and the ephemerality of their putative conversions: "But, alas, as a general truth, in a few weeks after the noise of the multitude, and the eloquence of the preachers died on the ears of the people, there remained on their memories and on the morals of society, no greater effect than was produced by a bygone thunder storm." Or as a group of slaves said on witnessing the revivalistic fervor of some slaveholders, "'I wonder if there is any religion in that noise? . . . Pshaw, no, . . . Religion is willing and doing good to others—this is only bawling.'"[24]

Slaves gained some solace from recognizing that masters responded to their ringing of plantation bells. One escaped slave recalled, "I used to try to see how near I could come to making it say, come to dinner." Bondpeople had their own affective aural worlds and attached precise and distinct meanings to some of the things they heard. Ex-slave Elsie Moreland, for example, considered hearing an owl screech, a dog howl, or a cow bellow after dark to mean death.[25] The essence of slave resistance lay in alternative constructions of what was noise and what was sound and, especially, in the manipulation of the physical world of sound.

By shaping the heard world of the plantation, slaves protected themselves. Screams when one was whipped or about to be sold, for example, reminded masters of slaves' humanity and so played on calloused ears to save delicate flesh. Hence slaves referred to receiving a "whooping" not least because they whooped their pain into the ears of masters. Inanimate objects, they told whip-happy masters, were dumb and silent. Both slave singing and shrieking reminded masters of bondpeople's humanity and their masters' excesses. As Frederick Douglass recalled, "If anyone wishes to be impressed with the soul-killing effects of slavery, let him go to Colonel Lloyd's plantation, and, on allowance day, place himself in the deep pine woods, and there let him, in silence, analyze the sounds that shall pass through the chambers of his soul." But if masters' ears remained insensitive to cries (as abolitionists liked to claim), some slaves fancied that the groans of bondage might reach the ears of others—if not friends to the north then perhaps a higher authority. As former slave Peter Randolph hoped, the day would come "when He shall say, 'I have heard the groanings of my people, and I will deliver them from the oppressor.'"[26] Vocalization of pain served multiple functions.

Slaves used not just singing and screaming but other audible devices to communicate. To contact his wife, for example, an escaped slave ran to her cabin "and gave his peculiar whistle under her window." Not unlike white women, female slaves developed strategies to avoid white men's wrath by limiting the noise of children. Laura Clark recalled that slave women on her Alabama plantation gave them candy "to keep us quiet."[27] Quietude is a difficult thing to impose, and slaves shaped the limits of their masters' paternalism. This applied to slaveholders' tolerance of slave noisiness, particularly at "hog killin' times," when as Susan Bradford Eppes explained, the slave "leader now and again would groan heavily and each groan called forth a louder, wilder, outcry from the participants." That slaves on her plantation turned an aural privilege into an aural right was clear to Eppes: "We were always glad when it ended and we have no pleasant recollections of 'hog killin' time.'"[28]

An equally effective tool of resistance for slaves was not making sound but, rather, learning how and when to be silent and thereby turning the very idea of the quiet bucolic life against their oppressors. Too much can be made of slave singing on plantations. Certainly the ring shout and various other forms of audible and spiritual expression were critical for sustaining slaves, lessening the terrible drudgery of labor in piercingly hot and humid fields, and enabling them to carve out a degree of autonomy under bondage. Yet the emphasis on the audibility of slave resistance tacitly undervalues the importance of silence to the continued viability of the slave community. It is well worth listening to Frederick Douglass on this matter for, as he remarked, "A still tongue makes a wise head." Nor was slave singing always as communally empowering as work on the subject suggests. Some African Americans found slave songs "wild" and failed to "understand the deep meaning of those rude and apparently incoherent songs."[29]

It is also easy to exaggerate the prevalence of slave singing on large southern plantations. Evidence from ex-slaves, for example, suggests that work was so hard, masters' insistence on quietude so strict, and their tools of enforcement sufficiently potent that singing was not as common as some historians imply. Easter Jones, ex-slave from Georgia, said of singing, "No'm, ain't sing nothin', too tired when dey git in fum wuk. We warn't 'lowed to make no noise nohow." Or as an Alabama former slave put it, "Dey didn't allow us to sing on our plantation 'cause if we did we just sing ourselves happy and get to shouting and dat would settle de work."[30] By the same token, songs could be coerced, with masters threatening to use the whip if slaves would not "sing and dance some mo'." Moreover, tight, usually temporal strictures were placed on the slaves' aural world and their contribution to it: "Den we sang all de songs dat we could think of, till nine o'clock, when Massa rang de bell, for all de noise to be cut out, soas all de white folks could go to sleep. Den we had to go to bed, or be awful quiet."[31]

Effective resistance to slavery resided in both the singing of songs (important information could be conveyed in words) and slaves' ability to control their own sounds. The ways in which the enslaved used sound and silence to resist slavery varied enormously, and some times were quieter than others. Following the Nat Turner massacre, silence was a slave's best friend. This did not mean that slaves abandoned aural practices that were important to them; rather, it meant they altered them. As one slave explained after the Turner insurrection, "On Sundays, I have seen the negroes in the country going away under large oakes, and in secret places, sitting in the woods with spelling book. . . . All the colored folks were afraid to pray in the time of the old Prophet Nat. There was no law about it; but the whites reported it round among themselves that, if a note was heard, we should have some dreadful punishment."[32]

Plainly, slaves thought it important to shout ("The Lawd done said you gotta shout if you want to be saved. That's in the Bible," recalled Elizabeth Ross Hite of Louisiana). Equally clear, though, they understood that a quiet shout was better than no shout at all, and they took steps to preserve worship by muting volume. Former slave Becky Ilsey recalled, "we'd have a meetin' at night, wuz mos' always 'way in de woods or de bushes some whar so de white folks couldn't hear, an' when dey'd sing a spiritual an' de spirit 'gin to shout some de elders would go 'mongst de folks an' put dey han' over dey mouf an' some times put a clof in dey mouf an say: 'Spirit don talk so loud or de patterol break us up.'" A variety of techniques were used to muffle and distort the sounds of the shouts. Some slaves prayed directly into dirt in an effort to let sound out and keep it from going anywhere but into the ground. Others hung wet rags and quilts and prayed behind them "to keep the sound of their voices from penetrating the air." Quiet shouts did nothing to lessen the spiritual worth of the exercise, and full-throated roars when patrollers were not within earshot could make up for temporary deficiencies in spiritual volume. No wonder bondpeople found evangelicalism so attractive. Hush arbors allowed slaves to moan and shout quietly and did not detract from the depth, meaning, or conviction of their beliefs. The religious practices of slaves were characterized by controlled volume as well as by unfettered, shouted passion or "the riot of sound."[33] Slaves understood that God's voice was often silent. His "voice every slave as distinctly hears, by his spiritual organs, as he hears thunder by his natural ones." Slaves used "a voice silent to sense, but audible to spirit and God."[34]

Sound itself was contested in the Old South, and southern bondpeople attempted to control it, especially where religion was concerned. One cause of tension over slaves' religious worship centered on the ownership and use of time. As patrollers told Minnie Fulkes, "Ef I ketch you here agin servin' God I'll beat you. You haven't time to serve God. We bought you to serve us." Little wonder, then, that slave mothers advised children to listen hard for patrollers. Fulkes recalled that her mother adopted the following strategy: "An' she sed dat dey use to have meetings an' sing and pray an' th' ol' paddy rollers would hear dem, so to keep th' sound from goin' out, slaves would put a gra' big iron pot at the door."[35] So, too, with Marriah Hines, born in 1835 in Southampton County, Virginia. "On our plantation we had general prayer meeting every Wednesday at church. 'Cause some of the masters didn't like the way we slaves carried on," she noted, "we would turn pots down, and tubs to keep the sound from going out." Listen also to former slave Oliver Bell of Sumter County, Alabama: "I 'members when us had de big prayer meetin's. Dey would shut de door so de voice won't git out, an' dey would turn de washpot down de door. Dat was to keep de voice inside, dey tol' me." Fannie Nicholson described a variation on the theme: "When de slaves got together an' had prayer an' sang, we put large tubs of water outside of de huts to catch de sounds so we wouldn't bother our master or missus."[36]

Secrecy was maintained through the slaves' control of silence and sound, and the role of upturned pots in killing "de noise" of prayer was important. Slaves put their heads in pots "to keep the echoes from gitton' back," as William Mathews, ex-slave from Louisiana termed it. Certainly, placing one's head in a pot muffled whatever was said or sang. But slaves spoke not simply about putting their heads in pots; they also mentioned turning pots upside down, gathering around them, and singing and praying in the belief that the pots somehow captured the sound. They were not far wrong. When propped up, the pots could mask the low-frequency sounds of consonants and fricatives. According to former slave William Lee, "We couldn't serve God unless we stole to de cabin or de woods. In our meetings we turn de pot down wid a stick on one side to drown sing-ing to keep de patterrollers from catching us."[37] Wholly upturned pots served no acoustic muffling function because they were "closed off," so some were upturned only in part, often propped against a wall or up by sticks, as William Lee noted. Some ex-slaves recalled that a number of large empty vessels were scattered around during worship with open ends up, which probably enhanced the reverberative properties of the space in-side the pot or vessel, especially in the bass frequency range. As a result of excitation by the sound of each new voice, the vessels stored the sound and released it slowly, thereby giving rise to an upward masking effect. While not reducing the level of sound that reached the remote listener, it helped distort the intelligibility of the speech.[38] This was resistance sotto voce, and many African American slave spirituals were probably quiet af-fairs. Remembered Angie Garrett of Alabama, "Sometimes us sing and have a little prayer meetin', but 'twas mighty easy and quiet like."[39]

Slaves' appreciation of the physical properties of the aural world, and the safety such properties afforded or denied, extended beyond pots and into slave cabins. The sound absorbency of a slave cabin, with the ex-ception of the earth floor, was low simply because sounds leaked out. Booker T. Washington remembered being born in a typical log cabin in Franklin County, Virginia. The cabin was "without glass windows; it had only openings in the side which let in the light, and also the cold. . . . There was a door to the cabin—that is, something that was called a door—but . . . the large cracks in it . . . made the room a very uncomfortable one." Other holes included several cat doors.[40]

In such an environment sounds made inside trickled out, and sounds from the outside seeped in (including the sound of the plantation bell). Descriptions of cabin acoustics by slaves are rare, but depictions by white settlers in similar cabins on the western frontier offer clues to the aural world of slaves when in the ostensible safety of their homes. Recalled Richard Cordley of his cabin in western Kansas in the late 1850s,

> The cabin was about fifteen feet square and of very simple construc-tion. There was no chinking between the logs and I could almost roll through the openings into the yard. I could look out and see the ponies and the pigs and the cattle, and could hear the chickens talking in their sleep. Now and then I could hear the bark of a prairie wolf, or the screech of an owl in the woods, or the yell of an Indian who was late getting home. All around the cabin the family lay on their shelves, and

were snoring in that peculiar piping key which none but an Indian larynx can produce. This music of the night was made all the more impressive by the deep bass snoring of my negro driver.

Such building materials (or lack of them) made the enslaved sensitive to sounds generated in their own cabins and aware of the sounds outside them.[41]

Other aspects of cabin construction afforded some incidental protection and allowed sounds slaves made inside to be absorbed and interiorized. Some slaves appreciated the acoustic absorptive capacity of dirt floors, which dampened sounds of stamping and shuffling. "Sometimes us slip off an' have a little prayer meetin' by use'ves in a ole house wid a dirt flo'," recalled George Young, explaining, "Dey'd git happy an' shout an' couldn't nobody hyar 'em 'caze dey didn't make no fuss on de dirt flo'." Just in case, though, "on stan' in de do' an' watch," and "some folks put dey head in de wash pot to pray, an' pray easy."[42]

If controlling sound provided slaves opportunities to maintain their dignity and spirituality on a day-to-day basis, their profound understanding of the plantation soundscape allowed them to commit more substantive acts of resistance against the peculiar institution. In short, it allowed them to leave the plantation temporarily, cavort at night, escape, and even revolt.

Slave children were trained in the use and abuse of sound. Sometimes parables were used to instruct slave children when not to betray presence acoustically. When former slave Josh Horn lost his dog, a voice in his head "say to me, jes' lak dis: *Josh, blow your horn!*" Josh did so: "Well I give three loud, long blows." Too late, the trick had been played, for another voice intervened, "whispering: *Josh, you out here in dese woods by yo'self. You blowed dat horn and your enemy heard you. You's a fool, you is.*" Josh barely managed to escape a patrol that arrived soon thereafter.[43]

Josh's lesson was important. Slaves understood that aural betrayal of misdemeanors was everywhere. "Our boss didn' keep no parrot to tell on d' niggers," remembered Horatio Williams, "but some did." Aural strategies were important to slaves and were employed for protective purposes. "De slaves," recalled Andy Williams, "'ud lie on de groun' an' put dier ear on de groun' ter lis'en an' see effen er pattyroller er comin'." The world of the slaves, then, was often deliberately onomatopoeic. They spoke of "paddlerollers" when the patrol beat them with paddles, and they spoke of the "patterrollers" because that was precisely what they meant, a point that seemed lost on the interviewers of former slaves in the 1930s. Similarly with slaves' wooden shoes. Bondpeople understood that "de wooden bottom shoes sho would make a loud noise" that might betray out-of-sight activity. Booker T. Washington agreed. "The first pair of shoes that I recall wearing were wooden ones," he noted, continuing, "When I walked they made a fearful noise." Slaves without shoes could walk more carefully barefoot; slaves who had them probably took them off for nocturnal activities.[44] Slaves' ability to quash sound and to impose silence was itself resistance against the master's will. It was also useful. Harriet Tubman certainly thought so: from her father she learned "how to walk soundlessly through the woods," and this proved handy, as posterity testifies. Slaves who hunted probably developed similar skills.[45]

Judging by the testimony of former slaves, antebellum bondpeople were extraordinarily sensitive to the aural world and the practical and qualitative meaning of particular sounds. As one former slave recalled of her North Carolina mistress, "You could tell by de way dat cane tap whether she was mad or glad. If it come tappin' sof' like, den she was glad, but if it come down de hall wid a rap, rap, rap, den you better be runnin' befor' she crack you 'side de head wid de steel end of dat stick." This ex-slave learned the benefits of silent walking: "She didn't hear me kaze I tipped so easy."[46]

Myriad aspects of slave life and resistance necessitated a highly tuned ear and the ability to control sound. When Charles Ball followed a man for whose crime he had been wrongly convicted, he did so at night with cultivated stealth and the skills of an aural sleuth. "Stealing quietly into the road," Ball followed "the wretch" into the South Carolina undergrowth. "It was easy for me to follow him," recalled Ball, "for I pursued by the noise he made, amongst these brushes; but it was not so easy for me to avoid, on my part, the making of a rustling, and agitation of the bushes, which might expose me to detection." Ball was "obliged to depend wholly on my ears," and he tracked the suspect by placing his ear to the ground.[47]

Notions of escape could be inspired by the soundscape that slave owners coveted. Like the fancy for pastoral sounds that attracted antebellum elites generally, slaves heard the "harmony of woods and fields, of birds and flowers," which in turn "gave birth to ideas that chimed" with freedom. The juxtaposition of the two southern soundscapes, the sweet purl of nature and the cruel timbre of pain, colluded to heighten slaves' desire to escape. One moment a slave could be "sauntering amid the green grass and blossoming fields, and regaling his senses with the music of birds and insects, and the outspreading beauties and harmonies of nature" that served to "announce the presence of an unseen God." Then, "when all was quiet within, and all beauty and harmony without," "a howl of agony. . . . Screeches, and screams, and cries for compassion" could rudely interrupt so that the "charms of nature in a moment vanished, and the voice of God was drowned by the cries of misery."[48]

The ability to control volume was critical to those slaves who struck at the heart of the peculiar institution. Runaways and conspirators alike used silence as their ultimate tool of resistance, and those who ran applied what they had learned to execute their escapes. J. W. Loguen, for example, correctly recognized that the Ohio "river was quiet in the embrace of winter" but that horses' hooves on the ice could generate dangerous noise. Escape from the plantation required that he move "swift[ly] and noiselessly as possible," and this demanded "a gentle whisper." Unnecessary decibels elicited a "Hush!" and the following advice from worried conspirators: "Hallo, friend!—a little too loud. There are things a colored man may think, but not speak above his breath, until his eyes and ears assure him that he is alone."[49]

Charles Ball understood the necessity of sensitivity to his aural environment. With his homemade (and quiet) moccasins on his feet, Ball prepared to flee. He also understood that successful escape depended on controlling sounds that might betray his activities. Accordingly he left his beloved dog behind because "I knew the success of my undertaking depended on secrecy and silence," which the dog could not be trusted to keep. With stealth he left his plantation, "seeking refuge in the deepest solitudes of the forest." The first night in the woods was unnerving. He "slept but little; for it seemed as if all the owls in the country had assembled in my neighborhood to perform a grand musical concert. . . . Although I had passed many nights in the woods at all seasons of the year, I had never before heard so clamorous and deafening a chorus of nocturnal music." At daybreak, acoustic sleuthing began in earnest with his "listening attentively to every noise that I heard in the trees or amongst the canebrakes." Ball had been a slave long enough to know the sounds of danger. When he caught "the voices of people in loud conversation" and "the cry of hounds," he adjusted his tactics. "The first resolution that I took was, that I would travel no more in the day-time." Knowledge of seasonal southern soundscapes informed his decision: "This was the season of hunting deer, and knowing that the hunters were under the necessity of being as silent as possible in the woods, I saw at a glance that they would be at least as likely to discover me in the forest, before I could see them." This was aural skirmishing with real stakes. Ball, like other escapees, knew that aural betrayal could be preempted by hard listening, which is why fugitives "lay out in de woods half asleep and listen for bloodhounds."[50]

The remainder of Ball's escape (on this occasion, a failed one) was navigated within the coordinates of the aural and visual worlds. Before he moved anywhere, he "examined everything around me, as well by the eye as by the ear." He kept away from dogs because he appreciated their sensitive hearing, but even then their olfactory advantage foiled him: "I presume he smelled me, for he could not hear me." And he used a technique to find his way that must not have been alien to slaves in the quarters: he listened to conversations. Finding himself on the wrong road in South Carolina, "I resolved to try to get information concerning the country I was in, by placing myself in some obscure place in the side of the road, and listening to the conversation of travelers as they passed me." It worked: "I heard one of the drivers call to another and tell him that it was sixty miles to Charleston." Ball was always mindful of the consequences of movement. He avoided eating chestnuts because he was "afraid to throw sticks or to shake the tree, lest hunters or other persons hearing the noise, might be drawn to the place."[51] Even when captured, Ball used his acoustic sense to escape again, this time from a jail. By accident he hit the wooden door to his cell, and "to my surprise, I discovered by its sound, to be a mere hollow shell." He split it open and absconded. Eventually Ball made it onto a ship and hid as a stowaway. Freedom's sweet tones reached his ears days later: "In a short time I hear much noise, and a multitude of sounds of various kinds. All this satisfied me that the ship was in some port; for I no longer heard the sound of the waves."[52]

Slaves' aural gambits were particularly effective in towns. Here they used the noise of city life to mask the sounds of their activities (on the plantation where sound carried farther with little distortion, such sounds could mean betrayal). In the Old South's urban environment, furtive activities were more difficult to hear, and slaves found more sounds to appropriate in southern cities.

Like plantation bondpeople, urban slaves were regulated by sound. They were subject to curfews, announced ordinarily by bells, and were arrested for being out after the appointed time.[53] But the city had its advantages, and as masters well knew, order could be lax in urban environments. Slaves exploited the situation and sometimes ignored aural summonses.

Of a Washington, D.C., hotel, Charles Dickens noted, "Whenever a servant is wanted, somebody beats on his triangle from one stroke up to seven, according to the number of the house in which he is required." Dickens thought the system less than effective because no one "takes the smallest notice of the triangle, which is tingling madly all the time."[54]

More sinister than willful deafness to the commanding sounds of the master class was slaves' careful manipulation of them. Urban slaves used city bells to create confusion that enabled them to conspire, insurrect, escape, or carve a few hours of uninterrupted freedom from surveillance. Well aware that the safety of the southern city was contingent on the ringing of the bell to warn inhabitants of fire or social discord, bondpeople recognized the benefits of creating aural confusion. So suggests an 1853 city statute from St. Augustine, Florida. To make best use of a new bell, the first part of the ordinance refined previous statutes dealing with what amounted to a city curfew for blacks. It was the duty of the "City Marshal to Ring at 9 Oclock each night the Bell lately placed on the Beef Market as a warning to the negroes in the City to retire to their respective homes." Fines and stripes were punishments for "any negro or mulatto [without a pass] caught in the Street . . . more than thirty minutes after said ringing of the Bell." Part 2 of the ordinance suggests how St. Augustine's African American community may have used that same bell to resist bondage. The statute stipulated "that the City Marshall after ringing said Bell, at 9 P.M. as aforesaid, Shall lock the trap door which secures the Bell Rope from reach of unauthorized hands, and turn the Key of the lock over to the Captain of the patrol to the end that he may Ring the Bell in case of Fire or any other sufficient cause of night alarm to the Community." To make the point clear to those who read posted copies of the statute "in the Market & other public places in the City," it was "further Ordained that if any person or persons, not expressly authorized as aforesaid to ring said Bell shall ring, or in any other maner [sic] cause the same to be Sounded, or shall wantonly injure and do any damage to said Bell or to anything connected therewith, . . . shall on being convicted . . . be fined if white or whipped if black." The reasoning was sensible and as revealing of white fears as it was of black hopes. Authority over public bells was a power recognized by contemporaries, slave and free. For the former, access to a sound that could be heard throughout the city and was obeyed by whites could be the key to successful insurrection; to the latter, false aural cues and misleading alarms meant confusion, a collective dulling of authority's senses to the sound of danger, and most worryingly, potential revolt. Access to authoritative sounds was among an antebellum southern city's most closely controlled resources.[55]

Because slaveholders were successful in imposing a social quietude on their slaves, they ultimately failed to secure the order and safety they sought. The source of their failure lay in the nature of bondpeople's resistance. Slaves' ability to control sound and, especially, learn the art of silence rendered the master class skittish. Certainly southern elites had always feared the silence of the ostensibly weak. Native Americans and slaves in the colonial period had been—and were still—worryingly silent. In the context of a mature slave society under increasing attack from northern advocates of free wage labor, additional imperatives applied. Every rustle, every murmur reminded slaveholders that slavery was possibly being undermined from within in response to agitation from without. Slave noise, to be sure, was unpleasant and could, to some extent, be controlled in this relatively closed society. But slave stealth, grave silence, and misleading aural cues were feared even as quietude was respected, courted, and insisted on by masters. Mary Boykin Chesnut knew as much: "If they want to kill us, they can do it when they please, they are noiseless as panthers." The master class had only themselves to blame in creating a people who knew how to move "as silently as a cat."[56]

In her retrospective of life on her antebellum Florida plantation, Susan Bradford Eppes agreed, also invoking the feline simile. Aunt Dinah, matriarch of the day nursery, "moved noiselessly as a cat and oftentimes we would not know that she was anywhere about until," concluded Eppes nervously and with some relief, "we would hear her cheery 'Good mornin' ladies.'" Slaves' ability to negotiate plantation space silently and quietly was obviously a source of great worry to masters and mistresses and of satisfaction to slaves who aimed to escape owners' eyes by depriving and fooling their ears. Bondpeople knew the power of their stealth and recognized the discomfort their silence caused masters. "A silent slave," remarked Frederick Douglass, "is not liked by masters or overseers."[57]

Travelers recognized the effect of the dynamic. Harriet Martineau detected the jittery and highly strung nature of southern urban life in the 1830s. "I was never in Southern society," she commented, "without perceiving that its characteristic is a want of repose. . . . It is never content; never in a state of calm."[58] Good reasoning grounded slaveholders' fears. Silent slaves had silent allies in abolitionists who infiltrated the South.

These northerners listened. Although slaveholders certainly exaggerated the numbers of such infiltrators, they did exist. One abolitionist, John Ball Jr., responded to critics of his "silence policy" in the *Liberator* in 1854. Ball had lived in the South and resented the charge that all who were against slavery should vocalize their sentiment. Such critics, he maintained, do "not seem to be aware that there is a masterly as well as a slavish silence." Ball's silence while in the South, he argued, enabled him to hear things his blustering would have prevented. "What has been the result of my silence policy?" he asked. "I have listened to confidential lamentations and curses—excuse the phrase—uttered by prominent pro-slavery men. . . . I have learned facts which will yet crush many of the false assertions that at present are so boldly advanced as truths by pro-slavery preachers and politicians." No wonder masters were so jumpy. Not only must they listen for slaves but also guard against being listened to, which meant that they, not unlike their chattel, had to regulate their own sounds.[59]

Planters also knew that antislavery writers spoke a truth when they fictionalized accounts of slaves running away. A tale depicting slave runaways in Madagascar, reprinted in Toronto's *Provincial Freeman* in 1855, had all the elements of an Old South slave escape or, worse, insurrection. Escape, suggested the writer, happened in the context of silence: "On the sugar-estates a dead silence reigned, and the harsh voice of the terrible overseer was no longer heard: the cock had given his first nocturnal crow—the signal of repose—and but for the piercing cry of the tec-tec, or night-lark, as it hopped from sprig to spray, not a sound could now be heard." Success meant not disrupting the delicate skein of quietude. Four runaways-cum-insurrectionists recognized as much: "They grouped themselves against the huge trunk, each taking up his position as he arrived, but not saying a word to break the dead silence."[60]

In fact abolitionists were good at gauging aspects of the slaveholders' world, and they were not far off the mark in their depictions of southern fears about slave insurrections. Some antislavery writers captured the aural dimensions of such fear with wonderful clarity. The "appalling terror of a slave revolt," wrote "A Southron" for the *Liberty Bell* in 1848, "pervades the community" so that southern words were "whispered along the streets." Slave silence underscored the gravitas. Masters who found slaves keeping mum worried: "They were as silent as the grave. Even 'Momma,' privileged to say and do what she pleased, and who would be heard amid the laughter and tongue-clatter of the rest," was ominously quiet.[61]

Just how fearsome were rumors of a slave revolt? Chilling enough, suggested antislaveryite John C. Vaughan, to be quieter than plague and louder than gory death. "All other perils are understood," explained Vaughan. "Pestilence walking abroad in the city, making the sultry air noisome and heavy, hushing the busy throng, aweing [*sic*] into silence heated avarice. . . . But the appalling terror of a slave revolt, made instinct with life, and stunning as it pervades the community—the undescribed and indescribable horror which fills and sways every bosom as the word is whispered along the streets . . . must be *felt* and *seen* to be realized."[62] Vaughan was wrong on one point only: the terror could be heard, too, in silence.

So edgy were slaveholders that any sound discordant with the rhythm of plantation labor sent shivers up spines. Sounds made by slaves at the right time of day or year gave southerners in a particular locale a certain reassuring cadence, indicating that all was right with their world. But sounds out of time caused alarm. When the rush to get the crop in deferred a celebration on a plantation for a few days, the "neighbours, hearing the noise on a quiet working day, had jumped to the conclusion that it was an insurrection."[63]

Suspicions of slave insurrections and the impending dislocation of order that they represented encouraged painfully close listening by slaveholders. Natural sounds that customarily promised refuge from the noise of the world suddenly assumed violent sonority so that "the sound of the wind among the trees was like the unsheathing of swords." Fear of slave revolt, remarked one Alabamian in 1840, "greatly disturbs the quiet & repose of many of our citizens."[64]

Antebellum southerners, especially those in frontier regions where Indians and slaves posed a palpable threat to social order, found both worryingly silent. Of the two, though, slaves were the stealthiest, and it was their silence that was most feared. Micanossa, the "head chief" of the Seminole nation in 1835, for example, was believed to be "under the most complete control of an artful, shrewd, and cunning negro" who had "been silently at work for the last two years in exciting the Indians to disaffection and resistance."[65] When the two stealthiest groups in southern society conspired against the master class, slaveholders had much to fear.

The silence of the slaves and the heightened sensitivity to the heard world that it encouraged in masters meant that those most accustomed to speaking their minds sometimes had to remain conscious of their own sounds. Slaveholders had to keep mum on certain issues in order to maintain plantation discipline and order. "I wanted to sell him softly, that is without his Consent and Knowledge," revealed one owner, inadvertently testifying that to avoid disruption the master class itself accepted self-imposed silence. Charles Ball similarly recalled that his overseer "called my young master aside, and spoke to him in a tone of voice too low to be heard by me." But there were tensions here, especially in the three decades preceding civil war. Fear that secrets might be discovered encouraged masters to control their own audibility, and yet the growing attacks on southern identity from without demanded that slaveholders counter abolitionist charges.[66] The years leading up to the Civil War saw the playing out of these two imperatives and affected how the master class fashioned its defense.

(((PART II
Keynotes Old and New
Listening to Northern Soundscapes

Like their southern counterparts, northern elites invested profound meaning in shaping their aural worlds and all that they stood for. North or south, American elites ruled "as a class . . . in its whole range, hence . . . they rule as thinkers, as producers of ideas, and regulate the production and distribution of the ideas of their age."[1] Some northerners' ideas of an ideal antebellum soundscape differed utterly from those of the master class and contributed powerfully to the construction of an aural sectional consciousness. Because northern elites were fairly successful at managing noises unleashed by the very modernity they courted, many of them applauded the keynotes of industrial capitalism, particularly when northern politicians contrasted modernity's throbbing registers with the enfeebling quietude and loud cruelty of southern slavery. Like that of their southern counterparts, northern elites' understanding of their world was shaped both by an idealization of what it should sound like and by the limits imposed by working-class resistance. The working class's use of democratic voice—their shaping of the northern soundscape—encouraged northern patricians to embrace the reality of democratic political participation, just as the silence of southern bondpeople made masters fret about the safety of their own society.

Aural boundaries in the North were relatively more elastic, as they had to be in a society that was becoming more urban, industrial, and capitalist. Although injunctions against public and private noise remained on statute books in the North, northern elites accommodated the sounds of free labor and industrialism and the muttering of virtuous, uncorrupted democrats. While they continued to define riotous mobs as clamorous (what ruling class has not?) and while they disassociated democracy and mobocracy, of necessity they proved willing to construe the sound of democratic politics as necessary and even desirable.[2]

But as a broadly constituted group sharing similar concerns about order, progress, and danger, northern elites built their heard worlds in ways remarkably similar to those of the southern master class. Like their southern counterparts, northern elites had their aural limits. Sounds made by northern factory workers that were not part of the production process or deemed deleterious to it were perceived as noises precisely because they threatened the regulation of labor, which in turn undermined order, production, and profit. The fitful evolution of democracy and urban-industrial capitalism in the antebellum North necessitated a reconceptualization of which sounds were desirable and which were not. Sounds that threatened economic production and the functioning of a virtuous republic were considered inappropriate; sounds that testified to the health of northern political democracy and northern economic productivity, even when loud, were accepted by advocates of northern modernity as happy humming.

Much may be done by a human being
with a pure will and amid a quiet life.
— "Renovation," *Godey's Lady's Book,* September 1853

(((4. Northern Echoes

Like the South, the North had predictable and reassuring
aural signatures. Christmas bells were particularly important for reasons
sacred and secular. Rung throughout the North, they marked Christ's
birth and served to punctuate the northern civil year.[1] Bells and their distinctive sounds anchored people to place and time, and emotion was invested in campanological soundmarks. Wherever he traveled, one man
always remembered the bells of his village. "He had heard grander bells,
louder bells, more famous bells, but none so dear and touching as his
own."[2]

National sounds were also secular, and their commonality was important for helping Americans imagine their country. As in the South, July
Fourth was ritualized aurally in the North. Although the meaning of the
day was contested (it was an occasion for local electioneering and, increasingly, for scathing abolitionist critiques of slavery), the commonality
of bell ringing and cannon firing on July Fourth helped join Americans
in time and space through real and imagined sound. Independence Day

was supposed to be loud. When it was not, people commented. "I hear no cannon's roar. . . . I hear no cheering sound of bells hailing the bright morn of American glory," noted a concerned Hartford, Connecticut, listener in 1837, worrying, "All are still & silent." So disconcerting was the quiet the following year that he committed the matter to verse:

Of our independence the glorious day.
In this modish town pass'd listless away.—
At morn the bells rang & a few guns were fired
But the powder was soon gone & the ball men soon tired.
The City soon sunk into quiet repose
Nor for mirth or hilarity at even'g arose.

The annual sound of political independence and patriotism was meaningful to Americans, and its absence was vaguely troubling.[3]

The heard world was also temporalized. Like southern elites, northern patricians believed in "a sort of Sabbath-quiet." Saturday night, they fancied, was "prelude to more pure, more holy, more heavenly associations." "'Tis then the din of busy life ceases" and the quietude of Sunday began. Sometimes, in fact, God's sanction was sufficient to quiet northern cities. "It was a Sabbath morning," wrote Henry Ingersoll Bowditch in 1843, and "men, women, and children [were] on their way to the temples of *Christian* worship! Every thing was so quiet; the hum of the great populace was so entirely hushed, that my own tread sounded loudly and coldly on my ear."[4]

As in southern cities, northern urbanization and industrialization threatened to mask and disrupt such moments of quietude. "A stranger who had been brought up where the observance of the Sabbath is rightly attended to—where the noise of coaches and the prancing of horses are prohibited on that day, would be surprised" at the noise in some of New York City's churches. Churches near railroad lines found the rumble of modernity drowning out their voices and, by implication, rendering garbled what God's ears should have heard clearly. The courts sometimes stepped in to protect the sounds of the old from the noises of the new. In the late 1840s the First Baptist Church in Schenectady, New York, charged the local railroad company with disrupting their worship with rumbling railroad cars, screaming steam, and ringing bells. The court agreed for two reasons: first, there was the matter of property (vibration from the railroad cars damaged the church) and second, the noises disrupted religious worship.[5]

Urbanization heightened tensions, and the battle escalated to preserve the serenity of night (so important to sleep) and Sabbath quiet (so important to God). Genteel folk had to share streets with increasing numbers of politically active, noisy plebs. "Censor," writing in 1811, believed roaming gangs of young men disturbed New York City's streets at night "to the utter annoyance of all quiet [people], both male and female." When lower orders were the source, even "humming, or whistling" was deemed inappropriate. Other elites agreed. New York City's "gangs of lawless ruffians," noted one correspondent in 1811, threatened "the peace and order of society" not simply by making noise but by actively seeking it out. The "villainy . . . sally forth in the night in quest of tumult, riot and disorder." Nocturnal encounters with groups of "whistling, hooting" street boys and "not being within call of the watch" made elite men indignant (and fearful) at being "followed by their shouts, hoots and hisses." The noise per se was doubtless annoying; that it acted as potential political subversion was also worrying.[6]

Like his southern counterpart, the northern gentleman, while able and allowed to be loud, was defined by his sobriety and by being "thoroughly quiet in his manners." Although ostentation was deemed noisy, ill breeding was the common cause of noise presumably because it betrayed an inability to control the passion of self. As the "Lounger" moaned in *Harper's Weekly* in 1862 about an annoying hotel guest, "Why not learn how to shut a door? Why not place your boots quietly?" Clearly, training in aural etiquette was the problem. Besides, continued Lounger, not only was the noise gratuitous, but the sleep of others was of great importance: "Why should you wake up those to whom sleep may be peculiarly necessary by a perfectly unnecessary noise?" Poor manners and ignorance of the serenity and solemnity of darkness were to blame: "Have you ever asked yourself how an ill-bred man would behave if he were going to bed at such an hour and in such a place? Have you ever reflected that a man is honorably upon his good behavior always?" Lounger knew the essential cause of noise, though. His noisy adversaries were "strangers," as was increasingly the case in a modernizing North.[7]

Even sacred quietude was not immune to acoustic rudeness. In 1811 a man of "uncouth manners" was removed by police for disrupting a Sunday service in New York City by "rasing his voice." Quietude in polite northern society was tied to gentility and godliness, and elites were often reminded of the connection lest they slip. "The Gentleman at Church," reported New York's *Colored American* in 1838, "comes in good season, so as neither to interrupt the pastor or the congregation by a late arrival. . . . Opens and shuts the door gently, and walks lightly . . . and gets his seat quietly. . . . Does not cargo in conversation. . . . Does not whisper, or laugh," and "does not rush out of the church like a trampling horse, the moment the benediction is pronounced, but retires slowly, in a noise-less, quiet manner."[8] Bodily sounds, especially when heard in public, were also noisy to elites, north and south. *Harper's Weekly* printed a letter from a man in New Orleans in 1859 on the noise of public kissing. "I have often been very much shocked," he wrote, "by the manner in which some people kiss each other; that is, in giving one of those unearthly smacks that sound like the explosion of a Paixhan gun or something similar."[9]

Quietude was not simply about aesthetics and decorum. The sick, it was believed with some justification, were best served with "repose and quiet." Ill in Savannah in 1851, Henry L. Cathell of New York City thought serenity his best cure. To combat chronic chest spasms, Cathell "spent the day in a quiet manner—hoping to wear the pain off." Others, particularly in cities, tried to silence passing cartwheels by strewing tan-bark in streets adjoining their homes. Even the hum of cottage indus-try could jar the nervously disposed. Industrious daughters caring for ill mothers were caught in a paradox. To support the sick economically, women did weaving at home, so that the "whirr of her shuttle and the stroke of her lathe were heard from morning till night." But invalids had little appreciation for the sound of such activity. Some reached the point where they "could no longer bear the noise of the loom," and they feared that the "slightest noise" would induce conniptions.[10]

Old age also was supposed to be experienced in "calm, peaceful" qui-etude and constituted retirement from a buzzing life. Death itself offered utter silence. The deceased had earned respite from the noise of the world, and their final resting places were, by design and respect, quiet. A hillside cemetery near "noisy and grand" West Point was "so intensely quiet, that no sound is to be heard but the plash of oars from below and the hum of insects around, except when the evening gun booms over the heights, or the summer storm reverberates among the mountains."[11]

The invention of print helped interiorize thought. Contemplation, it was believed, had to occur in quietude if thought were to be uninter-rupted. There was a certain silence to productive mental leisure not found in the noisy "tumult" of working-class pastimes such as gaming. The *vita contemplativa* could operate only in relative quietude, and this belief tes-tified to the recovery of positive silence in the eighteenth and nineteenth centuries.[12] It was understood, "How noiseless is thought! No rolling of drum, no tramp of squadrons, or immeasurable tumult of baggage wagons attend its movements." Hence libraries were quiet sanctuaries, sometimes too much so. One library patron in 1863 explained the system of library silence. He was "requested, almost inaudibly, to write his name, address, etc., upon a slip of paper, and, upon inquiry, was 'curtly answered' that communication between the librarian and the reader was to be held by signs that there might be no noise of conversation." Considering himself a man of refined aural sensibilities, the visitor was affronted. He had "sufficiently good manners not to disturb others"; he was not a "young and heedless" person who did "not know the value of silence." When elites were shushed, their honor and political independence were challenged.[13]

If there was a class basis for who was considered noisy and who produced sound, the same may be said of hearing or the ability to consume, with discretion and judgment, the heard world. The cultivated heard better than the laboring classes in part because, according to one 1863 formulation, there were physiological differences between rich and poor. "In the higher animals," it was maintained with accuracy, "the organ of hearing is formed of three parts, an external, middle, and internal portion; but in birds the external ear is wanting"; fishes lacked "the external and middle parts"; mollusks had nothing but a simple "organ [that] can distinguish noises only." "Notes, tones, melodies, harmonies" were beyond the mollusk. What separated animals was paralleled in humans. As everyone knew, there was a "great difference between the disciplined . . . ear of a cultured person, and the undisciplined, and therefore less differentiated ear of a boor." No wonder workers were becoming ever noisier.[14]

If the poor were noisy to elite ears, especially so were the newcomers who increasingly swelled their ranks, specifically plebeian immigrants. In the 1830s, for example, Frederick Marryat spoke of "the turbulence of the Irish lower classes."[15] Drunkenness heightened their noise, as the Temperance movement made clear. Although some northerners denied that "nature has prescribed bounds to the minds of race or color," they nonetheless believed that different nationalities had different oral and aural traits. "The Spaniards are essentially different from the Americans in their dispositions," opined one Philadelphian in 1843, explaining, "on the contrary, he loves to assemble nightly among his friends [and] . . . join in warbling forth a melody accompanied with his guitar. In fact, rather than be silent and alone, he will go talk Spanish to his horse."[16]

Melodic strains helped combat plebeian social noise. Good music preserved "great order" during its performance. Exceptionally fine music was celestial and allowed one to hear "heaven-born sounds" drowning out the noise of the secular. Music was also socially functional. "The utility of music," opined one observer, "is almost equal to that of poetry. . . . Leading armies to battle by the sound of music, has a tendency to dispel fear, to preserve order, and inspire heroism." Music, in short, "dissipates every care from the troubled mind." The Aeolian harp, commented one northerner in 1811, could make one "deaf to all other noises and sounds." In a world of jarring noises, "music calms the mind, and unloads it of the cares of life."[17]

Conflict ensued when elites found their attempted consumption of sublime melodies disturbed by persons they considered too uncouth to appreciate good music. In a particularly snobbish piece titled "Casting Pearls Before ———," written after a recent concert, "Urania Rus" complained in *Harper's Weekly* in 1858, "I *am* going to complain, as loud as you will let me, that people who *do* like . . . and who came far to hear it, were not allowed to hear it by those who do *not* like it, and who lounged in from some neighboring street. We sat in the balcony; and such scuffling, talking, rustling, tapping, beating time inanely with the feet, and general indifference and indecorous riot, I have never before encountered." What was "every real lover of music" to do?[18]

Class muddied gendered dimensions of sound. Lower-class women who joined male relatives in Revolutionary military encampments, for example, offended officers' sensibilities concerning women's place and were heard as "yelling in sluttish shrills as they went spitting in the gutters." While their passion and weaker discipline of the self made all women prone to noisiness, lower-class women seemed more predisposed than most. Street women were heard negatively and often.[19] By contrast, the antebellum northern middle-class woman (not unlike her southern counterpart), though given to emotional outbursts, was a lady not least because she was quiet and submissive. Entrusted as she was to raise sons of a "sound and virtuous character," woman "was to work in silence" in this endeavor, bear indignation "quietly," "suffer and be silent," and so help cultivate a "quiet life" for her family.[20]

Representations of domestic serenity and women's quieter nature seeped into all sorts of discourse. Pleasant music was as "soft as a virgins whisper," and presumably, a virtuous woman was like soft, whispered music. A silent wife was a good wife, or so Canadian native Frank Hatheway wrote from Florida in March 1846: "On the 18th, I am to be married, or, in the language of the times, change the form. I hope I shall have a *silent* partner at any rate." Ideally the northern home was always a "quiet home," and the custodians of domestic silence, those responsible for the front-line defense against public and private noise, were women.[21] Home was comforting because it was quiet, the defense against the boisterous world of the public, a chamber of quietude in which men found repose and sustenance to face another day of noisy battle. Thus even as they celebrated the aural confirmation of men's public, bustling activity, New England poets portrayed women who supported such endeavors as domestic and silent. In 1850 John Greenleaf Whittier applauded the public, masculine, and republican world in "The Drovers" using spouses as an aural counterpoint: "quiet wives are knitting" while loud men are working. There were benefits for women, too. Lydia H. Sigourney, popular promoter of domesticity north and south, advised mothers to create a "season of quietness" while lactating. Northern women's public and private duty also included protecting husbands and men generally from the noise of infants. Steaming on the Hudson, Harriet Martineau noted that "the greater number of ladies on board remained in the close cabin among the crying babies," while the men surveyed the fine scenery from the deck.[22]

Obituaries stressed that good women kept quiet homes, thereby civilizing northern society. According to the obituary of Rebecca Gilman, who died in 1823, "She lived in Ohio from the time when it was the abode of savages, until it had become a civilized and powerful state. She saw wilderness bud and blossom . . . where the war whoop interrupted the sleep of the cradle" but later "became the quiet home of domestic enjoyment." The older the woman, the quieter she was. A "good old aunt," while "she loves to *talk*," had learned well her sex's silence because "she never interrupts me."[23] Women who failed to keep their tongues still were excoriated. As one "Epitaph on a Scolding Wife" of 1811 expressed it, "Here lies my wife, poor Molly! Let her lie, / She finds repose at last — and so do I." Gentlemen blasted putative ladies for their "love of *little tittle tattle*" and "*gossiping tales*" but nonetheless remained convinced that "doubtless, there are very many ladies, both in this city and elsewhere, who do not come under the denomination of *gossips,* who prefer *silence* to *scandal*."[24]

Children, north and south, were considered noisy creatures, perhaps because they had yet to learn the discipline and habit of silence. On his visit to an antebellum Florida plantation, for example, an abolitionist advised a slaveholding couple what they knew only too well, that "children should be seen and not heard." "When walking through the streets in the evening," commented one listener in New York City in 1811, "we are often astonished at the noise of children." But children per se were not to blame. The onus lay with parents, especially mothers, who, given their own training, should have known better.[25]

As in the South, aural disruptions of genteel religious worship were commonplace, and the culprits were predictable: the lower orders, but also children of any stripe and the mothers who were supposed to silence them. So disruptive did New York City men find children in church that they advised that no "moral woman . . . ought to be suffered to enter the threshold of a church, with an infant in her arms, unless for the purpose of having it christened." The logic here was clear: a minister's "audience are loath to hear the sound of a footstep, fearful that they may lose the train of his reasoning." All too often, though, "the ears of all are stunned, the imagination confused, and the soul tortured by cries and screeches of an infant." "The poor child," though, "is not to blame." Responsibility lay with "the mother who introduces it," believed a "Friend to Females." Such criticism did not go unchallenged. "Sarah Touchstone" made the reasonable point that if a poor mother's soul was to find regular sustenance and if such a mother could not afford a nanny, she had to bring the child to church. Besides, she continued, suckling babies were usually quiet precisely because they were suckling. "Friend to Females" remained unconvinced and fired back. "The screeches of a sucking babe," he maintained, "is a sound the most unpleasant to hear." Moreover, screaming was spiritually damaging because "the noise possesses the less harmony, and consequently, pierces the soul quicker. The cries of a babe operates upon the mind equally as bad as the filing of a saw. I therefore maintain, that *one* child can disturb a *whole* congregation."[26]

Northern genteel church members saw the control of vocal volume during worship as testimony to their secular and spiritual discipline. Episcopalian services, such as the one held on Christmas Eve in Hartford's Christ Church in 1836, impressed on congregants the celestial solemnity of silence through biblical injunction: "The quoire chanted the fultering sentence, 'The Lord is in his Holy Temple let all the earth keep silence before him,' which had a very great effect upon the crowded audience— & completely silenced them." It is not surprising that genteel northerners winced at the spontaneity and fervor of marginal faiths. Early national Shakers, for example, produced "a kind of wild plaintive tune" and sang "doleful melodies," and even though they were considered "a sober, honest and industrious people," their singing "not disagreeable," the sound of their religion was "wild and something like that of our American savages."[27]

In many respects New England genteel clergymen were similar to their southern counterparts, for whatever their theological disputes and rancorous disagreements over slavery, they both aspired to rational learning. New England Unitarianism appealed to the North's cultural elite because it equated "Christian grace with capitalist effort." Moderate New Light evangelicalism was both rational and individualistic and stressed the importance of self-discipline. Like southern gentlemen theologians, New England clergy also excoriated passionate evangelical New Lights, although for slightly different reasons. The multitudes who found the fervor of antinomianism so attractive, complained Connecticut's Rev. Timothy Dwight in the early 1820s, were entirely "too talkative, too passionate," their camp meeting shouts hardly compatible with his vision of clerical authority and capitalist discipline. Revivalism per se was not the problem, since rational evangelicalism could promote worthwhile social reform. Rather, evangelical excess was the worry.[28]

Norms of genteel Protestant worship required that parishioners refrain from excessively emotional displays. Northern clergymen critiqued black spiritual expression in ways similar to those of their southern brethren. Methodist minister John F. Watson of Philadelphia, for example, blamed blacks in particular for the general excitability of evangelical fervor. In 1819 he lamented, "We have too, a growing evil, in the practise of singing in our places of public and society worship, merry airs, adapted from old songs, to hymns of our composing: often miserable as poetry, and senseless as matter, and most frequently composed and sung by the illiterate blacks of the society." The religious sounds made by Philadelphia's black Methodists, maintained Watson, testified not simply to Calvinist declension as he heard it among the black community. Blacks' "audible sound of the feet at every step," the noise they generated by vigorous thigh-slapping, and the general volubility of black services had "already visibly affected the religious manners of some whites." "What in the name of religion," he asked, "can countenance or tolerate such gross perversions of true religion!"[29]

As in the Old South, Native Americans occupied an ambiguous place in northern aurality. Fictional histories of Indian pasts depicted "the murmur of voices" and "peals of laughter" echoing in northern woods. Their "old women who were for ever whispering like the forest leaves" were peaceful creatures. But quiet belied coming storms. Indians attacked white settlers using "the terrible war-whoop" that, in turn, produced a "wild shriek of woman's terror." White victory, though, ensured that all "was hushed and still as though the voice of pain or discord had never echoed through the wilderness." Abolitionist literature echoed the ambiguity. A piece in the *Liberty Bell* in 1841 charged Florida's Indians with disrupting the state's "dim and leafy aisles" using the "yell of fiercer men," "the din of war," the "shout, the groan, the fierce hurrah," and the "fearful war-whoop."[30]

Native American sensitivity to the heard environment helped shape those images. A profound awareness of the heard world, after all, was indispensable to successful hunting. During the mid-nineteenth century the Ojibway and other northern tribes managed to retain the inestimable skill of "shooting without a light," or night hunting. "Many were so expert, and possessed such an accuracy in hearing," it was said, "that they could shoot successfully in the dark, with no other guide than the noise of the deer in the water." Little wonder, then, they were considered "The Silent Enemy."[31]

Not unlike slaves who had learned to walk silently and shape their aural worlds, Native Americans trod noiselessly because they had been so trained from childhood. One Sioux, for example, recalled how his uncle taught him to walk in the woods at night and that such an exercise required consummate stealth lest the sound of his footsteps alert "wild beasts" or "hostile bands of Indians" to his presence. He was "always careful to make as little noise as a cat." Charles A. Eastman, who learned to hunt with the Santee Sioux in British Columbia in the 1860s, explained that an Indian's "moccasined foot fell like the velvet paw of a cat—noiselessly." Southern slaveholders understood the analogy only too well. According to Chief Kahkewaquonaby of the Ojibway, sound battle tactics mitigated against "meeting the enemy upon an open plain face to face, to be shot at like dogs." Instead Ojibway warriors aimed "to surprise the enemy by darting upon them in an unexpected moment, or in the dead of night," which involved "creep[ing] up slowly and stealthily like panthers. . . . When sufficiently near, they simultaneously raise the war whoop, and before the enemy awake or have time to defend themselves, the tomahawk is rattling over their heads." These aural skills of silent approach and loud surprise had been honed by tribal warfare long before the arrival of Europeans.[32]

Native Americans could not win the battle for image because to elite ears their silence and their noise were two sides of their savageness. In recounting an Indian raid on Buffalo, New York, during the War of 1812, Harriet Martineau anchored their barbarism in their quietude and noise. On one hand, "these savages" kept their white enemies in a state of prolonged nervous excitement through their worrying silence: "They listened, but they knew that if the streets had been quiet as death, the stealthy tread of the savages would have been inaudible. There was a bustle in the town. Was the fight beginning? No. It was an express sent by the scouts to say that it was a false alarm. The worn-out ladies composed their spirits, and sank to sleep again. At four they were once more awakened by the horrid drum." This was the quiet before the storm: "A host of savages . . . like so many kangaroos . . . came crashing in . . . yelling." But none of this meant that Native Americans were beyond the reach of bourgeois redemption. Separating them from "the white population, and . . . settling them in a quiet home" was a "wise policy."[33]

Tied to the idea of nationalism and the acoustic construction of class and northern bourgeois aesthetics was the sound of civilized economic development. Not all sounds conveying progress were hefty or resoundingly metallic. As in the South, preindustrial sounds connoted industriousness. The chinking hammer of the busy blacksmith was ubiquitous throughout a country still tied principally to horsepower. Each tap produced its own sound, depending on the manipulation of the iron by the smith's assistant and the force of the strike. The sounds could be regular or irregular, and they could be loud, sometimes over 100 decibels, which meant that during harvest seasons, when scythes had to be flattened regularly and tools repaired frequently, sounds of the blacksmith were heard often. Even as the North industrialized, the sound of hammering did not disappear; rather, it was enveloped and joined by new sounds.[34]

Before the North industrialized, Americans generally embraced sounds of progress, not least because they had always heard and coveted industriousness. All good men, it was maintained, should urge the modernization of their country regardless of their region, and they quoted approvingly one another's views on how modernization would be heard. An article on the sound of progress, virtue, and industry from Virginia's *Richmond*

Enquirer, for example, was reprinted with approbation in New York's *Independent Mechanic* in 1811. The author recounted a trip to Switzerland in 1772 and was "struck . . . with the picture of national happiness" that apparently was not exclusive to town or country. "If I entered a town," the traveler explained, "I heard, on every hand, the rattling of the hammer, and the clinking of the trowel," which in his estimation bore "witness to the progress of wealth and population." Preindustrial sounds of progress were heard elsewhere, albeit in slightly different form: "If I sauntered into the country, I heard the rosy daughters of industry, singing aloud to their spinning wheels." These "primeval" sounds echoed in the "shepherd-master," who "with his flaglet to his lips and peace and gladness in his heart, poured from the echoing mountains into the valley that smiled below, the simply wild and touching notes of his favourite air, the *rans des vaches.*" The rattling of urban hammers and the sound of rural trowels, songs, and spinning wheels were not in tension because they were fundamentally preindustrial. They were aural testimony to "the effect of *industry* . . . virtue and health"—in one word, activity. North and south, a man's industry and character were measured by the sounds of his industry. "My neighbor, Samuel Steady," commented one writer, is "an industrious man." How did he know? "His hammer is heard at the cock's crowing."[35]

Even the canal was heralded as aural progress. "Ungovernable" rivers, believed Benjamin Franklin, would be silenced by "quiet and very manageable" canals. The aural cost of quiet canals was the noisy work and play of the canal builders, whose ill discipline in turn would be tamed by the work "bell [that] must be rung every morning ¼ before Six OClock" to get the men to work. No matter that they resisted such discipline. The "continual buzz" of building the waterways, commented a correspondent for the *Detroit Advertiser* in 1854, the industriousness heard in the "noise" of the blasting of rock, "the voices of the overseers," and the "ceaseless music" of man mastering nature were sufficient to convince observers that canals were indeed a hallmark of industry.[36]

With urbanization early-nineteenth-century northerners became especially sensitive not simply to the difference between rural and urban soundscapes but to the ways that urbanites and country dwellers perceived those soundscapes. An anecdote from 1812 captured the essence of those differences. "A huntsman," according to the account, "invited a city friend out to his country residence, to a fox hunt." The tale continued, "The morning came, and the friends rode out together. As they ascended the hill, the voice of the dogs broke on their ear. The huntsman, in an ecstasy of delight, exclaimed, 'Hark, my dear fellow, do you not hear that music?' The citizen listened—'Music, (cried he) no—I don't hear a note of it, the cursed dogs raise such a yell.'" For a man who lived in a city where dogs barking disturbed peace, the response was hardly surprising. Because courtesy demanded the "citizen" return the favor, he invited the huntsman to the city and took him to the theater. Asked what he thought of the orchestra's "heavenly strains," the huntsman replied, "I can't hear them for my life" because "those noisy fellows in the cellar make such a damnable racket." Whatever separated the acoustemology of the two men, they agreed that sounds of activity, progress, and industry were neither utterly urban nor utterly rural. "To rise . . . and to hear the hammer of industry resounding through the village" was agreeable to both because progress was linked inextricably to a particular construction of the sound of progress rooted in a long past.[37]

Northern elites were similar to southern masters in other ways, and they were not any less nostalgic about the heard past. Postbellum northerners listened back and heard the progress of their history. The quiet wilderness of early Massachusetts, cooed the mayor of a city on its 250th anniversary in 1886, was replaced by "busy manufacturing towns . . . filling all these valleys with the cheerful music of busy and prosperous industries." "How complete the change!" he continued. "Nature's sounds then delighted the ear. . . . Now, two thousand children are ready to chant the hymn of peace and success. Their sweet notes take the place of the savage cry." Yes, chimed in others, and Connecticut residents boasted, "The hum of their industries is heard . . . around the world."[38]

America's modernization was contingent on mastering a wilderness that was both howling and silent. By the late 1840s northern frontiers were still places of quietude, and wilderness was still home to "noiseless fish" and "the stealthy leopard." Settlers found in the wilderness a "sweet quiet" distinct from the "hurries and perplexities of 'woeful Europe.'" "In due time, the mind, devoted to better accommodation, seeks for its permanent settlement. Then the busy, bustling era begins!" But mastering the wilderness took time, so that "although the astonished tenants of the forest thus feel and fear the busy stir of man throughout the day . . . they were not immediately to be driven from their favorite haunts. . . . It was therefore no strange thing with the primitive population to hear occasionally at safe distances,—'the fox's bark, or wolf's lugubrious howl.'"[39]

The association of sound with progress and utilitarianism, with the mastering of the silent and howling wilderness, reached maturity in the heady days of the expansive antebellum period when persons charged with ensuring northern progress injected aurality into the market revolution and early westward expansion. Lonesome travelers west heard pockets of civilization before they saw them, and the sounds of activity came as a relief from the quiet of the wilderness. As John Hamilton Cornish noted after he rode into Jonesville, Michigan, at night on August 30, 1836, "The shout of the boys the rattling of the cow-bells the cries & prattling of the children the sound of the ax the busy hum of the women cooking & preparing comfortable lodgins for the night rendered the scene lively and animating beyond description." To a man emerging from days in the woods, these sounds represented civilization, acoustic islands of safety, a heard archipelago in a quiet, howling wilderness. In Cincinnati in the 1850s, "where the busy din of commerce and the rolling carriages are heard from morning till night," one could also hear the sound of progress.[40]

The islands were becoming joined in a growing American soundscape. As the United States pushed west, the rhythmic, ordered sound of progress quashed the sporadic, unpredictable noise of the as yet unconquered. Hezekiah Niles put it admirably in 1815: "Everywhere the sound of the axe is heard opening the forest to the sun, and claiming for agriculture the range of the buffalo. . . . The busy hum of ten thousand wheels fills our seaports, and the sound of the spindle and loom succeeds the yell of the savage or the screech of the night owl in the late wilderness of the interior." Certain sounds heralded progress. Rhythmic and predictable, they regulated, controlled, and it was hoped, silenced the visceral and undisciplined "yell of the savage." Each push westward was an aural victory of white America over red, the silencing of the Native American testifying to U.S. victory. As John F. Watson remembered in 1891, "The print of the moccasin is so soon followed by the tread of the engineer and his attendants, and the light trail of red man is effaced by the *road of iron;* hardly have the echoes ceased to repeat through the woods, the Indian's hunter cry, before it is followed by the angry rush of the steam engine, urged forward! still forward! by the restless pursuer of the fated race."[41]

There was a price to pay for such expansion. While the soundscape of the wilderness could be refashioned, the initial settling produced keynotes antithetical to the northern bourgeois aesthetic. New western towns were especially prone to noise, not least because police and ordinances were, at first, few and far between. Early Chicago, for example, was cacophonous, as a poem composed in 1837–38 suggested:

Was you ever at Chicago? No sir, no!
Then, sir, I'd advise you never to go,

.   .   .   .

When night invites sleep, you can take no ease
For the luxuriant mud breeds abundance of fleas
The Fort is on one side, and Indians at t'other
A dog yells in one place, a mongrel in another
Now speaks a guard['s] gun,—the poney bells gingle—
Now cats, dogs—and bells in loud chorus mingle
So, stranger, retire, if weary, sleep sound if you please
Mid this charming music and the biting of flees.[42]

Northern elites, like southern, did not always revel in registers of progress, but they nevertheless recognized them as necessary and manageable through legislation. Antebellum municipal statutes targeted noise producers with increasing precision. False sounds, such as those produced by guns and fireworks, were among the first identified as disruptive of the functioning of urban centers that relied on sound to regulate life. Market commerce was similarly regulated. The sale of perishable goods, for example, had to be announced aurally lest the goods spoil. Hence in early nineteenth-century Philadelphia "a small bell was rung . . . to inform the chief part of the town that the fish were come." Northern markets, like southern, were opened and closed by the sound of bells.[43]

Statutes were just one way in which northern elites tried to ensure urban tranquility and order. Like southern elites, affluent northerners had relatively quiet houses because rooms and buildings were insulated with glass windows, rugs, carpets, wallpaper, and a host of other materials that absorbed sound. The poor, by contrast, lived in flimsier, more populous accommodations that let noise creep through cracks in walls. The rich could also stay in hotels that used plaster and horsehair "to secure for the inmates of the establishment exemption from noisy interruption, and to lessen the contrast which generally exists between the bustle and confusion of a public establishment and the quiet retirement of a private residence." Although antebellum architects were sometimes only dimly aware of the absorptive tendencies of different materials, it seems probable that the northern middle class stumbled on and became appreciative of materials that absorbed sound in their efforts to keep homes warm. Wallpaper, curtains, and glass windows not only kept wind out and warmth in but also absorbed sounds that were made within rooms. While the historical actors themselves were sometimes unaware of the physical, acoustical properties of their internal decor, this did not necessarily mean that there was no incidental, experiential awareness.[44]

If the sound of the market preceded the sound of the market revolution, what did the revolution itself sound like? One indication of the extent to which the sound of the market had infiltrated the northern soundscape may be found in the region's moments of silence. As in the South, economic silence testified to the extent to which the sound of market commerce had insinuated itself into people's ears. Within a year of Thomas Jefferson's signing of the Embargo Act in 1807, New England's ports were listless because of the suspension of trade. Following the breakdown of the banking system in the mid-Atlantic states in late 1814, New England mill building declined so sharply that the sound of growth was no longer heard, for a while at least. Similarly, economic silence reigned over Philadelphia following the recession in 1819. "The enlivening sound of the spindle," noted one anxious listener, "the loom, and the hammer has in many places almost ceased to be heard."[45] Conversely, the market revolution was heard positively when recession did not loom. American ears adjusted to the sound of steamboats, for example, and within a generation the superior speed and efficiency of steamboat travel converted potentially noisy vessels into silent gliders. New York City's Henry L. Cathell wrote in 1851 during a steamboat trip, "We are now on our way, under 16″ of steam, quietly gliding up the Savannah river."[46] Steamboats were loud in an objective sense, but their speed, efficiency, and importance to the market revolution made them decidedly quiet.

As with the South, the northern social fabric and the coordination needed to hold the strands together were indebted to sound, especially bells. Their indispensability and relative inexpense made them commonplace. "School Houses Without Bells," ran an advertisement by a New York firm in 1860, "can be supplied with IRON AMALGAM BELLS, one-third the cost of other metals and quite as good." This applied "also, [to] hotels, factories, boats, engines, and all purposes needing a bell." The biggest bell produced by this company weighed 150 pounds and cost $20; the smallest weighed 50 pounds and cost $6. In more rural areas "cheap Church[,] School and Farm Bells" were sold at lower cost. Hedges, Free & Co. of Cincinnati advertised a bell with a hanging weight of 170 pounds for $18 in 1860.[47]

Bells sounded pleasant because they facilitated everyday coordination of everyday activities and because their tones were aesthetically pleasing. Aluminum bells were rated highly in this regard. "No metal, or combination of metals," waxed a writer for *Harper's Weekly* in 1857, "yields a tone so musically sweet when struck as aluminum." Good bells also shared the same qualities as sound social order. The best makers produced bells with a "homogenous character" deemed "essential for producing evenness of tone, freedom of vibration, and strength of cohesion."[48]

All antebellum Americans, regardless of their origins, understood the summoning function of the bell. When Yankees moved south or southerners moved north, the sound of the bell provided the same basic function and meaning. The bell's ability to alert, summon, and coordinate behavior was so embedded in Western culture that foreigners in the United States experienced smooth, seamless transitions in their understanding of and response to local bell ringings. Frank Hatheway of New Brunswick, Canada, for example, understood that bells in Tallahassee, Florida, called him to church in 1845 as New Brunswick bells had presumably done before his move in the late 1830s.[49]

Sound and bell ringing could be both arbitrary, as in the case of fire alarms, and predictable, as with the sound of time. Urban growth affected aural systems of alarm in northern cities just as it did in southern towns. Until the mid-1830s, for example, Lynn, Massachusetts, had "but two bells in town; one on the 'Old Tunnel,' as it was then called . . . and the other in the belfry of the First Methodist church." Lynn's denizens attempted to use the two bells as best they could in their growing city. "If a fire occurred in Woodend . . . it would take some time to get the alarm to West Lynn," remembered David Johnson in 1880. "Vociferous yelling, pitched to all sorts of keys, from the shrill tenor of the small boy, to the deep, if not sonorous bass of the full-grown man, was heard all over town, as fast as the alarm spread, till somebody got hold of the bell-rope, and the people were thoroughly aroused." Then came the debate about the fire's location: "'I heard the 'Old Tunnel' strike first.' 'No, the Methodist struck first.'" Confusion often reigned, and the numerous false alarms, caused by shouts of "fire" on Saturday nights especially, lasted until the 1870s, when the telegraph supplanted voice and bell. Quieter and more reliable, the telegraph marked a distinct departure from the noise of the old system. Civil authorities in aspiring northern towns also echoed their southern brethren by encouraging the sound of time to facilitate commerce. Inhabitants of Albany, New York, for example, complained in 1830 that the two existing town clocks "were almost useless to the business and laboring part of the community, from the circumstance of their not being heard throughout the city." [50]

Northern elites submitted to the bell's regulatory power and used it to shape the behavior and labor of others. The bell per se did not constitute the essence of power. Rather, it was a proxy for all sorts of relations. For myriad aspects of antebellum northern society to work properly, some coordinating functions of the bell were deliberately neutral and even reversed the traditional master-servant relationship. Elites, like everyone in fact, acceded to the bell, which summoned them to church and warned them of fire or imminent danger. They also allowed themselves to be regulated by house bells. Diary entries made by antebellum patricians, rife with testimony on this point, illustrate the ambiguous nature of the sound of the bell and reveal the importance of who rang and who responded. As Robert Habersham, a privileged southerner who studied at Harvard, noted of an everyday event at home with his family in Saratoga in 1831, "I wish I could have time to write some poetry for the 1st dinner bell. . . . There is the bell, I must carry the girls down." Sound and time were so braided that contemporaries frequently did not bother to disassociate the two and allowed both to regulate the self. In Connecticut in 1834, for example, John Hamilton Cornish noted simply in his diary that "the clock long struck nine and I break of[f] in the midst of a story at this late hour." The sound of time was a powerful force in instilling punctuality among the elite. Indeed, familiarity with local bells and clocks could make one ahead of time. In 1837 Cornish, for example, "got to the college just as the bell was ringing for 3½ oc. Recitation." Such wonderful examples of time obedience were, of course, encouraged.[51]

Elites readily allowed themselves to be regulated by a bell that was rung by others, not because doing so appreciably lessened their authority, but because to exploit labor power, they necessarily had to allow servants and civil authorities to summon them by bells. The servant's act of calling his or her employer or master to the dinner table did not empower the servant or slave because it did not alter the fundamental relations of power inherent in the relationship itself. Servants and slaves were obliged to ring bells as directed, just as they were required to respond to the bells rung in society at large and at work specifically. By virtue of their authority and control of the means of production, elites governed workers' bodies via bells. Even before the widespread use of electric bells, customers of refined hotels could (theoretically) summon servants. In the 1820s a Washington,

D.C., hotel was equipped with "rows of large bells hanging by circular springs on the wall, each with a bullet-shaped tongue, which continued to vibrate for some minutes after being pulled, thus showing to which room it belonged."[52] Servants could ignore the call, but the price they paid usually ensured that they did not do it often.

Bells and the times they sounded had remained unchanged for years, and they proved as effective in regulating antebellum cities as medieval ones. Bells retained the pedigree of nostalgia and helped ease the transition from old to new. In the American context the bell was more than a courier of religious and secular time because it also became an icon of nationalism and freedom. Yet the bell of liberty was subject to qualification, and its ambiguous legacy was contingent on class. Not only did it regulate the hours of commerce and the city, but it also regulated hours of labor, north and south. Northern mills invariably used bells to call workers to labor, as contemporary paintings and sketches make clear. Mill managers' diaries also record the extensive use of bells in northern factories. N. B. Gordon, a manager in Mansfield, Massachusetts, noted on July 6, 1829, "H. Kingman new hung Bell." For fairly obvious reasons, managers liked bells. Factory workers in all industrializing societies were, managers believed, "bound punctually to obey the summons of the factory bell." They were not alone in their subjection to the sound of the work bell. Artisans regulated their apprentices with the town clock and bell. Ben Johnson, "a brick-layer by trade, and a rigid observer of early hours," for example, obliged his apprentices "to be at home every evening before the clock struck ten," presumably to ensure that they could rise early.[53] In regulating labor by the bell, masters of capital shared much.

Northern industrialists and southern slaveholders also agreed that the aural regulation of workers should ideally extend beyond the time and place of actual labor. Like southern masters who essayed to regulate the religious expression of bondpeople and southern industrialists such as William Gregg who enforced temperance in his Graniteville, South Carolina, mill village, northern advocates of industrial management exhorted workers to make particular sounds on Sundays. S. V. S. Wilder of the Ware Manufacturing Company advocated the building of a church for mill workers in the late 1820s in an effort to replace "Sabbath-breaking revelry" with "sheltering groves, destined ere long to become vocal with songs of praise."[54] Zachariah Allen critiqued English industrialization in the 1820s and 1830s by condemning the "proprietor of a manufactory in Manchester" who "rarely troubles himself with investigations of [the workers'] conduct whilst they are outside the walls of his premises, provided they are reported to be regular at their labor whilst within them." Allen believed morally continent workers were workers controlled at work and play. Others agreed. In his 1836 memoir of Samuel Slater, George S. White maintained that "manufacturing establishments exert a powerful and permanent influence in their immediate neighborhoods, and time, if not already, will teach the lesson, that they will stamp indelible traits upon our moral and national character." This was a moral and political matter. The "spirit of enterprise" and civil, orderly conduct applied to workers outside as well as inside the factory and avoided "creating discord and producing civil commotions." Industrialization, after all, courted social noise: "Manufacturing, instead of going on quietly and single-handed in private families . . . grows into large establishments, which employ and bring into association, masses of population." But by regulating working behavior off the job as well as on, factory owners "have it in their power to change the current vice from its filthy and offensive channel, and make peace, order, and comfort among those they employ." Workers who were thriftless could not be trusted with high wages they invariably frittered away in noisy entertainment. It was far preferable to avoid "the mists of error, discord, and confusion" by prohibiting gambling, grog guzzling, raucous leisure activities, and "vicious amusements" by teaching workers "sound morals."[55] Northern capitalists had only marginally more faith in the assiduity of their laboring class than southern masters had in theirs.

If the myriad attempts to control and discipline this "canting and shrieking, very wretched generation or ours" failed, northern elites retained a trump card not unlike that held by slaveholders: the state's ability to try to enforce quietude. The basis for the criminalization of noise was already well established by the late antebellum period. "For making a noise while in his cell," for example, one prisoner in Massachusetts in 1845 was given a day in solitary confinement.[56] The control of sound was the essence of discipline in antebellum northern prisons. In the workshop of a South Boston penitentiary, for example, Charles Dickens noted, "The treadmill is conducted with little or no noise; five hundred men may pick oakum in the same room, without a sound; and both kinds of labour admit of such keen and vigilant superintendence, as will render even a word or personal communication amongst the prisoners almost impossible." Although the "noise of the loom, the forge, the carpenter's hammer" in other parts of the prison "greatly favour those opportunities of intercourse," even these were necessarily "hurried and brief," and the aim of the penitentiary was to promote penance through silence. The women "did their work in silence like the men," and Dickens thought escape impossible, "for even in the event of his forcing the iron door of his cell without noise," the escapee would be rendered visible by the prison's spatial arrangement. Philadelphia's penitentiary was even more attuned to disciplining the prisoner with sound, or rather, its deprivation: "There is no sound, but other prisoners may be near for all that. . . . The cells were so constructed that the prisoners could not hear each other, though the officers could hear them."[57]

Disciplining through imposed silence grew increasingly common in the early nineteenth century. The "silent system" proved popular in England because, reformers believed, enforced silence encouraged prisoners' quiet contemplation and permitted the creation of an almost monastic space of quietude where penance would flourish. A similar mentality inspired the adoption of the silent system in the antebellum North. Harriet Martineau visited the Philadelphia prison in the late 1830s but, like so many other antebellum elites, found that silence was impossible to enforce, even when prisoners were in solitary confinement. "The convicts," she observed, "converse with . . . ease, through the air-pipes or otherwise, at night, as they do by speaking behind their teeth, without moving the lips, while at work in the day." If the men could not be kept quiet, there was even less hope for women: "The attempt to enforce silence was soon given up as hopeless; and the gabble of tongues . . . was enough to paralyze any matron." Martineau thought brutish a restraining chair with a muzzle to keep "refractory female prisoners quiet."[58] Several other northern prisons refined the silent system to some effect. At Sing Sing in New York, officers, presumably remembering the silence of the Indian warrior, wore moccasins, which allowed them to approach cells without convicts' knowledge. There, rules of silence were, according to contemporaries, effective; rarely were prisoners heard talking after lockup.[59]

Occasional whispers aside, the ability to maintain silence, northern reformers maintained, not unlike southern masters, betokened regularity and sobriety. In prison convicts learned when to speak and when not to, and the bell was their timepiece (just as it was for those who had industrial jobs). Cells were therefore designed with thick walls. "No communication whatever between the persons in the different cells can be effected," boasted reformers in Philadelphia in the 1790s. "The most agreeable order and quietness" resulted. The silent system incorporated penalties for chatter and swearing in an effort to prevent the spread of vice through the tongue. Moreover, "solitary confinement effectively dissolved networks of collective support, forcing the inmates to confront authority alone." Inmates would return to society with a sober quietude. Reform of the individual and the preservation of peace in society were the aims here. Criminals' quiet cunning, their mastery of "prowl" and "lurking," and their ability to be as stealthy as Indians rendered them social adversaries of the worst kind; their threat lay in both their noise and their silence.[60] Despair followed silence, and the noise of the impoverished mob followed the silence of despair. A prison riot in 1820 in Philadelphia disrupted the aural serenity not just of the prison but of "a quiet city like ours," argued elites. This was a veritable "mob," "tumultuous" and "thundering" in its noise and deadly in its intent.[61]

From the mid-1790s on, penal reform in Philadelphia attempted to shroud prisoners in silence, marking the intersection of "coercive, institutionalized labor with monastic traditions of penitence and self-transformation." Benjamin Rush, perhaps the best known of such reformers, urged the precise mapping of space and time onto prisoners' bodies through the construction of prisons focused on quieting the imprisoned. The place should produce sounds that penetrated the body and led to penitence: "Let its doors be of iron; and let the grating, occasioned by opening and shutting them, be encreased by an echo from a neighboring mountain, that shall extend and continue a sound that shall deeply pierce the soul." Ensconced in solitude, inmates had time and quietude enough to hear their souls. Deafened to political and civic registers, they would hear only the sounds of their immorality.[62]

Anxieties about the nature of working-class life during and after the 1820s encouraged Philadelphia patricians and reformers to project the discipline of the prison onto society at large. Values about order, silence, and discipline radiated from the prison system and were used to tame the excess passions of a working class prone to rowdy protest. To be sure, there was growing criticism of the solitary system as cruel and expensive. Furthermore, some reformers feared that solitary confinement infantilized inmates and dispossessed them of qualities necessary for participation in the body politic. In voicing these worries, northern reformers were perhaps thinking of the silence of slavery. Still, such qualms did little to temper beliefs that the laboring classes could benefit from a dose of enforced quiet solitude. Consequently asylums for disruptive youths, the poor, and prostitutes embraced "privacy and quiet" as key features of their reformative measures.[63]

The economic cost of maintaining single cells helped end widespread use of the silent system and helps explain why it was less common in the fewer southern penitentiaries. Southern leaders undoubtedly preferred the enforced silence of criminals—and the moral introspection that such quietude afforded—but simply put, southern states could ill afford to build many such facilities. Southern prisons, as visitors often remarked, were noisy places. Not that it mattered much. Slaveholders, after all, preferred to deal with their most refractory populations on their plantations and in ways that were sometimes not far removed from the silent system of northern prisons.[64]

Whatever the salutary effects of the sounds of progress, northerners at times preferred the sounds of nature, its purrs offsetting the noise of modern urban life. "The tradesmen, the attorney, comes out of the din and craft of the street, and sees the sky and the woods, and is a man again," reckoned Ralph Waldo Emerson. Holidays from the buzz of the modern were often sought in nature's soothing trills. "At last I have escaped from the dust and turmoil, the thronging crowds, the ceaseless din . . . and ten thousand annoyances of the great city of New York!" cheered one New Yorker in 1847. He had found the countryside, where "omnibuses no longer deafen me with their unending clatter."[65] Tuckerton, New Jersey, "about fifty-two miles from Philadelphia," was a "delightful refuge . . . from a noisy world. You get rid of hubbub and humbug the moment you leave Camden. No locomotive shakes you to a jelly; no turnpike keeps up an incessant din in your ears. You travel silently, slowly, softly through the warm, gushing sands and still pines, everything . . . disposing thought to repose."[66]

The sound of American progress sometimes seemed decidedly discordant with nature. This was the crux of the growing Transcendentalist critique of the market, replete with pastoral conceits and focused on the untinctured authenticity of romantic repose. But as loud as it was, this critique was insufficient to drown out the growing boosterism of northern industrial capitalism. Self-imposed exile, such as Henry David Thoreau's at Walden Pond, while giving him reprieve from the noise of northern modernity, was an ephemeral escape from a society that had already gone a long way down the road toward liberal, capitalist progress. Such romantic refuges were just that: mere escapes for northerners, islands of rural serenity to be consumed for brief periods, places where they could, perhaps for the first time, hear the curious warbles of birds and wildlife before they returned to their progressive, bustling society. To southerners they were a way of life. By the 1830s that way of life, while similar to the northern in several key respects, was becoming increasingly distinct, premodern, and threatening to many northern ears.[67]

The whites now . . . cause the peal of the church-going
bell to resound where formerly, with rude ceremonies,
the Great Spirit was invoked; . . . cause the busy hum of
industry to be heard over the land; . . . and fill the once
untrodden wilds with the resounding fame of literature,
arts and arms.

—Wilbur M. Hayward, "Black Hawk's War,"
Frederick Douglass's Paper, November 19, 1852

(((5. Sounds Modern

       Although colonial patterns of sound and noise lasted well
into the nineteenth century, changes in technology and society required
some regions to redefine and reconceptualize legitimate and illegitimate
sound. Northern industrialization, urbanization, and democratization (in
essence, the rise of liberal capitalism) required configuring new tones and
increased decibel levels and, in effect, merging such registers into modernism itself. Northern elites listened to the growing volume of "Chants
Democratic" and promoted a more boisterous social order that applauded the voice of democracy and industrial capitalism, even though
the sounds produced by such developments sometimes set them on edge.[1]

    There was a good deal of exaggeration and distortion in the construc-
tion of sectional differences, especially where democracy was concerned.
By the 1830s most southern states were reforming or had reformed their
voting provisions as thoroughly as their northern counterparts. While
state legislatures in both regions often denied the vote to free blacks,
property requirements for the vote were abolished. Moreover, elections,

campaigns, and political meetings had a similar tenor in both regions. Yet there was a difference, and northern abolitionists in particular pointed to it in an effort to cast the North as democratic and the South as archaic and aristocratic. That difference, they believed with some justification, was heard principally in the silent political voices of roughly 4 million southern slaves and the supposed quiescence of poor whites. Abolitionists, free soilers, and later many Republicans certainly exaggerated the differences in southern and northern democracy, and they, too, harbored doubts about excesses of democratic participation. But that they exaggerated these and other distinctions for political purposes does not deny the material basis of those differences.[2]

It is worthwhile elaborating on the structural basis of the diverging northern and southern social and economic soundscapes. The South and the North sounded increasingly different (qualitatively and quantitatively) because their competing modes of production produced different material and cognitive soundscapes, altering how leaders in each region heard not only themselves but also one another. Treating the South and North as uniform without much comment on the varieties of soundscapes that plainly existed in each region is something of an oversimplification. Structurally there were pockets of industrial and urban noise in the South, just as there were acres of rural quietude in the North by 1860. But like contemporaries, we necessarily deal in aggregates and stereotypes that were used to characterize each section. It is also worthwhile noting that while there were faint strains of free wage labor in the South, heard most obviously in southern factories and towns by the 1830s, there were no longer sounds of slavery in the North. The acoustic dimensions of sectionalism evolved and gained currency in the 1840s and 1850s especially when leaders in each section listened to and heard one another in binary terms.

The North was increasingly more urban, industrial, and populous than the South. Between 1820 and 1860 New England's urban population more than tripled. The proportion of its population residing in towns and cities climbed from 10.5 percent to just over 36 percent. The Old South's South Atlantic urban population doubled in the same period, to be sure, but from a measly 5.5 percent in 1820 to 11.5 percent by 1860.[3] Southern towns and the South generally did not attract nearly as many immigrants as the North. The percentage of the Old South's population that was foreign-born was less than one-quarter of the North's; as such, the South's white population was ethnically more homogenous. In 1800 82 percent of the southern labor force worked in agriculture; the figure for the free states was 68 percent. Sixty years later 40 percent of the northern population worked in agriculture, while 84 percent of the southern population were rural laborers. Northern agriculture also became increasingly capital-intensive and mechanized. By the eve of the Civil War the free states had almost twice the value of farm machinery per acre per farmworker as did the slave states. The proportion of the South's capital invested in manufacturing declined from 31 percent in 1810, to 20 percent in 1840, to 16 percent in 1860. The North had roughly 2.5 times the amount of capital invested in manufacturing in 1810 and 3.5 times as much by 1860. Moreover, the West was beginning to take on the characteristics of the North, not the South. It was urbanizing and industrializing faster than both regions and, by 1860, was becoming more northern than southern.[4]

Consider maps 1 and 2, bearing in mind that denser, urban populations are louder than less dense ones; that more ethnically diverse populations, by virtue of the different sounds created by different tongues and accents as well as cultural styles, can sound more discordant than more homogenous populations; that steam power is louder and sounds qualitatively different from muscle power; that industrial-urban societies are characterized by increased decibel levels; and critically, that slavery had a timbre different from that of freedom.[5] Leaving aside for the moment the cognitive, perceptual values given to these decibels (loud machinery and towns are not necessarily bad to all ears), material changes in the North made it sound louder than and qualitatively different from the South. The North's higher rates and levels of urbanization and industrialization gave the region a denser, more cluttered soundscape. The transition to such a soundscape took time, but there was no mistake that it was occurring in the rapidly industrializing and urbanizing antebellum North.[6] New England's cotton mills, for example, altered the region's soundscape. From the erection of four such mills in 1805, the figure climbed to twenty in 1813 and accelerated thereafter. David Wilkinson recalled that in 1791 his father built a small air furnace or "reverbatory," which doubtless made the sound of industry palpable and physical. Such machinery tended to add to the ambient noise level in a community and increased the unintentional radiation of sound. Unlike horns or bells, which were intended to generate sound, factories and machinery had noise as their by-product that added to the North's ambient noise levels.[7]

The rural North must have sounded quieter and qualitatively different from the region's industrializing areas, just as the border states of the

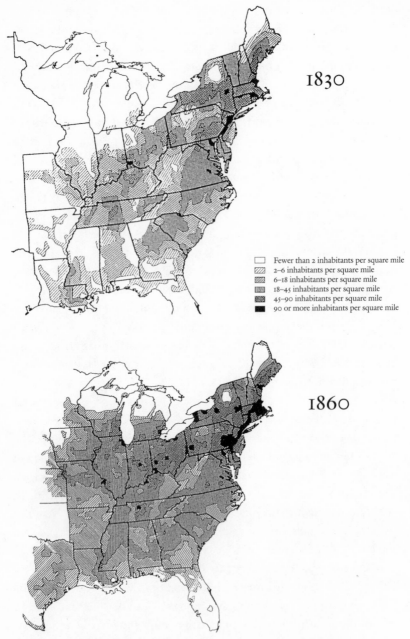

1830

Fewer than 2 inhabitants per square mile
2–6 inhabitants per square mile
6–18 inhabitants per square mile
18–45 inhabitants per square mile
45–90 inhabitants per square mile
90 or more inhabitants per square mile

1860

MAP 1. *Population Density, 1830 and 1860 (Paullin,* Atlas of the Historical Geography of the United States, *76–77; reprinted courtesy of the Carnegie Institution of Washington and the American Geographical Society of New York)*

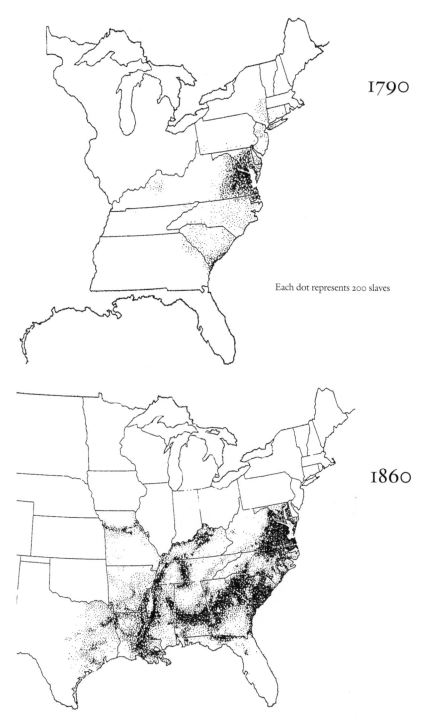

1790

Each dot represents 200 slaves

1860

MAP 2. *Number of Slaves, 1790 and 1860 (from Fogel and Engerman,* Time on the Cross, *fig. 11, p. 45; © 1974 by Robert William Fogel and Stanley L. Engerman; used by permission of W. W. Norton & Co., Inc.)*

South and that region's few cities sounded louder than the vast majority of the slaveholding, rural South. But northern industrialization, growing use of wage labor, and the absence of slavery made the South sound different to contemporaries. For example, the rural Ohio soundscape was not composed of the sounds of slave voices, slave sales, slave punishment, or plantation sounds generally. Certainly southern border states were more industrial and rather more urban than states in the Deep South, more so toward the end of the antebellum period. In 1850 Virginia, for example, ranked eighth in capital invested in manufacturing establishments, behind (in order) New York, Pennsylvania, Massachusetts, Ohio, Connecticut, New Jersey, and New Hampshire. Ten years later Virginia ranked sixth nationally in number of manufacturing establishments, bested only by New York, Pennsylvania, Ohio, California, and Massachusetts. Yet the difference between Virginia and New York was far greater than, say, the difference between Virginia and South Carolina. In 1850 South Carolina ranked eighteenth in the nation measured in capital invested in manufacturing establishments (totaling $6,056,865); eighth-ranked Virginia had three times as much capital invested in manufacturing ($18,109,993). But the figure for New York was more than five times greater than Virginia's at a colossal $99,904,405. A decade later, this time measured in number of manufacturing establishments, Virginia still echoed South Carolina more than it did New York. New York had 22,624 manufacturing establishments compared with Virginia's 5,385 and South Carolina's 1,230. In other words, New York was home to four times as many manufacturing establishments as Virginia, the numerical difference being 17,239. While Virginia had four times as many establishments as South Carolina, the actual difference in number was 4,155.[8]

Such relative differences help explain why contemporaries thought that even the more industrialized states of the border South sounded more southern than northern. As abolitionist Ebenezer Davis said while steaming along the Kentucky-Ohio border by boat in 1840, "Upon the left bank of the stream the population is rare; from time to time one descries a troop of slaves loitering in the half-desert fields. . . . Society seems to be asleep, man to be idle." From the Ohio side "a confused hum is heard, which proclaims the presence of industry . . . and man appears to be in the enjoyment of that wealth and contentment which are the reward of labor." For Davis the difference in regional sounds had moral and economic dimensions, with the registers of premodern slavery taking center stage. Southern ears heard the same sounds differently. A correspondent for *De Bow's Review* in the 1850s hears "activity" and "bustle" in the North, but when he turns "his face southward, and breathes the milder air of Virginia and Carolina, a far different prospect meets his eyes"—and his ears, too, for "the sounds of the steam engine and the manufactory rarely falls upon his ears." As for Davis, the sounds were moral proxies for this listener: "The sufferings of the slave are nothing in comparison to what I have often witnessed in the cities and poor-houses of the North." Southern quietude for this listener was good not least because its social and economic underpinnings kept wage labor immiseration at bay.[9] In short, the border South's commitment to slavery, while allowing urban-industrial growth on the margins, nevertheless ensured that the drone of bondage would muffle the sounds of modernism.

More than the North, the South resisted the incursion of the industrial-capitalist soundscape and all that those sounds represented, and in this respect the North followed a relatively traditional trajectory. The northern states behaved as had most industrializing societies. Despite the increase in noise generated by the new machines, opposition was relatively rare because the new sounds were considered congruous with the prevailing political and economic ideology emerging in the North. During the ante-bellum period the North passed little formal legislation to lessen or resist the sounds of industrialism. Some voices protested the rumbling of the machines, the literati figuring prominently among them. But, that much said, even poets liked some of the new sounds.[10]

Slavery did not sound like freedom to northern capitalists or abolitionists. The social and economic basis of each section grounded the aural images of sectional otherness in material reality even as those representations became politically contorted, clumsy, and exaggerated. The differences between North and South operated subjectively and objectively, the latter shaping the former until cognitive constructions took on a saliency of their own. But clumsy though they were, the projections of sectional otherness never became so abstracted from material reality that they lost authenticity. Once structural differences were cast in metaphoric and symbolic terms, the images themselves took on a reality that confirmed even as it distorted the differences between North and South and undermined the real structural similarities in class formation, ethnic heritage, common national memory, levels of economic and social inequality, and legal and religious traditions shared by both sections. This process helps explain why contemporaries commented on the profound and increasing differences.[11] As a South Carolinian told London *Times* correspondent William

Howard Russell in 1861, "We are an agricultural people, pursuing our own system . . . breeding up women and men with some other purpose than to make them vulgar, fanatical, cheating Yankees." Or as Louis Wigfall of Texas advised Russell, "We are an agricultural people. . . . We have no cities—we don't want them. . . . We want no manufactures: we desire no trading, no mechanical or manufacturing classes." Alternative sentiments emanated from northern free wage labor advocates and reached southern ears through print or in person. For Theodore Parker the South was "the foe to Northern Industry—to our mines, our manufactures, and our commerce. . . . She is the foe to our institutions—to our democratic politics in the State, our democratic culture in the school, our democratic work in the community, our democratic equality in the family." [12] Politicians' use—through printed and heard words—of powerful imagery to portray each section as aurally distinctive convinced many contemporaries that they belonged to a sectional soundscape even if their specific geographic locale did not always echo precisely the structural sounds of their section writ large. The antebellum revolution in print culture (combined with word of mouth for the ears of the illiterate) helped people hear the pulse and keynotes of their societies. In this way a northern farmer who was far removed from city and industry and a southern yeoman who, particularly in a non- or small slaveholding region, rarely heard the actual registers of plantation slavery became embedded in sectional soundscapes that applauded particular soundmarks even as the individuals heard only the faintest literal strains of that acoustic environment.

Qualitative evidence points to the growing volume and changing timbre of northern society, particularly in the last thirty or so years before the Civil War. As Isabella Lucy Bird observed in 1856, immigrants to New York City from rural Europe "found themselves in the chaotic confusion of this million-peopled city" and were bewildered "by cries of 'Cheap hacks!' 'All aboard.'" [13] By virtue of the rapid economic and industrial development in the North, the region was filled with registers loud and alien to immigrant ears. Isaac M. Wise from Bohemia found New York City in 1846 a commercial cacophony. Recalled Wise, "I had never witnessed anywhere such rushing, hurrying, chasing, running. In addition to this, there was the crying, blowing, clamoring, and other noises of the fishmongers, milkmen, ragpickers, newsboys, dealers in popcorn, etc.—ear-splitting noises, which were often drowned in the rumblings of the wagons and the cries of the street gamins. All this shocked my aesthetic sense beyond expression." [14]

Many northerners, though, accommodated the sounds of industrialization. They did this in part by reconceptualizing the new within remembered intellectual frameworks. The sounds of insects, for example, more than any other in nature, give the impression of being "steady-state or flat-line sounds." Although this is an auditory illusion because many insect sounds are modulated, the subtlety of that modulation sometimes escapes the human ear and gives the sound of insects a monotonous, continuous quality. Because the first man-made flat-line sounds came with the industrial revolution, one metaphor used for progress was the sound of insects, which helps explain why many northerners, even those who grew up in the countryside, accommodated urban-industrial registers quickly. James Hamilton Cornish, who moved from rural Michigan to urban Connecticut in the 1830s, heard the hum of northeastern cities much as he heard nature, the harmony of the two blending in his ears: "The musical Grasshopper is merryly tuneing his pipes, while the little birds are chirping upon the trees under my window. How calm and tranquil all, the distant rumbling of the heavy drays, the constant din of the City, the occasional peals of the bells fall not unpleasantly upon my ear." Even the rabidly antimodernist Henry David Thoreau sometimes heard nature in modernity and modernity in nature. He believed that "the hum of insects" testified to "nature's health or *sound* state." During his week on the Concord and Merrimack Rivers in the late 1840s, Thoreau heard the noise of modernism. "Nashua," he lamented, "now resounds with the din of a manufacturing town." But many of his metaphors for northern progress were natural, and much of his appreciation of the natural world was refracted through his modern ears. He noted how the "honey-bee hummed," compared the alarms issued by squirrels to "the winding up of some strong clock," and was sometimes lulled by the pulse of the modern. Near Boston he remarked, "Instead of the scream of a fish-hawk scaring the fishes, is heard the whistle of the steam-engine, arousing a country to its progress," and he heard the telegraph as "faint music in the air like an Aeolian harp." Some understandable calls for occasional breaks from the new modernist soundscape notwithstanding, northerners generally found comfort in reconciling the hum of industrial modernity with the purrs of mastered nature.[15]

For northern industrialists the sound of industriousness served as precursor to the sound of industrialism. A new foundry in Delaware County, Pennsylvania, in 1849 allowed anxious moderns to hear activity: "This establishment is now in successful operation, the hum of industry being heard in all its departments."[16] Although elites north and south hankered to hear the hum of industriousness, northerners especially embraced the hum of industrialism. They listened for the unmistakable sound of capitalism in places that had abandoned bondage and modernized. The building of London's Great Exhibition in 1851, readers of the *National Era* learned, "gets on finely. There are now two thousand men employed upon it, and the music of their labor is the music of the Park."[17]

Incorporating the sound of industrialism into the New England soundscape was not particularly difficult. Although industrialization demanded the regulation of laboring bodies, some legacies of old proved good enough to lend assistance. Just as it had during the colonial period, the antebellum church bell retained the quality of "solemn tones" as it rolled "along the valley and echoing among the hills." But bells of liberty also regulated the hours of commerce, cities, and increasingly, northern wage labor. To the northern bourgeoisie, factory bells could be aesthetically pleasing. A "new bell at the Cotton mill" for someone who did not work there could have a "very clear musical sound" in 1849. Machines that made loud noise because of their various moving parts were "harmony itself" because they moved in sync. That silent farms were replaced with bustling cities, that the boom of urban and industrial capitalism drowned out quiet preindustrialism, was a source of admiration, not condemnation, for northern boosters. "Louder and louder, on each succeeding day, waxes the tumultuous hum of this bustling Babel," cooed an observer of modernization in New York City in 1850. "In all directions," he continued, "the great city of the New World is stretching forth her restless hands, converting silent fields into crowded streets, and invading all the old peaceful retreats with thousands of her fast-growing, heterogeneous population." This aural assessment was not ambivalent in any sense. The westward expansion of the sound of the modern and the enveloping of the silence of the old by the sounds of the new called for celebration, not lamentation in northern ears: "All along the noble rivers which embrace the city in their loving arms, how rapidly swells the great tide of life and industry! Miles and miles of ships . . . propelled by steam; and what a ceaseless chorus of human voices, and rattling of vehicles, and clanking of machinery, and ringing of bells, broken in upon, ever and anon, by the far-resounding echoes of ponderous hammers, fashioning into shape the iron skeleton of some new monster of the deep." All this was to the good, for it betokened "more and more of a metropolitan air" and "civilization, refinement, and wealth." Progress was heard as much as seen, and its meaning was invested with the future of a country. Little wonder that words describing industrial machinery were onomatopoeic. "Throstle frames" did what they suggested and gave the impression of industrial activity. The industrial process itself, the moving frames of power looms, were reckoned in terms of "beats per minute" in 1813, thus giving aural rhythm to manufacturing efficiency.[18]

Men of character and capital were responsible for this new soundscape. Dilapidated mills bought by Samuel Bancroft in Pennsylvania in 1857 were sure to get back on aural track. Although these "mills have not been in operation for some time," read an editorial, "we presume, from the energy of the purchaser, that the hum of industry will ere long animate the place." Refurbished mills, after all, meant wage-paying jobs, the stuff of which national futures were made. The health of factories near Chester Creek, Pennsylvania, in 1851 was evident because of wage labor ("over ten thousand dollars a month are paid in wages to the hands there employed") and because these "extensive cotton mills, paper mills, woolen mills, and various manufactories of edged tools, spades" and the like "teemed with the hum of industry, receiving their stock and supplies from Philadelphia, and returning their fabrics here for market." The registers of wage labor, industry, and the market were in harmony.[19]

Some factory entrepreneurs were better than others. James Campbell, proprietor of a Pennsylvania mill, was an industrial maestro. He, after all, had banished the deafening silence of economic dislocation. "The sound of the Pioneer [mill] bell, reminds us," chortled an 1858 newspaper report, "that the factory of our indefatigable friend and neighbor JAMES CAMPBELL, is again in operation, after remaining silent for so many months." Campbell and men like him were conductors of industrial orchestras: "It is gratifying once more to hear the sound of the shuttle and the hum of industry which resounds from the Pioneer establishment, when the looms are in full chorus."[20]

Setbacks happened, but even fires that silenced factories did not stop industrialists from listening to the future. Damaged plants could be rebuilt, "and, in a short time, the busy hum of industry will be heard upon the spot where the discolored walls now stand as a monument of the destruction which fire can produce." Similarly, a Pennsylvania cotton factory consumed by fire in 1859 was on the mend even as it smoldered. "Scarcely had the flames subsided" when "the bricklayers, masons and carpenters" were "busily engaged in reconstructing the work; in four weeks from today, it is probable, the busy hum of industry will again be heard within the walls of the building." A new railroad in Lynn, Massachusetts, beginning in 1836 animated the town and put it on the road to economic progress. Building it introduced "bustle and stir" that "broke the monotony of the dull season following the panic of 1837." Rock blasting, bell-regulated railroad labor, and the trill of drills only confirmed that Lynn was on its way to industrial greatness. Industry was booming elsewhere, too. "All the cotton mills in our Borough are now in operation," crowed the *Delaware County Republican* in September 1861, adding, "It sounds pleasant once more to hear the shrill whistle of the steam engine, at early dawn, after so long a suspension of business."[21] Industrial capitalism did not preclude a commitment to democratic freedom. Frederick Douglass said as much in 1848 when he described England. "Calm, dignified," England embraced economic modernity and its political handmaiden, and "her toiling sons, from the buzz and din of the factory and workshop" were not deafened to "the joyful sound of 'Liberty-Equality-Fraternity'."[22]

The sound of industrialization was not necessarily the same as the register of urbanization. The Waltham System, with its industrial villages located away from urban conurbations, managed to combine the tick of the pastoral with the buzz of progress—so much so that boosters believed northern urbanites would welcome the change in the soundscape. When the railroad linking Rockdale and Philadelphia was completed in the mid-1850s, one newspaper correspondent thought the industrial timbre of the "village, being almost within an hour's ride of Philadelphia," would prove alluring to the citizens of the big city. Should they choose Rockdale "as the place of the summer sojourn," ventured the writer, "they will enjoy the bracing and invigorating air of our numerous hills, and hear with pleasure the busy hum of industry in our valleys." Industrial soundscapes were now marketed as vacation soundscapes.[23]

A commitment to the registers of modernity did not preclude northerners' love of pastoral melodies, just as southerners' penchant for the sounds of slave-based agriculture did not deafen them to some sweet-sounding modern tones. Like their southern counterparts, northern elites listened to the present of natural sounds, chuckling streams, and sighing trees to remind them of what they perceived as a uniform historical past. A northern visitor to Niagara Falls in 1848, for example, heard history in rushing water: "Now, as for ages past, boiling, gushing, foaming ever, rolling on continually, like the car of Time," echoed the falls, at once uniting historical present with imagined past. Crashing water, warbling birds in "the far-resounding woods echo and re-echo the joyous melody, till it dies away in softly soothing murmurs, in the far distance." Here "the voice of God seems to whisper in the ear of listening nature—Peace! peace! peace to the troubled breast, rest for the wicked, hope for the oppressed." Adjacent "verdant valleys" and forests were quiet, too. But the scene was changing, even as the visitor listened. It was "not many years since, as appears by the stumps which still stand firm and sound," that the forest "reigned in undisturbed and silent solitude, over this entire district of country." The intrusion of human sound caused some angst because it threatened sublimity: "I can scarcely find in my heart to find fault with the man who, it is said, from motives of miserly selfishness, preserves the rapids from the clatter of the mill and the manufactory." Ultimately such motives were selfish, for this was an age of progress, an age when the "eye is not satisfied with seeing, nor the ear with hearing." But Buffalo's soundscape was applauded, even though its volume drowned God's whisper, cascading water, and warbling birds: "The steam engine is rapidly hewing a way through these wild forests, implanting civilization and commerce, and the hum of a teeming population." Moreover, this "change would be produced even more rapidly, were it not for the short-sighted and selfish speculators" whose purchase of large tracts of land inhibited their improvement and industrial development. Sacrifices were necessary because the sounds of progress were considered accurate proxies for capitalist modernity. Here nature's quietude lost to the sounds of the modern. It was aesthetically pleasing, to be sure, but not nearly as important as the American future. Anyway, the hum of industry was almost sublime, was it not? The "busy din and constant whirl of swift machinery," after all, was what "makes a town" and successful industry.[24]

Certainly there was no persuading northern youth of rural aural delights, for many of them had been reared on a dry diet of pastoral tinglings. No, they wanted the sound of invigorating modernity, and they were ready to add their decibels to the growing volume of northern urban democracy and capitalism. One agricultural writer suggested as much in 1851. "Another cause operating against the farm," he maintained, was the sound of the city. The "young man [hears] . . . not the green fields, with trees and flowers, and the singing of birds and the murmur of bees . . . and all the beautiful sights and sounds of his daily life; but his imagination dwells on paved streets, and brick walls, and muddy streams of gutters, and the rattle and din where men herd and jostle one another." There was a recognition that the price for national progress was a gradual abandonment of the quiet life, and increasingly men left the "silent retreats" of the rural North for the excitement and bustle of its urban and industrial centers.[25]

While the northern bourgeoisie was busily resurrecting old sounds and incorporating the growing hum of industrial capitalism into its cognitive soundscapes, workers both accommodated and resisted the new sounds and the uses to which they were put. In turn, their resistance encouraged northern elites to make concessions as to what was noise and what was sound and sometimes led them to reconstitute the faintest thuds of popular politics as fitting registers of democracy. Northern ruling classes found embedded in that resistance an implicit requirement to practice what they preached by reconfiguring the voice of the virtuous crowd as the happy throb of democracy. Other reconfigurations were also at work. While it took little time for the first factory noises to blend with the sounds of the prefactory northern soundscape in elite ears, the process whereby factory sound became legitimate to workers was a little more protracted.[26]

Although industrialization proceeded unevenly, clear benchmarks existed for the industrial revolution in the antebellum North. The constant negotiations that free wage labor industrialism inaugurated between workers and managers are apparent if we listen to what contemporaries heard.[27] A good deal of the North's earliest industrial activity was performed in homes, not in factories, as with most of Philadelphia's hand-loom weavers. But their activities gave the city's working-class districts a distinctive sound: "Throughout parts of the city, especially that formerly known as Kensington, the sound of . . . looms may be heard at all hours — in garrets, cellars, and out-houses, as well as in the weavers' apartments." The concentration of the weavers in certain streets doubtless increased the volume and stamped parts of Kensington as aurally distinctive, which in turn may well have helped delimit that part of the city as especially Irish.[28]

Workers who labored at home and who were equally responsible for unleashing the enormous productive capacity of industrial capitalism could be accused of making noise, and they found themselves in the impossible position of having to generate decibels to survive in private lodgings where landlords defined legitimate sound. Carl Conrad, a tailor in New York City, put the problem into graphic relief. "Before we had sewing machines," he recalled, "we worked piecework. . . . We had no trouble then with neighbors, nor with the landlord, because it was a very still business, very quiet; but in 1854 or 1855, and later, the sewing machine was invented and introduced . . . and the bosses said 'We want you to use the sewing machine.'" Conrad found his need to produce limited by the machine he was required to use. His landlord told him that "the machine makes too much noise in the place, and the neighbors want to sleep," and he found that "we have to stop sewing earlier; so we have to work faster."[29] Circles like these became vicious. As wage labor slowly undermined the old apprenticeship system, young workers found that they "had no quiet place to read or relax," and many, for want of brief moments of the *vita contemplativa,* turned to gangs and taverns.[30]

Outside home and urban work, industrialization in the North took hold in particular regions that less than a generation earlier had sounded like the countryside. Pawtucket, Rhode Island, which in 1790 became the site of the first mechanized textile factory in the United States, reveals the complexities of the industrialization process and offers an aural richness testifying to that development. Unlike many other factory towns in New England, Pawtucket was a settled village "with a substantial artisan population" before it industrialized fitfully and incrementally. As such, sounds of the traditional coexisted with the tones of the modern.[31]

Sounds of industry and industriousness were heard in Pawtucket village before the coming of the factory. In the mid-eighteenth century "over three-quarters of households . . . owned spinning wheels and looms." Household production had already introduced the throb of industrialization and was undoubtedly a necessary evil for families that had to hear production within their homes. Acoustic privacy was a luxury early textile workers in homes could not afford, and in this sense the reallocation of production to a central location—such as a factory—probably had the effect of quieting the laboring household even as it concentrated industrial noise in one location.[32]

When it occurred, industrialization in Pawtucket disrupted seasonal soundscapes. Workers were tied to the factory for much of the year, and the sounds of the summer harvests and of autumnal repairs to equipment were gradually displaced by a throb that demanded workers do the same thing at all times of the year. Industrialization and the disciplining sounds of the factory bell introduced qualitatively new sounds of progress.[33]

The transition, while fitful, happened quickly. Textile manufacturers mechanized speedily, and sounds of old industry harmonized with the keynotes of new factories. By 1790 textile manufacturers had been established in Boston, Worcester, Beverly, and Haverhill, Massachusetts; New Haven and Hartford, Connecticut; Portsmouth, New Hampshire; Philadelphia; Baltimore; and New York.[34] Although the machines stopped frequently for lack of cotton, parts, water, and labor, when operating they were "constantly in motion, emitting a low and continuous hum."[35]

Once industrialization was under way, northern capitalists turned to old sounds to regulate new processes. Bells figured prominently. The application of the bell to the factory was, from the perspective of capitalists, seamless and logical.[36] Inside the factory, workers were regulated by aural signals and were required, not unlike southern bondpeople, to perform even peripheral work noiselessly, which suggested efficiency and order to masters of capital and labor. At the Linwood factory in Delaware, for example, workers operated a new steam-driven pump bought to quench fires. At "a certain signal" the hands scurried, and "each individual was at his post, and with far less noise and confusion than attends the efforts of well organized fire companies." Here were the ideal workers: efficient, smooth, sober, and quiet. Similarly, from 1805 to 1813 workers in some Pawtucket mills were fined for talking to fellow workers and for "moving the wheel which belongs to the bell and by that means making short bell ringings." In both instances managers charged that worker shenanigans altered the soundscape of the mills and limited the factory's productive capacity.[37]

What workers considered the noise of the pitiful, managers often heard as positive. Self-styled management expert Samuel Ogden insisted on labor discipline on the factory floor. In 1815 he described with relish how quality control could be enforced in a cotton mill: "The inspector takes every fleece (one by one) and holds them to the light, so that he can see through them, and if one fleece is defective, the whole is returned, to be looked over, and picked better where it is not clean." To the working women who were paid by the piece (Ogden's preferred system), the process was agonizing—returned fleeces meant less money. For Ogden, though, the sound made by the women during the fleece inspection held a curious fascination that identified in aural terms the crux and one-sided balance of labor relations in New England textile mills: "During the time of looking it over again, a mortifying bleating is uttered from the pickers, in various parts of the room, and that, together with the trouble, brings them into the way of doing their work well."[38]

Resistance to such exploitation took several forms. When trained textile workers were scarce (and demand for their talents still high), skilled weavers on both sides of the Atlantic could keep to their own pace, heard in the rhythm of the loom: "Plen'ty of time, plen'ty of time" at the beginning of the week; "A day t'lat, a day t'lat" at the end."[39] Although they often remained quiet while at work and, in fact, were sometimes made to work "in complete silence," as one foreign observer noted of an artillery factory in Boston in 1826, workers resisted by voicing their concerns during their time off.[40] In this sense, leisure time was hardly that, since some of it was taken up with resisting work-time exploitation. For a working class perpetually admonished to keep quiet, evangelical Christianity and the revivals that went with it offered a release whereby their social noise was transmuted legitimately into God's golden sound. At meetings they noised with relative impunity. Workers embraced the most expressive aspects of evangelicalism because "religious enthusiasm . . . served as counterpoint to the deadly tedium of machine tending [and] . . . as a counsel of quietude." Mill owners sometimes frowned on the excesses of evangelicalism because rather than promoting quietude it encouraged expression of the plebeian self. Even as managers championed the virtues of sobriety and thrift that were an important part of northern revivalism, too much evangelical enthusiasm, tied as it was to excessive and irresponsible democratic passion, worried them, and they tried to stop it. The worry was real. When workers organized, they marked their solidarity aurally, and capitalists heard "tumultuous crowd[s] . . . shouting." Workers using their voice—one of the few weapons in their arsenal—to protest their exploitation worried mill owners to the point of distraction. Consequently, operatives at the Pawtucket mills were instructed in disciplining aspects of evangelical religion in an effort to give "explicit religious sanction to the virtues of sobriety" and the industry that sobriety encouraged. Industrial morality emphasized order, inner discipline, and productive quietude.[41] Penitentiary silence was resonating in northern society beyond prison walls.

Skilled workers in particular offered alternative values to elite sounds. Artisans, who preferred the feel and the sound of hard money, critiqued the paper money policy of the 1830s by recalling their delight at receiving the "kind of money that would '*jingle*.'" "Such is the power of coin—or rather such *was* the power of coin," one man noted. Artisans also had something to say about the putative music of industry in a tightening and deskilling labor market. Listening back to the antebellum period, David Johnson, a Lynn shoemaker, recalled the traditional grinding of knives by skilled workers on Saturday afternoons. "Kepler," he remarked, "might indulge his fine theories about the 'music of the spheres'" of the sound of metal rasping stone, but the grinder did not: "His eye rolled, but it was not because he was entranced with the 'concord of sweet sounds,' . . . it rolled because his back ached." The analogy was not altogether inaccurate: "Any quarry slave, any galley slave, *any* bondman, whether 'hereditary,' or otherwise, was a king on a throne compared to the unfortunate victim bound to such miserable toil on a Saturday afternoon." Similarly, the behavior of Philadelphia's artisan handloom weavers on the night of August 11, 1828, not only reveals their ability to use sound as a tool of resistance but also explains why so many antebellum northern town statutes attempted to regulate who was allowed to make particular sounds at night. Not unlike southern bondpeople, northern workers found their production of the urban soundscape hedged by statute. Affronted by a city watchman's insult to "bloody Irish transports," Kensington's weavers administered a healthy, thirty-minute beating to a group of watchmen. Fife and drum warned scabs to stop working in 1839 when Kensington handloom weavers marched through the streets attempting to intimidate those who still wove.[42]

Workers' resistance to the appropriation of their labor was understandable because they found the sounds of the new machines alienating. Lucy Larcom recalled a Lowell dressing-frame she had to tend: "I felt as if the half live creature with its great groaning joints and wizzing fan, was aware of my incapacity to manage it, and had a fiendish spite against me."[43] Female factory workers of the 1830s and 1840s located the essence of their exploitation firmly in the sound of work time and its insidious ability to infiltrate the sanctity of preindustrial household relations. They also indicted the noise of factory time for killing its female operatives:

> The father roused his child;
> Her daily morsel bringing,
> The darksome room he paced,
> And cried "the bell is ringing,
> My hapless darling, haste!"
>
> .   .   .   .   .   .
>
> Again the factory's ringing,
> Her last perceptions tried,
> When, from her straw bed springing,
> "'T is time!" she shrieked, and died.

Some put it more simply still:

> The factory bell begins to ring,
> And we must all obey,
> And to our old employment go,
> Or else be turned away.[44]

Factory workers did not condemn the factory bell unequivocally. Instead they excoriated its power to usher in work and applauded its inauguration of all-too-brief breaks. As one verse from 1844 had it,

> Loud the morning bell is ringing,
> Up, up sleepers, haste away;
> Yonder sits the redbreast singing,
> But to list we must not stay.
>
> .   .   .   .   .   .
>
> Sisters, haste, the bell is tolling,
> Soon will close the dreadful gate;
>
> .   .   .   .   .   .
>
> Now the sun is upward climbing,
> And the breakfast hour has come;
> Ding, dong, ding, the bell is chiming,
> hasten sisters, hasten home.
> Quickly now we take our ration,
> For the bell will babble soon;
> Each must hurry to her station,
> There to toil till weary noon.
> Mid-day sun in heaven is shining,
> Merrily now the clear bell rings,
> And the grateful hour of dining,
> To us weary sisters brings.
>
> .   .   .   .   .   .
>
> Ding dong ding,—our toil is ended,
> Joyus bell, good night, good night.[45]

Plainly, though, the bell as the harbinger of work was noise; the sounds of rest—birds, breaks, and the "joyous bell"—were positive, the opposite of managers' aural constructions.

For most factory operatives the tocsin became toxin and the aim was to be "far from the factory's deaf'ning sound, From all its noise and strife."

Worship offered some refuge, provided workers were given quiet space and time in which to pray. Factory owners required the women to attend public worship but, according to workers, did so at inopportune moments when the noise of progress and work disrupted the solemnity of God's divine silence: "Will those who are obliged to hear the noise and confusion caused by some fifty or more men, with teams of oxen, and all the noise consequent on such occasions, together with the splitting and blasting of rocks, to their great annoyance while *in* their places of worship—will these be deceived by such hypocritical pretensions of piety, and love to the moral interests of the community in which they live?" The religious instruction of the working class, and the silence appropriate to such occasions, was compromised by the sound of progress, and workers knew it.[46]

Some factory owners agreed. Thomas Mann, a member of a family who owned a cotton mill in Pawtucket, made the tyranny of the factory bell and the noise of factory production his prime target for criticism in his overwrought 1833 poem, "Picture of a Factory Village." For Mann the factory bell, and the wage labor system that it represented, was an aural scourge that, by virtue of its ability to penetrate the mind through the ear, rendered factory workers automatons:

> Hark! Don't you hear the fact'ry bell?
> Of wit and learning 'tis the knell.
> It rings them out, it rings them in,
> Where girls they weave, and men they spin.

The noise inside the factory had a similar effect, making workers senseless:

> Hark! Don't you hear the Picker hum?
> It would a deaf and dumb man stun!
> Sounds like the wailing of the damn'd,
> Who in the lowest Hell are cramm'd.

The factory's yes-men were also culprits, their sycophancy sounding like sheep:

> Now, see! Their lordling strut along,
> Follow'd by his bleating throng.

To Mann's thinking, the noise of factory life, rather than representing the aural progress so fancied by boosters, had instead reduced operatives to animals:

Hark! Hear the looms, shuttles rattle,
The girls they stand like frightened cattle

and

Hark! Hear the breakfast bell's loud call,
Like sheep begin to scatter all.[47]

Industrial workers themselves recognized the profound ambiguity of the sound of factory time and spoke eloquently on the duality of the bell as liberty and tyranny. As "Ada" put it in 1846,

Hark! the liberty bell is pealing
Loud and clear its accents fall;
Tyranny beneath its chime is reeling, —
Arouse! enslaved ones, at its call![48]

Combined, these sounds and their ideological and actual manipulation by factory owners urged women workers to contribute their printed voice to Lowell's *Voice of Industry* in an effort to combat "the discordant sounds of human degradation and woe" inaugurated by the industrial, wage labor age.[49]

More substantive resistance to the sound of factory time occurred. In many mill villages the principal source and record of public time was the mill owner's factory bell. Abuse riddled this system since workers had little choice but to obey the sound of time as defined by managers. Managers' diaries are clear on this point. N. B. Gordon recorded the irregularity in the ringing of his Mansfield factory bell in 1830. On September 20 it was rung at 5:00 A.M., as it was on the first two days in October. As dawn came later, so did the bell's tone: On October 5, "Bell rung ¼ past 5"; on October 8, "Bell rung 20m past 5." How he justified ringing "the bell 10m after 5" on the even darker morning of November 10 must have eluded his workforce. Conflict ensued. In 1828 Pawtucket workers broke the factory's aural and temporal monopoly by establishing a public clock and bell. "A time-piece which can be depended upon as a regulator," commented the *Pawtucket Chronicle,* "located in so central and public a situation as the new Congregational Church, will be of great utility in this village." Not only would the central location ensure that the bell was heard, but placing it in God's house gave the workers a moral edge in defining whose time and sound was legitimate: "All are aware of the vexatious confusion occasioned by the difference of time in the ringing of the factory bells at the time, and which can only be remedied by erecting a clock that will always give '*the time of day*.'" Ultimately this episode reveals not so much laborers' radicalism as their profound conservatism. Workers fought not against aural time itself but, rather, its accuracy, a point that factory owners in their contemplative moments doubtless took as representing broad agreement on the legitimacy of heard, timed labor.[50]

Whatever their critiques of progress, working men and women shared some modern values. Like managers, they came to accept their new and evolving acoustic environment not least because they heard new sounds within the context of comforting registers of the past. David Johnson, who spent most of the antebellum period as a skilled shoemaker in Lynn, Massachusetts, heard the changes of 1830–80 from memory. Johnson rightly noted that Lynn was relatively small in 1830 (with about 6,000 inhabitants) and only on the cusp of manufacturing greatness. It was, he remembered, "a quiet town with its industry divided between farming and shoemaking." Skilled artisans still worked independently in small shops. Their culture was largely oral ("Books were comparatively scarce in those days"). Because the Bible was the most common text available, "as a consequence, religious discussions were often heard in these shoemakers' shops." Then, slowly but surely, industrialization hit. The sewing machine was first used in Lynn in 1852. "The peculiar rattle of the machine," remarked Johnson, "made a new music."[51]

Even the first generation of factory workers testified to the modernizing association of industrial sound and progress. Lowell workers, for example, appear to have been converted to the sound of progress even though their labor was increasingly hedged by it. One recalled, "At first the hours seemed very long, but I was so interested in learning that I endured it very well; and when I went out at night the sound of the mill was in my ears, as of crickets, frogs, and jewharps, all mingled together in strange discord. After that it seemed as though cotton-wool was in my ears, but now I do not mind it at all." In her effort to promote the benefits of factory life, Lucy Larcom drew on bourgeois notions of sound and progress. She maintained that she missed the refrain of progress even as it polluted her soundscape and organized her work: "When I returned I found that I enjoyed even the familiar, unremitting clatter of the mill, because it indicated that something was going on."[52] Larcom and some of the other women probably described early stages of sociocusis, tinnitus, or boilermakers' disease. By becoming accustomed to the noise, they underwent physiological and psychological changes. Physiologically they experienced aural fatigue whereby the sensory cells and fibers in their ears were damaged by constant noise. But they, like northern boosters, adapted cognitively to the sound of their environment. Associating loudness with power led elites and workers in the antebellum North to interpret clanking engines as indicative of power and efficiency. The sound of industry was aural testimony to northern progress, and members of all northern classes increasingly shared that assessment.[53]

Even operatives who warned of the alienating tendencies of the sound of factory time succumbed to the instrumentalism of the factory bell. Dismayed that Lowell women assembled at the gates before "the ringing of the second bell," "Juliana" argued that instead of "waiting at the gates for the bell to strike, the *gates* should wait for *them* after the bell gives the summons!" Since the sound of time was tied to wage labor, Lowell women were preempting that sound and thereby depriving themselves of "ten minutes twice a day."[54]

Workers' qualified embrace of the hum of industry did not necessarily mean their wholesale endorsement of the noise of their exploitation. If managers failed to listen to workers' demands, laboring people reasoned they would use the mouthpiece of democracy to sensitize patricians' cochleas. The Boston Working Men's Party, for example, identified a number of "actual evils" in 1830 that they deemed injurious to workers of that city. If their delineation of these evils fell on deaf ears, the party did not hesitate in "loudly demanding correction." When the law prohibited strikes by journeymen tailors—as it did in New York City in June 1836— artisans proclaimed, "The Freemen of the North are now on a level with the slaves of the South!" Should that not inspire workers, the journeymen took a leaf from the abolitionists' book by accepting their construction of how slavery sounded: "Go! Go! Go! every Freeman, every Workingman, and hear the hollow and the melancholy sound of the earth on the Coffin of Equality!" Not all laborers were so loud. Skilled workers especially were careful to couch their demands within the idiom of political sobriety and quietude. In its claim for more common schools and fewer colleges, for example, Philadelphia's *Mechanics' Free Press* argued in 1830 that it was not workers' "object . . . to raise the hue and cry against colleges" but merely to argue for the civic benefit of a more democratic education in sensible, constrained terms.[55]

Elites heard factory workers as both rowdy and disconcertingly quiet. Economic recession and the subsequent labor troubles in Springfield, Massachusetts, in June 1847, in which workers gathered to protest their impoverishment, were heard clearly because of the aural vacuum created by economic recession. The gathering of this "mob" was sufficiently disturbing for town authorities to ring church bells for two hours, and the "sounds of strife" were silenced only by the intervention of a "military company."[56] In other words, elite fear of economic recessions was well founded. Not only would the pause in the sound of industry testify to economic retardation, but perhaps more worrying still, that silence could be replaced with the discordant timbre of worker resistance.

Workers' use of democratic voice as protest exposed some basic contradictions in the constructed heard world of northern elites. When that voice threatened social order, property, and life, northern elites heard the activities of crowds as the noises of the mob. Antislaveryites who attempted to rescue fugitive slave Anthony Burns from Boston's Faneuil Hall in the summer of 1854 threatened social order and property, and even Toronto's sympathetic abolitionist newspaper, the *Provincial Freeman,* called them "pell mell," causing a "melee." Police attempting to control this "evil disposed mob" were "saluted with hisses, groans, and other marks of derision." "The noise of the mob," the editorial concluded, that created "much alarm" was hardly consonant with northern democracy. This was behavior unbecoming recipients of the vote and the sort of thing extending the franchise was supposed to prevent. Such fears did not undermine elite commitment to democracy. They considered riotous mob action not a fault of their vision of rational, democratic debate but, rather, a reflection of a corrupted portion of the citizenry who, thankfully, could always be reformed and checked through the power of the state.[57]

Even Republican John P. Hale feared that aspects of northern democracy were perilously close to "riotous or tumultuous assemblages" that were especially dangerous when they threatened property, as they did in the District of Columbia during the late 1840s. Willie Person Mangum of North Carolina smiled knowingly and told him that where there was northern democracy, mobocracy was not far behind. "Most of the time in North Carolina," he advised, "I have never seen anything in that State approximating even to a spirit of popular tumult." The implication was clear: northern democrats only had themselves to blame if suffrage sounded like the noise of a property-destroying mob.[58]

Given their preference, the northern bourgeoisie and patrician elite opted for a quiet, orderly democracy, its primal urges held at bay. While elites recognized that elections were shaped by "the one who yelled the loudest" and acknowledged that the politics of class and labor was aural, when workers dared critique their conditions of labor, boosters warned "those who have heeded the ranting assertions of an unprincipled, or unreasonable portion of the work-people" to ignore such noise.[59] Equally disturbing was the noise of multitudes. According to the profactory Pawtucket *Manufacturers' and Farmers' Journal,* in 1824 striking textile laborers had placed the town "in a state of excitement and disorder," which "reminds us of the accounts we frequently read of the tumults of manufacturing places in England." More disconcerting still, the strikers used the sound of labor to coordinate their noisy protest: "And when the bell rang to call them to their employment, they assembled" instead "in great numbers." A "tumultuous crowd filled the streets" and "made an excessive noise" by visiting "successively the houses of the manufacturers, shouting, exclaiming" within deliberate earshot. The crowd used sound to intimidate the factory owners. When workers wanted considered, thoughtful deliberation on whether or not to strike in the first place, they demonstrated the oral and aural self-control prized by elites and managers. As the *Journal* reported, "The female weavers assembled in parliament to the number, it is stated, of *one hundred and two*—one of the most active, and most talkative, was placed in the chair, and the meeting, it is understood, was conducted, however strange it may appear, without noise, or scarcely a single speech." The result was "a resolution to abandon their looms." Elites were surprised and a little concerned by such self-restraint. Noiseless crowds were possibly thoughtful and might be even more troublesome. Laborers were at their most dangerous when they could generate either noise or eerie silence. According to the manufacturers, who were acutely aware of market vagaries and the precariousness of new industrial endeavors in an early capitalist economy wracked by periodic dislocation, the price to pay for such tumult was the silence of economic ruin for the entire community. The day after the riot "the manufacturers shut their gates and the mills have not run since." Predictably, "a comparative stillness now reigns." The moral was clear: "The prosperity of a community can never be promoted by riot and tumult." Noisy labor meant silent ruin; quiet workers meant the hum of ever bountiful industry.[60]

In trying to persuade Philadelphia Irishmen to vote against Andrew Jackson in the 1828 election, the city's protectionist *Democratic Press* used identical tactics. "A little reflection," pondered the editor, "would teach them that the greatest calamity which would befall them, in the United

States, would be the election of General Jackson. Then would their looms be silent, and their spades, shovels and wheel-barrows be no more in demand."[61] Indeed, the frequency with which management and probusiness sources threatened striking workers with the silence of industry suggests the degree to which the working class understood the silence of economic dislocation. That they heard poverty approaching is significant testimony that workers and capitalists increasingly heard the economic world in similar terms.

Basically, then, cultural and economic elites tried to marry the cadence of industrial, republican activity and productivity to what they thought democracy should sound like. Listen to an early effort to blend the two in 1816:

> Care's busy notes, and art's responsive sound,
> In cheerful concert through the vale resound.
> The breaker-lads, with voices shrill and clear,
> In treble accents greet the musing ear;
> The buzzing spindles fancy's tenor trace,
> And angry pickers thunder on the bass.
> My heart, enchanted, to this music beats,
> My willing hand the joyful duty meets!

When one did not have to perform the labor, textile manufacturing could doubtless take on a musical timbre even if there was occasional clanging discord in the process:

> Hydraulic movements call the watchman's eare,
> Harmonious discord fills the laboring air.
>
> .   .   .   .
>
> And clattering engines intervolve my way.

Should discord get the better of harmony, managers could always reassert aural order:

> The master's voice directs the movements all,
> And drives disorder from the busy hall.[62]

Because robust democracy and industrial capitalism increasingly distinguished the North, the region's elites necessarily embraced the sounds of both with ever greater tenacity as sectional tensions increased. Newspapers described an 1847 New York mayoral race aurally: "A very democratic crowd . . . rushed along Broadway. . . . Here and there along the bustling thoroughfares a hoarse voice shouted. . . . The cry was re-echoed. . . . And so with great shouts and blasphemies and laughter . . . King Caucus made his proclamation, announcing that he had at length chosen a chief magistrate for this mighty Christian city!" This political sound was in perfect keeping with "the loud hum and smoke and dust of the city." "Noise and confusion," after all, according to a northern thinker in 1848 in an article titled "Suffrage," were "incident to an electioneering campaign." In breathless prose Walt Whitman agreed:

> Americanos! Conqueros! marches humanitarian!
> Foremost! Century marches! Libertad! Masses!
> For you a programme of chants.[63]

It would be a mistake to assume that northern elites thought democracy should be passionately loud and boisterous, for they worried almost as much as their southern counterparts about the noisy rabble and the threat to social order political cries foreboded. Some northerners who witnessed southern elections believed democracy sounded similar in both regions. Henry Cathell of New York City, for example, thought the "rows" he heard at an election in Savannah in 1851 similar to those in New York City, the main differences being stylistic not substantive. "This all appeared to me very natural," he commented, "with exception, that here, Bowie Knives Revolvers, and Clubs were used, and with us fists only."[64] But there were important differences, and they resided principally in the nature of what many northern democrats demanded. The silence of 4 million slaves echoed loudly in the North, and reformers maintained that they had a moral and republican obligation to give their democratic voice to those who did not have one. It was through the idiom of a loud, heard, but sober democracy that northern reformers such as Henry Ward Beecher, defender of "the oppressed and crushed," tried to arouse "an apathetic world." This carefully delimited democratic throat, with its "clarion voice" and its "democratic spirit" in defense of "slaves of the cotton-field, the servant-girls in lordly mansions, the poor city seam-stresses, or the struggling apostles of liberty in far-off Hungary," enabled reformers to cause "the ears of the rich to tingle."[65] Sober hubbub testified to the health of northern democracy. "The various speeches" delivered at the Free Soil convention in August 1848 "were received with loud cheers, and excited great enthusiasm among the vast assemblage." Then, "Mr. Chase mounted the box and called the meeting to order, amid such a din as was never heard. A scene of such noise and confusion would have perfectly annihilated General Cass." All this commotion was good, for it betokened the sound of democracy trying to speak for silent slaves. Resolutions were proposed in support of free soil in the West, and the "response was a deafening 'Aye,' which burst forth with entire unanimity from the throats of twenty thousand men, and which resembled a terrific clap of thunder," or more presciently, "the roar of a park of artillery." That the "entire mass of men . . . yelled and howled liked tigers in a fray" was simple, aural eloquence.[66]

(((PART III
Aural Sectionalism
The Politics of Hearing and
the Hearing of Politics

Antebellum workers understood the power of silence and sound as tools of effective resistance to their exploitation. Ruling classes in the North and the South were not deaf to this development. Elites heard social order at the everyday level of interaction and, simultaneously, as an abstraction, often with the former reinforcing the latter. What they heard either reassured or frightened them and helped shape their political discourse during the sectional crises of the 1840s and 1850s. Aural images of sectionalism exerted a powerful influence on leaders and acquired authenticity by virtue of their insertion into local and national political discussions. The acoustic construction of otherness enabled many slaveholders who had never seen industrial capitalism and northern elites who had never witnessed slavery to become increasingly exercised about the other and the legitimacy of their own societies. Oral and published reports by travelers often confirmed worst fears and reaffirmed to elites in both regions that the country was developing two competing, antagonistic societies. The increasingly combative choices of aural descriptions used by elites to critique and construct one another proved critical in heightening and shaping sectional consciousness and in foreclosing, little by little, efforts to find common ground.

Aural sectional projections proved clumsy because when actual sounds gained symbolic, metonymic, or metaphoric status, they lost a good deal of their precision. Even as they failed to convey accurately the entire range of sounds heard and produced in each section, the representations became real and tangible to those who constructed and listened to them. Southerners heard northern liberalism and capitalism as noisy; many northerners heard slavery as disturbingly silent and cacophonous. Obviously these raw aural representations did not and, by their nature, could not reflect the subtleties of actual sounds in either region. By the middle decades of the antebellum period, however, few were interested in their precision. Listening elites—their hearing pricked by the day-to-day shaping, management, and defense of their worlds—used aural imagery to capture the ideological and material differences between North and South. In doing so they gave emotional, visceral depth to increasingly pronounced sectionalism. Aural constructions of "the North" and "the South" gained wide currency so that northerners and southerners heard one another in profoundly and emotionally divisive ways. The heard world, imagined and distorted though it was in part, was real to those who did the selective listening.

Listening to these descriptions—themselves grounded in a material reality—shows that the mobocratic tendencies of urbanism, industrialism, and free wage labor especially were nothing but noise to the master class. True, northerners also at times cast the mob as noisy, but given the nature of their democratic and industrializing society, they tended to deem such sounds as integral to their society. Southern elites proved less forgiving. To them the North was filled with the cacophony of the modern, the noise of the laboring mob, and the aural detritus of industrialism (it was the "ism" that they heard and rejected most forcefully). In contrast, proslavery southerners fancied they heard the sounds of conservative economic, political, and social relations on their plantations and sober, republican democracy in southern society at large. They delighted that these sounds reflected and reaffirmed an industrious, honest, pastoral southern life.

Abolitionists, free soilers, and later, Republicans heard emanating from the South the noise of immoral slavery and the retarding effects of a silent institution whose lack of industrial and wage labor registers testified to its economic backwardness. An important method for critiquing that wretched system was their sustained and deliberate application of precise and cutting aural projections. Having heard the South via slaves' ears, they in turn attempted to make slaves hear the free wage labor critique of slavery and, in the process, inaugurated a relentless chorus of criticism on masters' highly sensitive ears. Masters responded to this aural onslaught the way they did because they were not deaf to their actual and metaphorical aural worlds and the authority they invested in them.

While listening to the heard worlds of increasing sectionalism can no more explain the specific timing of the outbreak of hostilities between the North and the South than can other, more conventional analyses of the coming of the Civil War, an understanding of aural sectionalism helps to explain how and why tensions and passions ran so high and why elites in each section perceived the differences between themselves and their counterparts as beyond compromise by 1861.

The wails of bondmen are on my ear, and their heavy
sorrows weigh down my heart.

    —Frederick Douglass, "Frederick Douglass' Address,"
     North Star, August 4, 1848

(((6. Listening to Bondage

Politics had always been clamorous. Of American revolution-
aries, the English Tory Samuel Johnson asked how it was that "we hear
the loudest *yelps* for liberty among the drivers of negroes?"[1] In the early
national period, individuals such as New York's "industrious mechanic,"
seeking to disassociate themselves from "*Party politicians,*" claimed to be
above "the filthy sloughs of party declamation" and denounced party politics on the grounds that virtuous men were "sick of clamor, invective,
[and] slander." It was also understood that irresponsible political rhetoric
could so perplex "the public mind . . . that it will be difficult to quiet it."
Political discourse, after all, was public by definition, and keeping political passions in check was among the most trying disciplines of the self.
Even with the decline of viva voce voting in the antebellum period, discussions in the polis were aural because disagreement was a function of
burgeoning democracy and clashing views necessarily upped the volume
of debate. Hence political noise could "assail the eyes and ears" of antebel-

lum politicians. While party managers, floor speakers, and handlers were there "provoking calmness," and the "Speaker's hammer" could quash "the din and tumult," there was something sublime and healthy about heard debate that preceded harmony. In the spoken political word was heard democracy.[2]

The noise of politics was not simply a sectional matter. The two-party system and the slew of minor parties that operated within each section throughout the antebellum period ensured as much. Some heard their own politicians negatively. Robert Raymond Reid of Florida heard the forces of nullification within the South as disruptive. Reid, who rejected Calhounism, embraced slavery, and coveted the Union, believed strident nullifiers could fracture the nation in 1833. "I think we shall have a *row*," he wrote, but concluded, "the *Union* will weather it—no thanks to the South Carolina aspirants." Thirty or so years later, southern Unionists felt the awesome power of secessionist sentiment and complained of campaigns to "browbeat & bully into silence those whom they cannot persuade to go with them."[3]

Intrasectional and interparty metaphoric noise sometimes became actual, especially during political debate between those who would expand slavery into new lands and those who would not. The 1849 debate in Missouri between fire-eaters and followers of Thomas Hart Benton, a southern Democrat, is a case in point. Trouble was heard even before resolutions were offered: "Mr. Hoit re-appeared upon the stage again, who, with tenacious perseverance, stood it out amid all kinds of noise and confusion, including violent thumping upon a ten quart tin pan, which one of the strikers had, on the anti-Benton side of the house, for the purpose of aiding in drowning the voices of Free-Soilers, until the tumult subsided." Speech was drowned out by noisy southern politicians who, in effect, silenced a countryman.[4]

Southern Whigs and Democrats also injected aural imagery into their political discourse. Southern Democrats heard Whigs as old fashioned and backward. "Mattie is so cute," wrote a child diarist in 1860, "because her Father is a Whig she says she is a Whig too, she shakes her golden curls and turns up her pretty little nose when cousin Rob and I sing Democratic songs, one especially, sung to the tune of 'Benny Havens, Oh.'" Little wonder that Mattie was so incensed. The song cast Whigs who became conditional Unionists on the eve of war as sounding slavishly old and noisy:

Hark from the tomb a doleful sound,
We hear a mournful yell,
Old fogies shout discordant notes
For Everett and Bell.[5]

Conversely, William D. Valentine of North Carolina sometimes worried about the true Whiggishness of southern Whigs. John Kerr, North Carolina's gubernatorial Whig candidate in 1852, sounded too democratic for Valentine's taste, not so much for what he said but for the way he said it. He described one of Kerr's speeches as betraying "much boisterous action and dramatic effect. Though pleasing to the crowd there was perhaps rather too much rant." A good Whig, conversely, was one who, like a speaker at an October 1852 convention, "had the silent close attention of the mass."[6]

Coveting social order and quietude, Whigs tended to respect the throb of industrial capitalism, and they were not unknown to accuse Democrats of courting the noisy mob. Whigs wanted to hear the gentle pulse of economic progress, not the cacophony of democracy. Valentine blamed northern agitators and southern demagogues for increasing the political volume of the 1830s and 1840s and for rendering what should have been sensible, sober, quiet political discussion noisy and worryingly mobocratic. As he wrote in 1840, "Men are made up of humbugs. The present excitement about the abolition of slavery is a proof of this. The cry of abolition! is frightful indeed." Valentine found the aural dimension of the southern response equally disconcerting. "When the question was first agitated," he continued, "it had but little effect than mere speculation and abuse. But now it is a political hobby, and it is used for little other purpose than to affect the presidential question. No matter what office a man is a candidate for, the charge of abolition is thundered against him most deafeningly." "Any man," he ended, "can be a big bear to the mass of men with this word abolition."[7]

Some southern Whigs pushed hard for calm, sober politics. Again, the observations and language of William Valentine are revealing. "The people of the United States, the Presidential election being over," he wrote in November 1844, now needed "rest from politics and elections." Valentine believed Jacksonian Democrats were the cause: "These for the last ten years have kept the Republic in a constant ferment." Had he known of the noise yet to come, his aural perception of the political past would have been even more sentimental: "This state of [affairs] began at the close of Mr Monroe's Administration, which unopposed, quiet and harmonious may thus far be considered the halcyon period of the Republic." Valentine's quest for the restoration of political peace and order proved unrequited. A year after the Compromise of 1850, he lamented, "I thought after the said measures passed and Congress adjourned, quiet and order would return." Not so, however. "The general aspect of things have been more threatening and alarming since the adjournment than during the session." The political soundscape was, in his opinion, a veritable "tempest" whose "waves of dissatisfaction will lash other portions of the land" outside the nation's capital.[8]

Southerners sometimes showed equal contempt for politicians in both sections and, in effect, condemned the noise of partyism. "There are men, South as well as North, who used Slavery as a trumpet to excite sedition," remarked William J. Grayson in his 1862 autobiography. "The noisy and clamorous always the least fit to hold office are commonly the most likely to obtain it," he sighed. William F. Samford, one of Alabama's hardcore secessionists, detested blind allegiance to a political party and considered party hacks, north and south, noisemakers of the highest order. "To be bawling in the ears of freemen, 'the party'—'the party' like the town-crier," he railed in 1858, "is enough to disgust and insult the intelligence of free electors."[9]

Interparty competition in the North also had aural dimensions. Northern Democrats in the 1830s described Whig adversaries as "pant[ing] for monopoly," while some northern third parties were sometimes met with sibilance. Joel W. Jones recalled that his pro-Masonic toast at a July Fourth political celebration in upstate New York was received with "hisses and groans." Charles Francis Adams condemned opportunistic abolitionists who joined the Know-Nothing Party for wrongly fighting southern silence with a form of savage secrecy and thereby echoing southern masters and a dangerous red-skinned foe. Northern Nativists used aural analogies to denounce immigrants whose slums, "violations of the Sabbath, public drunkenness, pauperism, and crime" rendered the sober democratic process clamorous. For their part, free soilers were suspicious of the established parties who spoke out against slavery. As Indiana's free soilers proclaimed in 1849, "It is true that the Whigs make the loudest of professions, just now, in favor of Free Soil; but noise is not always an evidence of sound sincerity and faith."[10]

Northern Whigs in the late 1830s found visiting southern Democratic politicians nearly as noisy as their northern supporters. John Hamilton

Cornish heard as much in Hartford, Connecticut, in 1837. The "Hon. Mr Bell from the South," noted Cornish, "harranged in the City Hall, & poured out his invectives against the administration party—& roared and belched about two hours in the evening, to the infinite gratification of the rabble." Northern Democrats were similarly critical of political adversaries in both sections. Cincinnati's *Daily Enquirer* applauded the passage of the Kansas-Nebraska bill even though it "will, of course, send up a howl of rage," excite "the fanatical disunionists," and cause abolitionists to "fret and foam." The "noble band of Democrats from the North and West," the paper advised, must remain "unmoved by the clamors of fanatics."[11]

Southern Unionists worried that aural representations of the South had assumed a tone in northern ears that threatened to drown out their voices. In response they essayed their own acoustic engineering. Do not speak of "the voice of the South," recommended one southerner in 1848. He told northern antislavery sympathizers that the "great din" made by fire-eaters over slavery should not be taken as the sole southern tongue. No, he reassured them, there were reasonable men down there, men whose voices were being silenced not only by the volume of southern radicals but by the selective hearing of the abolitionists. Efforts to reassure northern capitalists of southern progressiveness were similarly aural. John Perkins of Louisiana argued in 1854 that northerners had nothing to fear from slavery's spread westward because the expansion to date had silenced neither the old Southeast or the new Southwest. "The cities on our Atlantic coast have not the dismantled appearance either of a decaying population or of a diminished commerce," he reassured his northern brethren, concluding, "The hum of industry has not ceased in their workshops, not grass grown upon their wharves."[12]

There was a good deal of reassurance heard from the North, too. Some tried to allay southern fears by arguing that "there has been a small band of fanatics who have made so much noise that many people having imagined them to be greatly more numerous than they are." Other northerners found solace in how they imagined plantation life sounded. "These large plantations were like little worlds within themselves," and in their "quiet streams" and "placid lakes" could be heard "God's country." Quieter than the noisy North, the plantation South exuded aural happiness. A few northerners agreed: "Compare the happy negroes you will find there [in the South] with the poor of our own section; hear them sing and dance and frolic." There was no noise of class conflict here.[13]

Northern admirers of southern civil society likewise voiced their affinity for southern social relations. Men such as Fitzwilliam Byrdsall of New York and Condy Raguet of Philadelphia "accepted the South's position not only on economic questions but also on the superiority of southern society to northern in the preservation of liberal principles." Too acquisitive by half, northern society had promoted an irresponsible individualism and the immiseration of its laboring classes that was bound to lead to strife. Here is how Raguet understood the problem: Contrast the "healthful and sprightly" yeoman of the South with the northern industrial wretch whose "ears hear nothing but the noise of machinery, or the reproving voice of a task-master." Northern elites were not willing to compromise by listening to southern complaints. Explained a remarkably prescient Raguet in 1830, "Instead of listening to the appeals of their fellow citizens of the South, who feel aggrieved by palpable violations of the constitution, as common justice and patriotic sympathy demand they should do, they turn a deaf and indifferent ear." As for the abolitionists, they would not be happy "until they have toppled the Union . . . and filled this happy country with the din, and guilt, and terrors of fratricidal and fraternal warfare."[14] Such articulate men were simply too few to effect any real change in sectional trajectory in a society that relied on democratic majorities for its direction.

Although intrasectional political identities and parties were heard as much as seen, the aural imagery of the two-party system was nowhere near as pronounced or as emotionally powerful as the aurality of sectionalism. Nonsectional parties generally did not embrace aural metaphors as fully as sectionally conscious southerners and northerners in their construction of otherness, and interparty competition failed to command the aural authenticity that sectionalism could and did generate. Whigs and Democrats were becoming increasingly similar by the 1850s, and there was little firm basis on which to portray each party as sounding different or as championing different aural social and economic relations. This was not the case with sections, however. The North and the South sounded different—increasingly so—and the aural projection of sectional otherness gained an authenticity that eluded nonsectional parties.[15]

Northerners from different backgrounds came together in the closing decades of the antebellum period to critique slavery in part because they constructed the evils and shortcomings of the peculiar institution through the heard world and communicated that critique using candid images of how slavery and, by extension, the South sounded. They chose to hear slavery in part because ex-slaves and abolitionists sensitized their ears to the aurality of bondage and because humanitarians and capitalists then critiqued that same institution using common understandings of the southern soundscape and what it represented. When abolitionists listened to southern bondpeople, they championed a "new mode of hearing" that "sought out the inner world that was presumably reflected in the expressions of slaves" and was linked to the abolitionists' "new humanitarian pursuit of the inner world of distinctive and collectively classifiable subjects." In this context Frederick Douglass's admonition to antislavery sympathizers to listen closely to slave singing helped build on the existing if inchoate humanitarian sensibilities of a northern audience who devoured, at times voraciously, the published accounts of authentic slave songs. But antebellum slave narratives and their audiences listened to more than just songs and literary texts; they communicated and listened to myriad other sounds of slavery contained in slave and abolitionist narratives, including the gruesome noises and silences of premodern bondage.[16]

Essentially, abolitionists and supporters of free soil and free labor heard the South in two ways that combined to create a distinctive southern soundscape in their ears and minds. The slave South was at once worryingly silent and horribly piercing, cacophonous, and loud. The two ways of hearing the South were not incompatible. To northern ears the South was a noisy place not least because slavery relied on bodily punishment to coerce labor and not on the silent hand of free wage labor capitalism. In addition, the tyranny of slavery was deemed to mute criticism and democratic voice as well as usher in industrial backwardness, and for these reasons the South was overly quiet economically and disconcertingly silent politically.

Northern listening to southern soundscapes was not without its tensions and contradictions. After all, many antebellum northerners ventured not just to northern spas but also to the South in part to escape, albeit temporarily, the noise of industrialism and urbanism. Southern quietude and the region's soundscapes generally held attractions for antebellum northerners. Some, for example, went south for sport and described the experience of hunting in the South in glowing aural terms. Northerner Henry Cathell traveled to Florida in 1852 for rest and recreation. He was not disappointed. His "ears were regaled," he noted of a hunting trip, "with the music of the pack in full cry." And when exciting sports were over, where better than the rural South to "have rested quiet and easy all day"?[17]

Others recognized that they trod a fine line in their charges of the silencing tendencies of the southern Slave Power. Their own "mob had attempted to silence a press," and abolitionists' freedom of speech itself was threatened by excessively boisterous democracy. More than one antislavery speaker found his or her words "drowned by the loud cries" of opponents. Indeed, for zealous antislaveryites, the northern mob, infected as it was with "Color-Phobia," sounded unmistakably like the master class. Nathaniel Peabody Rogers, for example, hated the noise of the northern mob because in its voice he heard the caterwaul of color prejudice. It was a condition which "makes them sing out 'Nigger—nigger,' sometimes in their sleep. Sometimes they make a noise like this, 'Darkey —darkey—darkey.' Sometimes, 'Wully—wully—wully.'" It was a mob that "howled like bedlam," that "sneers and scowls at woman's speaking *in company*," and that "will stifle and strangle sympathy for the slave."[18]

Northern abolitionists and advocates of free wage labor nevertheless fashioned a workable and easily communicable aural description of the South. Northern visitors to the Old South heard the silence of enslavement, the clanking of cruelty, and the noise of poor southern whites whose rowdiness and racism affronted their bourgeois sensibilities. The "whole tone" of slavery, charged General J. Watson Webb in 1856, was one of "street fights, homicides, brawls, and acts of violence." Such observers heard poor southern whites as indolent and lacking the discipline to work hard and steadily. Their own proletariat might lapse occasionally, but on the whole the sounds made by northern workers were associated with the work ethic and industrial and democratic progress. The sound of poor southern whites, by contrast, was the keynote of the wild, a people yet to be tamed by market sobriety.[19]

However much northern patricians complained about the noise of their lower orders, no matter how many laws they passed to keep their cityscapes relatively quiet, the aural activities of the northern working class evidenced the efficacy and appropriateness of northern freedom and democracy. "Emancipation is a democratic revolution," William H. Seward declared in 1850. Tyranny, by contrast, was silent and affected subjects accordingly: "The quietness of the people proceeds from a slavish acquiescence in the measures of government."[20]

Even prior to audible industrialization in the North, abolitionists heard slavery in general negatively, not least because the soundmark of free labor preceded industrialization. In describing slavery in the West Indies in a series of essays in 1811, one writer cast bondage generally in aural terms. He asked "countrymen" to "hear me" and to hear bondage: "The groans of slavery are heard among you, which sound more painful and horrible to the ears of humanity and justice, than the dying shrieks of an infant to its mother or the deep-toned thunder of hell's dark abyss to a damned spirit." This early abolitionist tried to stir his audience by recognizing it as one sharing common humanitarian and aural sensibilities. Abolitionists heard the world in terms of discord, with the quest for right and good as necessarily and comfortingly loud. "The history of the human race," argued a writer for the *Colored American* in 1841, "is but one continued struggle for rights; and though there is constantly associated with it the jarring sounds, the deep din and clamorous uproar of contending elements," such was the cost of the quest.[21]

Nonsoutherners were convinced that slaveholders heard and listened differently. Fannie Kemble found masters' commands to slaves repugnant to "listen to," and she rightly wondered "if the Southern slaveholders hear it with the same ear that I do."[22] Northerners had kinder, more refined ears. Although they resented the din of the proletariat, during periods of crushing unemployment reformers willingly heard suffering. "The cry of humanity is echoing from street to street," it was noted of northeastern cities in 1854. Such pain "cannot be unheeded." Indeed, for abolitionists, "it rises above the din of commerce." Thomas Clarkson was remembered for raising "a clear voice for the dumb," which included both slaves and New England's workers:

> Beside the forge, and at the loom,
> Amid the factory's din,
> Where little children weave their doom,
> His lineage looks in.

And when they wanted to critique the excesses of their own society, humanitarians made northern transgressions sound like southern slavery:

> I hear the hammer's sound!—I see
> The furnace-smoke of slavery!
>
> .   .   .   .   .
>
> Fetters are forged for human limbs,
> While Midnight wraps their work in gloom,

And makes the strokes of her sullen bell
Chime in with the strokes of the work of hell![23]

Northern humanitarians liked to listen to their own decency and slave-holders' cruelty.

"Whoso stoppeth his ears at the cry of the poor," maintained reformer Arnold Buffum, quoting Proverbs 12:13, "he also shall cry himself, but shall not be heard." Slaveholders, by this standard, would never be heard, for even though "They coldly hear the captive's moan / They bind anew his chain." Masters did not simply close their ears; they cruelly heard the pain and did nothing about it. In abolitionists' estimation, this was per-haps worse than not hearing per se, and as such, abolitionists felt obliged to open masters' ears.[24]

Humanitarianism and modernity were tightly braided for northern re-formers, and by listening to their own society, they distinguished slavery from freedom and, in the process, politicized the difference. Northern democracy was alive and well in the 1840s, and democrats made sure southerners heard it. Political songs filled the northern air, as did "shouts of the excited multitude." David Johnson, a Massachusetts shoemaker, re-membered the politics of the 1840 presidential campaign well and was sen-sitive to the aural sectional symbols and metaphors used to ignite demo-cratic passions. "Political meteorology was of the most startling kind that year. There were no balmy zephyrs or gentle showers." Indeed, John-son recalled, "There was nothing tamer than an earthquake during the campaign." "There was a good deal of thunder rolling over the prairies. . . . Gunnery, and especially artillery practice, was largely called upon for metaphors," he noted, adding, "As early as March was heard the 'First Gun from Illinois.'" Then came the key: "Voices echoed, torrents roared, the rushing sound of the Mississippi waters mingled with every breeze that 'swept over the plains of the South.' We were invited to listen to all kind of noises in every part of the land. Baltimore was the focus of all these noises. The North shouted to the South; the East called to the young West."[25] The sounds of democracy—the "shout . . . from count-less thousands, echoing back to earth, 'Liberty-Equality-Fraternity'"— as Frederick Douglass styled it, were not drowned by industrialism. But the masking effect of slavery could be more powerful than the volume of northern democracy and capitalism. "While our boast is loud and long of justice, freedom, and humanity," in the South the "slave-whip rings" deadened the timbre of freedom. "Three million slaves clank their galling fetters," and the result was the muting of communication between enslaved and would-be emancipator. "I am cut off from all communication" with them, remarked Douglass; "I cannot hear them, nor can they hear from me." A few sounds got through thanks to memory: "The wails of bondmen are on my ear."[26]

For Theodore Dwight Weld slavery was at once cacophonous and silently sinister. To prove his point, he picked on the way masters attempted to regulate slave behavior using sound and silence. For a man whose organization lauded the bell as the symbol of liberty, slaves "made to wear round their necks iron collars armed with prongs . . . and bells" was the height of perversion: That slaves supposedly wore "gags in their mouths for hours or days" was similarly horrendous, not only indicating the disturbing silence of a backward, Philistine South but also illustrating masters' designs for the rest of the nation. Here Weld exaggerated for political effect the extent to which masters used gags and iron collars, just as other abolitionists inflated the frequency of slave whippings.[27]

Other abolitionists also found the South generally, and the institution of southern slavery particularly, ominously taciturn. For advocates of colonization, such as Frank P. Blair Jr. of Missouri, southern quietude was economic and political. The master class, he argued in an address before the Mercantile Library Association of Boston in 1859, well recognized the economic "ruin around them" caused by slavery. Effete and backward, southerners lived in an "atmosphere that sighs around them." Irresponsible oligarchs, argued Blair, rather than remedying the problem by colonizing slaves, defended the increasingly peculiar institution for personal gain. Such men met the problem "in the spirit of daring men—political ambition whispers, and as from the beginning, its inaudible breathing swells the bosom of the proud and aspiring with the thought—'Evil, be thou my good.'" Abolitionists had to make the case audibly, Blair maintained, because as every good antislaveryite knew, "bondage is hoarse, and may not speak aloud." Silent slaves had to be given a democratic voice by those who had it.[28] There was "no free speech," for "the slave institution . . . only muzzles the mouths of these whites." Well, not quite. Darius Lyman conceded, albeit unwittingly, that poor whites did "shout and halloo for Slavery" but went on to suggest that the master class merely allowed such expression in an effort to divert attention from genuine democracy, the basis of which was education.[29]

Some antislaveryites were encouraged and thought they were making headway by the 1840s. The voice of freedom, reported the *Colored Ameri-*

can in 1840, "gathers strength daily. It is being heard above the din and confusion of battle, and is sinking deep into the hearts of those whose ears hear the thoughts it utters." They also heard pockets of freedom. Northern travelers to the South commented favorably on the sound of black church services. One correspondent for a Boston newspaper was "disappointed" by the singing of white Methodists in Fernandina, Florida, but found that "when the colored people sing their own songs . . . the effect is grand." These moments of liberty—confirmed in the freedom to produce and shape the heard world—held out some hope. Complacency was dangerous, though, for southern cunning took many aural forms. Beware the silent South, warned abolitionists. "Slaveholders and slave traders never betray greater indiscretion," offered an antislaveryite in 1848, "than when they venture to defend . . . their system of plunder." "The wise ones among the slaveholders know this. . . . They seem to fully understand, that their safety is in their silence." [30]

Aural politics intensified during the 1840s and 1850s. Northern congressmen, such as John A. Bingham from Ohio, interpreted gag acts, especially one in Kansas "which makes it a felony for a citizen to utter or publish in that Territory 'any sentiment calculated to induce slaves to escape from the service of their masters,'" a palpable threat to republican freedom and the westward expansion of capitalism. Bingham made the equation when he argued "that if free speech is tolerated and free labor protected by law, free labor might attain . . . such dignity . . . as would bring into disrepute the system of slave labor." Southern efforts to impose silence on slaves within their own states were translated as the Slave Power's effort to inscribe silence on free men nationally. This was in direct contradiction to northern practices. "We allow everybody to hold a public meeting that wants one. And he may say what he pleases," crowed Senator William Pitt Fessenden, neglecting to note that unauthorized public meetings and the noise they produced were as regulated by local ordinance in the North as they were in the South. Leonard Bacon went a little further. "If any individual has a right to be silent" on the subject of bondage, he maintained, "that individual is not the slaveholder. His silence respecting such an evil, is approbation." Those against slavery, argued Bacon, must be allowed to "speak out like freemen" and not be enslaved by masters' flinty silence. [31]

Southerners and northerners recognized the importance of free speech to republican government, but they disagreed precisely because speech could threaten that form of government. By the late 1830s much of the

South had quarantined antislavery expression because masters feared abolitionist tracts would only incite slaves to rebel. For the master classes, free
speech naturally stopped at the point of threatening civil order, and so
they insisted on national gag rules to impose silence. Public meetings proclaiming antislavery doctrines were dangerous and noisy because they disturbed the peace of the South, and thus some vocalizations were deemed
seditious and were criminalized. Such talk would not be tolerated "because our safety . . . our peace, the peace of our hearths, depends upon
the repression of such doctrines with us." Plainly, the North was essay-
ing to carry "discord, invasion, and anger" to the South. The argument
fell on listening ears and struck northern politicians as "a system of es-
pionage" reminiscent of the "darkness of the middle ages." Antislavery
Illinois congressman Owen Lovejoy contended that the North was not
noisy in its demands. Rather, antislavery demands were made "in a peaceful way, through the press." Alternatively it was the Slave Power that destroyed northern free speech and suffocated freedom. "The press has been
padlocked," it was maintained, which meant that "men's lips have been
sealed." [32]

Some northern politicians felt that slaveholders' insistence on plantation silence had been extended to the floor of the House of Representatives. Ohio representative Joshua Reed Giddings made the point in 1848
in the course of a congressional debate. "The gentleman [Mr. Haskell of
Tennessee] charges me with uttering sentiments on this floor, and out of
this House, which have tended 'to excite the slaves to rebellion,'" frothed
Giddings. Lest "Southern gentlemen suppose they can bring into this
body the practices which they pursue on their plantations," he reminded
southern representatives that although freedom of speech "was for years
trampled under foot by the slave power" and that he "sat here . . . in
degrading and humiliating silence," a new day had dawned. "Gentlemen
may play the tyrant on their plantations," he snorted, "but they shall not
impose silence upon Northern men, nor dictate the language we shall
use on this floor." Abolitionists and capitalists alike resented gag rules
because they effectively made silent slaves of northern men whose vocal
agitation was intended to free (indeed, was essential to freeing) southern
slaves and America's future from the unearthly hush of bondage and economic retardation. They believed gag acts were the Slave Power's attempt
to project the silent regime of the plantation onto the national body politic, and they interpreted this as a palpable threat to their freedom. By 1856

northern observers were complaining that they had long "submitted . . . in silence" and that it was now time to speak.[33]

Such hopes were sometimes denied. Unionists who found themselves south of the Mason-Dixon line in 1860–61 became acutely aware of the aural world and its relationship to politics and social safety. Calvin Robinson, originally from Boston but now a resident of Jacksonville, Florida, found that he could not win during the debates over secession. For the first two or three months of the secession crisis, Florida's Unionists "continued to utter their sentiments of opposition to the movement; but gradually the reign of terror gained full swing and the time came when for a northern man to utter openly his love for the Union would be almost suicide." It was "dangerous to talk 'Union talk' with anybody." Taciturnity, Robinson discovered, was the better part of valor, but "soon it came to the point, that silence was a crime. It was believed naturally enough that 'he who was not for secession was against it.'" Secessionists stepped up their aural surveillance and applied the strategies of slave control to new internal enemies. "Emissaries," recalled Robinson, "were even sent to watch us. . . . Eavesdropping was resorted to even, and towards the last when we did not dare talk, without first examining to see that there was no possibility of our words being picked up through some keyhole, door crack, or around a corner." Only Florida's hammocks provided aural security for voicing "Union talk freely," and Robinson compared his enforced silence to a "thralldom under which I was groaning." At the height of the secession crisis, silent slaves could be white as well as black. Robinson's only defense against the threats made by fire-eaters was aural. Fearing for his life, "I immediately proceeded to put my home in a more safe & defensible condition. I fixed a bell at the head of my bed and an arrangement so that no one could open either of the gates to my enclosure after nine o'clock at night without ringing that bell and instantly waking me."[34]

Abolitionists and capitalists fretted over the silence of the slave South because they associated southern quietude with indolence and economic backwardness—a regression that might well spread. Consequently, in their parables and stories they put the appropriate words into the mouths of reformed slaveholders. Pointing to the plantation, a mistress in Kentucky told her northern visitor, "It all looks bright, and peaceful, and happy, does it not?" The visitor nodded. "But, to us, who know what slavery is, this peace is the green corruption of a stagnant pool—the peace of death."[35]

Even when northerners heard sounds of progress and industriousness in the South, they considered them frustratingly modest and attributed them to northern initiative. "At the mouth of the St. Johns there is now erected a large steam sawmill," noted John C. Richard of Philadelphia on a visit to Florida in 1843, "whose puffs can be heard several miles out at sea, which gives at least some evidence of industry and enterprise." Once Richard reached Jacksonville, the explanation for such sound progress became clear: "The inhabitants are intelligent, and mirable dictu industrious; being composed chiefly of merchants and lawyers many of whom are from the northern states." The true South lay beyond Jacksonville. Just twenty miles inland Richard heard the "air of stillness" reminiscent of "a new and uninhabited country." Without northern enterprise the South was worryingly silent, a product of slavery, the indolence it induced, and in the case of Florida, the destructive capabilities of Native Americans. Now that red-skinned foes had been defeated and removed, there was hope for Florida, provided slavery was not reintroduced. Richard thought "the long protracted Indian warfare" a blessing. It had retarded the cancer of plantation slavery in Florida. In 1843 Floridians had the opportunity to make their world anew, the chance to modernize through freedom and introduce the soundscape that went with it. "Since all internal commotions have subsided," wrote Richard, "we find that under the cheering influence of a liberal and enlightened government the woods have been made to resound with the hearty strokes of the settlers' axes. The tide of immigration is diverted from the west and now flows with a rapid current to the rich lagoons of the luxuriant Territory. Innumerable settlements are already made by enterprising freemen, where lately the yell of the savage shook the foliage of the magnolia and startled the dove from her haunts in the Everglades." With the noise of savages, the silence of slaveholders, and the horrific whipping and clanking of slavery drowned out by northern-backed industrialism, the sound of progress and freedom now had a footing in the South.[36]

Northerners who had never been south nevertheless imagined the aurality of slavery because proliferating print culture, spread by market and transport revolutions, offered them aural metaphors and descriptions in abundance. Rumbling railroads, puffing steamships, and humming telegraphs carried print laced with aural sectional imagery that exploded in readers' ears through their eyes. Antebellum orators recognized this. The power of speakers' words on minds and hearts was complemented by the printing press, which captured those words and enabled readers to revisit and reexperience the emotive power of the orators' original lecture or speech.[37]

Sounds resounded in print. In response to one of John C. Calhoun's many denunciations of free wage labor, the *New York Plaindealer* ran an editorial in February 1837 contrasting societies free and slave. Contrary to Calhoun's claims about the efficacy of slave labor, the *Plaindealer* maintained that the shortcomings of slavery as an economic and social system could be seen and heard in "the languishing condition of agriculture[,] . . . paralyzed commerce," and desolate lands. Slavery produced inaction and, by default, was worryingly quiet. When bondage was heard, it was through the absence of human activity and in the unnerving screams of wilds that should have been conquered. Uncultivated fields and economic dislocation generally "howls in your ears through the tangled recesses of the Southern swamps and morasses." Silent or howling, the South sounded primitive, primordial, and backward. Should readers remain in doubt about the truth of such printed descriptions, the *Plaindealer* advised them to actually hear as well as see them: "If anyone desires an illustration of the opposite influences of slavery and freedom, let him look at the two sister states of Kentucky and Ohio." The contrast was not so much between an urbanizing Ohio and a rural Kentucky (although that difference was important) but, rather, between the strains of southern slavery and those of northern capitalism. In Ohio, "the air is vocal with the mingled tumult of a vast and prosperous population. . . . Every valley shelters a thriving village; the click of a busy mill drowns the prattle of every rivulet, and all the multitudinous sound[s] of business denote happy activity in every branch of social occupation." Villages hummed with activity, and the throb of mills and progress resounded everywhere. It had taken time, effort, and a firm commitment to northern principles to achieve this soniferous garden. After all, Ohio "but a few years ago, slept in the unbroken solitude of nature . . . and through the dusky vistas of the wood only savage beasts and more savage men prowled in quest of prey." Silent woods and savage screeches, though, had been "bowed before the woodman's axe." "This is Ohio, and this is what freedom has done for it." In Kentucky, on the other hand, laxity and indolence prevailed. "Lowing and hungry kine" echoed indolent planters and slavery's failure to produce agricultural abundance. Here men "lounge in the sunshine," and "languid slaves" were common. If clicks of axes sounded, they did not resound long: "The axe drops from the woodman's nerveless hand the moment his fire is scantily supplied with fuel." Slavery also rendered Kentuckians an irreligious bunch. "Whence comes it that 'the church-going bell' so seldom fills the air with its pleasant music, inviting the population to religious worship?"[38]

Whatever else they disagreed on, abolitionists and free soilers concurred in their aural assessments and characterizations of slavery. Certainly northern capitalists of all ilks preferred the gritty rasp of wage labor industrialism to the vaguely disconcerting cries of passionate democracy, and abolitionists hankered to hear more of the steady voice of a virtuous citizenry than of the occasionally deafening shrieks of working-class immiseration. But for the most part the sounds of political liberalism and economic capitalism were indistinguishable to antebellum bourgeois ears. By contrast, slavery's westward expansion and its attendant wars produced dangerous, degenerative noise. Free soilers, capitalists, and abolitionists heard "the fierce din of battle" during the Mexican War and commented, "the distant din of slavery's battles comes wafted on the soft South-western breezes to the walls of the Capitol." Even the noise of the ordinarily silent Slave Power was audible, particularly in Kansas during the 1850s. It was, after all, "a peaceful territory," whose free-wage hum was in danger of being silenced by the boisterous, ruffian emissaries of the South's master class. The passage of the 1854 bill "was a thunder-clap in a clear sky" in "a time of quiet" for the *Hartford Daily Courant.* If slavery spread, economic silence would reign. In 1854 Republican John P. Hale asked a New York audience whether expansion of slavery westward would jeopardize the progress of wage labor. "Shall the hum of busy industry be heard on its hills and in its valleys?" Not if slaveholders had anything to do with it, he suggested. Bleeding discouraged business, and free soilers' hopes to reproduce the echo of the East in the soundscape of the West, in the short term at least, were disappointed: "All growth had ceased, all business was depressed, and times were quiet enough for a hermit." Not only did slavery militate against the reassuring throb of industrialism, but the system also imposed "a general torpor" on enterprise. Free soilers and abolitionists wanting their own serenity restored argued that the prevention of slavery's expansion would "promote tranquility" because it would end acrimonious, noisy political debate and replace it with the song of free wage labor.[39]

Slavery and the economic stagnation that went with it, argued a writer for Boston's *Christian Reflector and Watchman* in 1849, posed a palpable threat to the economic future of the United States. "Slavery," remarked the author, "has greatly retarded the growth of our Commonwealth, and prevented the development of the resources with which she is so richly endowed." "Remove this incubus from her fair bosom," he continued in language and metaphor prefiguring Republican Party rhetoric, and as "a free State, [Kentucky] would resound from her centre to her extremities with the busy sounds of enterprise—her population would be doubled and trebled. . . . Works of internal improvement, facilitating transportation between different and distant points, would spring into existence. . . . Habits of activity would banish the languor that is now felt in every vein[,] . . . and the hum of industry would rise to heaven from every hillside and smiling valley, like an anthem of praise from a happy and thriving people." Sounds of southern industriousness fell on deaf ears precisely because they were a product of slavery.[40]

Foreign visitors authenticated such claims. "On passing from a country in which free institutions are established to one where they do not exist," commented Alexis de Tocqueville in 1835, "the traveler is struck by the change; in the former all is bustle and activity, in the latter everything is calm and motionless. In the one, melioration and progress are the general topics of inquiry; in the other, it seems as if the community only aspired to repose in the enjoyment of the advantages which it has acquired." His contrast was between democratic and aristocratic institutions, but it might as well have been between northern capitalism and southern slavery. When de Tocqueville arrived in the North, he heard its political economy: "No sooner do you set foot upon the American soil than you are stunned by a kind of tumult; a confused clamor is heard on every side; a thousand simultaneous voices demand the immediate satisfaction of their wants."[41]

Slavery threatened the nation's economic future. "The stream of immigration which is setting towards our shores from the Old World," exalted S. M. Janney in 1849, "spreads over the free States of the North and West, but shuns the deserted fields of old Virginia." Should doubters remain, Janney counseled them to listen hard: "Swarms of industrious and enterprising freemen from the prolific Northern hive of New England have built up great cities in the West, have covered her rivers with the floating palaces of commerce, and have broken the silence of her frosts with the hum of manufactories." Turn ears southward, he advised, and hear only the silence of slavery.[42]

Free soilers and abolitionists were so sure of their argument because of what they heard from within their own society. Northern towns veritably hummed. As one antislavery visitor commented during a visit to a New . York town in 1848, "This is a youthful town," adding, "Seven years ago, the place where it stands was a silent wilderness, now it contains an active population of nearly two thousand souls." A "good many colored persons engaged in the rafting business" had added to the hum of industry.[43]

Verse says much about social value; the following excerpt from Toronto's *Provincial Freeman* of 1854 struck chords deep in the hearts of northern boosters as well as abolitionists. Sounds preindustrial and industrial, new and old, mingled, the former reinforcing the latter as a legitimate, inexorable progression toward the modern:

> I love the clanging hammer,
> The whirling of the plane,
> The crushing of the busy saw,
> The creaking of the crane,
> The ringing of the anvil,
> The grating of the drill,
> The clattering of the turning lathe,
> The whirling of the mill;
> The buzzing of the spindles,
> The rattling of the loom;
> The clipping of the tailor's shears,
> The driving of the awl,
> The sounds of busy labor,
> I love, I love them all.

Slaveholders agreed with much of this, for these sounds connoted activity. But it was how these sounds were generated—the social and economic context of this aurality—that differentiated North and South. For as the poem continued, sounds of industry were "Not the toil and strife that groaneth, Beneath the tyrant's sway," but, rather, melodies produced by "willing heart."[44]

The counterpoint to the northern idiom of southern silence was the noisy South. The critique devolved on two principal categories of noise: political and economic. Combined, these generated a moral critique of the sound of southern social and economic relations. The economic noise of slavery did not reside in the South's few free labor factories. Plainly, northerners could not admonish the South for doing, albeit on a much more modest scale, what they applauded in their own society. Instead they reserved severe judgment for how slaveholders regulated their labor. They loathed the aurality of selling labor (not least because in their world, market allocation and supply and demand of free labor were silent), and they detested the sound of compulsion (whips and chains particularly) because their own invisible hand of wage labor incentives was also inaudible. The northern critique of corporal punishment of the 1820s and 1830s "was addressed simultaneously against older monarchical or aristocratic regimes and the contemporary slave regime." Whipping, which could be heard clearly in the slave South, no longer echoed in the liberal North, and for a good, simple reason: "Both penitential punishments and free labor aimed ultimately to create a productive and prudent work force. And they did so by denying that they acted on the body, insisting instead that they reformed and empowered the will." Thus northerners excoriated the sound of the whip because it was the sound of the old, the corrupt, the cruel, the unenlightened, and the economically premodern.[45]

From beginning to end, slavery sounded dreadful. J. E. Snodgrass told readers of the *National Era* in 1847 that after he "removed to Williamsport, a quiet little village on the Maryland side of the romantic Potomac," quietude was shattered by the noise of slavery's birth. He was awakened one morning by "a scream which violently cleft the cold air with its piercing agony. I instantly sprang to my feet, only to have my ears saluted by shrieks still more startling." A wicked medley of a primordial, savage past was Snodgrass's only comparative frame of reference: "So loud had the voice now become, that it seemed to startle from the bosom of the quiet river echoes such as perhaps had never been heard since the days when its glassy tide used to reflect the war-hoop of the Indian, and the scream of the panther." The cause of the scream was the enslavement by a master of a free woman of color.[46]

Once bondpeople were enslaved, the coercion of their labor produced a sickening din that echoed in humanitarian ears and made for potent propaganda. Elias Hicks winced by listening selectively: "We hear them summoned to their daily task, by the clashing of cowskin scourges."[47] Early antislaveryites agreed. When Jared Sparks, editor of Boston's *North American Review,* ventured to Georgia in 1826 he commented, "Two evenings ago, while I was sitting in my room, my ear was assaulted with the cries and screechings of a person in distress. I immediately went to the piazza, & could distinctly hear the strokes of a whip, & the cries redoubled. I walked in the direction of the noise till I came to the market house, where I found a small crowd assembled, & a man employed with his coat off whipping a negro." Such sounds were "shocking" and struck at the heart of northern sentiment concerning freedom. "Never in my life," frothed Sparks, "have I felt my indignation rise so high."[48] The emotional response to such sounds was potent and would become more so as northerners considered what they wanted for the future of their own society.

The sound of the slave market was especially offensive to abolitionists and free soilers. Slavery and the slave trade had been heard for years. In the mid-eighteenth century, for example, Quaker Anthony Benezet asked, "Will not God hear their Cry?" From early claims of noisy slavery, abolitionists and antislaveryites of the 1830s narrowed their critique to southern slavery. Of the St. Louis market, for example, British abolitionist Ebenezer Davies said, "The clock was striking twelve; and, before it had finished, the vast dome reverberated with the noise of half-a-dozen man-sellers brawling at once, disposing of God's images to the highest bidders. It was a terrible din." As modern as southern soundscapes were becoming, the aural world of the slave market was tumultuous and hardly in keeping with the disciplined timbre of the world of merchant capitalism.[49] Moreover, the noise of the South's mechanism for reallocating labor was far too near northern ears. That Washington, D.C., "is the only capital west of Constantinople" where the slave trade was in open operation was a national shame. The "noise and confusion," the "din and turmoil," and the "cries and groans" of the slaves produced by the hellish trade were an affront to the aural sensibilities of all decent men and women. A few were optimistic. "The day is not distant," opined a writer for the *National Era* in 1849, "when the soil of this District shall be consecrated to Freedom, and the deliberations of the Congress of the Republic be conducted, not amidst the discordant sights and sounds of slavery and slave dealing, but amidst institutions in perfect harmony with the Declaration of Independence and the Spirit of the Age." But even after the Compromise of 1850 had silenced the trade in the district, echoes lingered. In 1856 General J. Watson Webb said, "This is a city in slave District; its tone is the tone and sentiment of Slavery."[50]

Selling and buying chattel was, indeed, a noisy affair, especially when contrasted with the dulcet workings of the capitalist market for labor power. The first Monday of every month, for example, was undoubtedly a loud day in antebellum South Carolina, when most slave sales took place at public auction. Even northerners who were sympathetic to the South found such sale days wrenching. Boston minister Nehemiah Adams said of sales, "Such shrieks, such unearthly noises, as resounded . . . can not be described." "The sad and moaning coffles" were the "soundings of human misery." Slaves "mourned aloud, and their sighs and sobs, mingling with infant's screams, the crack of whips, and the curses of the drivers, made as discordant and infernal sounds as ever shocked the ear of night."[51]

Slavery was distinctly un-American. The noise, clamor, and discord generated by slavery and the people who defended it, maintained Frederick Douglass, belied the notion that "we, the people," aimed for "domestic tranquility." Nor was the sound of slavery consistent with America's God. In the beginning, God's world was harmonious, a place where there was "not a note to mar the general plan," a world where "Earth's concord with the lyres of Heaven" sounded "wo! wo!" But such "World-Harmonies" had been disrupted by the noise of slavery so that, abolitionists lamented,

> How harsh, hateful on undefended ears
> Rings the loud cry of Tyranny and Wrong
> Drowning the sweetness of the one great song.

Should good men doubt the heard evil of southern bondage, advised New York's *Colored American* in 1840, "go stand upon the Allegheny mountains and throw your eyes over the cotton plantations and rice fields of the South. Hear the groans of the father in bondage. . . . Can we do nothing in hastening forward the day when the trump of Jubilee will be sounded in this land?" Horace Greeley thought so and looked to men like John Brown to accelerate the process "until the not distant day when no slave shall clank his chains in the shades of Monticello or the graves of Mount Vernon."[52]

Questions remained, though. Now that northerners had been sensitized to the sounds of slavery and its political and economic implications, how were they to communicate the dreadful silences and noises of southern bondage and replace them with the melody of northern institutions? Perhaps a more important question was how slaveholders would respond to this rasping critique of their world.

I do not wish to think, or speak, or write, with moderation.
. . . I am in earnest—I will not equivocate—I will not excuse—
I will not retreat a single inch—AND I WILL BE HEARD.
—William Lloyd Garrison, *Liberator,* January 1, 1831

(((7. Northern Shouts and Southern Ears

The aural tactics used by abolitionists and northern capitalists
to critique the South depended less on gentle, whispered persuasion than
on sheer shouted force. To help cleanse the country of slavery and guarantee its free wage labor future, northern elites of all stripes lauded the
virtues of their own soundscape (and in the process minimized its shortcomings) and tried to force southern masters to listen to the advantages of
freedom and the enormities of bondage. Indeed, what united otherwise
disparate northerners was their common understanding of how slavery
and its supporting authorities sounded and their agreement on the keynote of America's future. While abolitionists listened to the inhumanity
of slavery and while free wage labor advocates heard the worrying quietude of a premodern economic institution, both constituencies nevertheless heard two sides of the same coin. Certainly both groups heard the
silent tyranny of the Slave Power. Although they tended to emphasize
different aspects of slavery's sounds, northern capitalists and abolitionists
concurred that because slavery's terrible authenticity revealed itself to the

attentive ear, the future in time and space must not resound with southern soundmarks. This common loathing of slavery's soundscape helps explain why northern capitalists and moderate Republicans added their voices to the louder chorus of radical abolitionism. A shared revulsion of slavery's registers and a common commitment to aural descriptions deployed to critique the South allowed a variety of political and cultural leaders to coalesce around a unified opposition to slavery and drive to establish a cultural hegemony over U.S. space and time partly by deploying a form of aural imperialism.[1] The northern critique of slavery's soundscape was critical not simply because it united the North to end the institution but also because it helped convince the master class that their own future lay not just in quieting their slaves but also in silencing their increasingly noisy northern enemy. In effect, by shouting southward, abolitionists especially and northern supporters of free labor generally confirmed what masters long suspected: northern society was a veritable cacophony that would not desist in its attempt to export its soundscape—and all that it entailed—unless silenced. If such silencing necessitated using some of the same tactics they had deployed against, for example, southern Indians, so be it. Short-term noises of warfare for long-term social quietude and serenity seemed a reasonable compromise for the master class.

Several tactics and strategies were available to northerners. Abolitionists asked England to listen to southern bondage in the hope that it might help. "Oh, listen to the voice that is calling to thee," pleaded the *North Star* in 1848. Of the "clank of [the slave's] chain," England was asked, "let it not echo in vain." Abolitionists were confident that freedom would sound in the South because it had done so elsewhere, particularly in the British West Indies. E. C. Ellis counseled countrymen to train their ears on places where emancipation had already been accomplished.

Each balmy zephyr o'er the Indian Isles,
Is redolent with joy and freedom's smiles;
The slave no more a slave, uplifts his hand,
And shouts for liberty and ransomed land.[2]

West Indian sounds of jubilation were preceded by silent solemnity: "Not a sound escaped from negro lips, which could wound the ears of the most feverish planter." Congregations of freedpeople fell "upon their knees, and receive[d] the boon of freedom in silence." At midnight came the medley of freedom: "The slow notes of the clock fell upon the ears of the multitude; peel on peel, peel on peel, rolled over the prostrate throng.

. . . A moment of profoundest silent passed—then came the burst—they broke forth in prayer; they shouted, they sung, 'Glory' 'alleluia;' they clapped their hands." Responsibilities of freedom were revealed when the Wesleyan missionaries spoke, "explaining the nature of the freedom just received, and exhorting the freed people to be industrious, steady, obedient." The sounds of the clock would thereafter take on new meanings, constituting important notes in the music of wage labor.[3]

The appeal to the English experience was sensible since slavery could be heard across national boundaries; metaphors of bondage's register helped cement an antislavery imagination and community. Despite "all the noise and strife of political parties," commented one writer in 1841, international abolitionism could still hear the "wailings of the broken-hearted captive, the shrieks of the bereaved slave-mother, the clank of the chain, the sounding of the lash [and] the curses of the slave-dealer," which "resound[ed] within the halls of that very senate-house, which within is ringing the loud panegyric of universal freedom." Abolition meant that the "cry of violence ceases" and "shouts of jubilee are heard."[4]

How to quiet the bellowing of tyrants and replace slavery with free labor were serious questions, but the tactic mattered not for some abolitionists. They were, after all, dealing with insanity. A slaveholder was "a 'mad bull,' who, if you should attempt to reason with him, would be sure to bellow and toss his horns! If we would escape being gored to death, we have been warned to keep our distance and hold our tongues!" If unreasonable men bellowed, the more cunning were a "knot of snakes," their presence betrayed only by sibilance. "Slavery crawls," shuddered the *Albany Evening Journal* in 1854, "like a slimy reptile . . . to defile a second eden." Noisy or silent, though, the true abolitionist had to brave the zoo and "find access to the ears and hearts of those who live amidst the monuments of slavery." Only by making masters listen to their own cacophony would the battle be won: "The facts you state; the arguments you employ; . . . your loud warnings, your fervent entreaties, will force their way into ears. . . . And those ears will tingle. Tongues, which a thousand artifices had been employed in vain to tie, will be set in motion. Tyrants may roar, and stamp, and curse. But what then? Surely the noise and tumult in which they may give vent to their windy rage, will but ill promote the cause of *silence*. By the very act of swearing that a word shall not be spoken, their own oath they will violate!"[5]

If the image of clanking chains conveyed by the printed page failed to stir the heart, abolitionists readily traveled to towns such as Springfield, Massachusetts, and "exhibited on the platform the very chain with which John Brown was led for thirty miles in a hot sun after his capture. Its clankings touched a cord [*sic*], and the City Hall was thunderous with emotion." Words on pages worked well, but re-creating slavery's sounds worked even better and bolstered the aural imagery.[6]

Escaped slaves also helped shape the abolitionists' mode of attack, for they made humanitarians listen to slavery. Frederick Douglass, for example, had heard clearly (and authentically) the ring of the slave whip and the "clank" of slaves' chains. "The wails of bondmen," he declared in 1848, "are on my ear." Frederick Douglass heard so well precisely because he had been a slave who inhabited that soundscape. "In the deep still darkness of midnight," he recalled, "I have been often aroused by the dead heavy footsteps, and the piteous cries of the chained gangs that passed our door." He also knew elite southerners well enough to know that they, too, disliked some aspects of their southern soundscape: "I was often consoled, when speaking to my mistress in the morning, to hear her say that the custom was very wicked; that she hated to hear the rattle of the chains, and the heart-rending cries."[7] This particular caveat fell on deaf antislavery ears.

It was Douglass's actual hearing of slavery that gave his critique of the peculiar institution such authenticity to northerners. "O! Had I the ability, and could reach the nation's ear," he would "thunder" his denunciation of bondage. So he did, by making his audience listen to the sins of slavery. It was Douglass and others like him who helped refine abolitionists' hearing of slavery's dreadful sounds. Of slave traders ("human flesh-jobbers") he declared, "Hear his savage yells and his blood-chilling oaths, as he hurries on his affrightened captives!" Of the slave pens and markets, which stood in noisy contrast to the silent northern market for wage labor, he said, "Heat and sorrow have nearly consumed their strength; suddenly you hear a quick snap, like the discharge of a rifle; the fetters clank, and the chain rattles simultaneously; your ears are saluted with a scream, that seems to have torn its way to the centre of your soul! The crack you heard, was the sound of the slave-whip; the scream you heard, was from the woman you saw with the babe."[8]

Slaveholders could not win for trying. Even when they heard in the South sounds familiar, comforting, and wholly in keeping with their pastoral conceits of a romantic age, abolitionists remained unmoved. What was innocent and aurally romantic in the North, they believed, was wholly sinister in the South. The melody of the northern hunt, for example, was paralleled in the South by the noise of the hunt for the runaway. As John Greenleaf Whittier put it in "The Hunters of Men,"

As the fox-hunter follows the sound of the horn:
Hark—the cheer and the hallo! The crack of the whip,
And the yell of the hound as he fastens his grip![9]

Similarly, Whittier's "A Sabbath Scene" was a direct challenge to northern churches' aid to slave catchers, one that described how the northern Sabbath soundscape had become polluted by the noise of the slave catchers.[10]

Because they understood the depth of slaveholders' convictions, abolitionists believed that emancipation would be loud. True, the ending of "property of man in man" in England was "silently and imperceptibly effaced," and "the noiseless operation of moral causes" was something to think about. However, although "certain moral causes 'noiselessly' abolished the system of villeinage it does not follow that the same causes will silently remove Slavery from this country." The abolitionists were right. Slaveholders gagged on the noise of free labor and refused to let its strains migrate south without a fight. Wage labor, argued proslavery thinker Henry Hughes, mixed with radical democracy was social vitriol because "the awful yell of blood or bread, shall ring from the throats of thirty thousand starvelings mustered in order and under arms in Broadway."[11]

Shall "we ourselves be free" was the question of the day, which led to the real issue: how should we know it? One common test was ascertaining what the silent/howling land to the west would sound like. In "An Appeal to the Women of the United States" in 1848, antislavery women answered and identified the political crux of the sectional crisis: "Shall the air of our mountains and our prairies, that hitherto has borne only the songs of the wild bird, the ring and echo of the pioneer's axe, and the busy hum of free labor, be burdened with the groan of the slave, the crack of the lash falling on woman's back, the mother's wail as her infant is torn from her embrace, the husband's muttered curse as he sees with fettered limb the wife of his bosom made the victim of lust?" No, was the answer. "But what, it may be asked, can woman do?" She could "raise her voice in eloquent appeal" and, presumably, be heard by virtue of her assigned quietude. When ordinarily silent women spoke, they would, perhaps through the sheer aural novelty, be heard more readily. And so the chorus of free labor and political liberty would grow louder, loud enough, perhaps, to be heard south of the Mason-Dixon line and loud enough to be heard in the cabins of the unfree.[12]

Antislavery men were not aural egalitarians. Some male abolitionists constructed female antislavery discourse as "cackling" and as different from and inferior to male political discourse ("crowing"). Many antislavery men aimed to silence woman's political voice by constructing it as sounding different. What they wanted was female silence on such matters.[13] Leaders of the antebellum northern women's rights movement ignored admonitions for silence and, through words written and spoken, tried to force American men to listen to their grievances in the hope that when men heard, they would acknowledge female political competency. Women's rights advocates well understood that men used sound as sentinel against disruptions of social order. Male newspaper editors were "already sounding notes of alarm, and calling upon their clan to prepare to resist the army of women that are coming against them in battle" from their "watch towers." When the men were too late to ring the bell, they attempted to quash women who did speak out in public by hissing, stamping their feet, and reducing their enemies to silence through cacophony.[14]

But women activists used the noise of slavery to empower themselves. How, some rightly asked, could the good woman remain quiet and submissive when her hearing and sensibilities were assailed by the horrors of bondage? "Talk not to me of a woman's sphere," wrote a Connecticut correspondent to the *Liberty Bell* in 1842,

> When the horrid sound I can almost hear
> Of the brutal lash, as it echoes by
> With the wail of the new-born infant's cry.

Women who had heard such sounds and who had learned the principles and tactics espoused by some of the same antislavery men who sought to silence them could now claim the right to answer negatively the question "Shall I close my lips, and in silence wait[?]"[15]

In addition to trumpets and bugles, bells figured prominently in the abolitionist arsenal to make masters listen.[16] The bell was both aggressive, punching its sound into the ears of the socially and spiritually deaf, and defensive, protecting liberty and calling citizens to its cause. Unlike horns, bells radiate sound uniformly in all directions (horns focus it specifically), and the bell's radiated consistency suggested aural nationalism to abolitionists. It was abolitionists' ideal symbol for historical and psychological reasons: historical, because U.S. freedom was symbolized in the bell of liberty; psychological, because of the bell's aggressiveness. The word derives from the Anglo-Saxon *bellam*—to bellow. It was also warlike (*bellum*), for bells had often been melted for cannon and then recast as bells, only to be melted again for war. In the battle against slavery the bell was the ideal image and metaphor for freedom.[17] Bells also defended liberty, especially when they prevented slaveholders from retrieving their property in free states. A quiet black man, regardless of his origins, could be kidnapped from northern states. Southern scoundrels would "drag him from his quiet home to a state of the most abject slavery." It was a crime all the more unpalatable because some African Americans had emulated the quietude and decorum of the northern middle class. But northern bells could preempt such kidnapping and so preserve liberty. Residents of New Bedford, Connecticut, in March 1851 received "intelligence that a steamer with 100 armed men and the Deputy U.S. Marshal left Charlestown Navy Yard on Saturday evening bound for this or some other port in the vicinity, with the avowed intention of claiming Fugitive Slaves. . . . The bell on Liberty Hall was immediately tolled and continued to sound the alarm for about one hour."[18]

Bells and their ringing, particularly on national occasions, had a greater metaphorical than physical aural range. Independence Day was heard locally when church bells were rung; but when all bells were rung, they resounded metaphorically throughout the United States and so contributed to the formation of an aural imagined community. Not everyone interpreted the bell simply as the sound of national identity. While workers and slaves undoubtedly heard the July Fourth bell in this way, they attached additional, less benign significance to the sound. It was simultaneously the sound of freedom and the sound of slavery, wage or otherwise. Once freed, however, slaves heard July Fourth bells as national symbols of freedom. As Frederick Douglass explained to listeners at the Rochester Ladies' Anti-Slavery Society on July 5, 1852, the "din of business, too, is hushed. Even Mammon seems to have quitted his grasp on this day." With business quieted, one could now hear the "ear-piercing fife and the stirring drum unite their accents with the ascending peal of a thousand church bells." Slaveholders doubtless heard the menace in Douglass's next words: "Prayers are made, hymns are sung . . . the quick martial tramp of a great and multitudinous nation, echoed back by all the hills, valleys and mountains of a vast continent." Yet, "Fellow-citizens; above your national, tumultuous joy, I hear the mournful wail of millions!" Douglass had, in effect, appropriated a national, aural symbol and recast it as sectional. The bell of freedom would soon reverberate throughout the vast continent, securing freedom for all by energizing those who heard it.[19]

The power of the bell had a long pedigree. The medieval notion that "the evil spirytes that ben in the regyon" were expelled "when they here the Bells rongen" lasted well into the nineteenth century. But the spirits had changed countenance. Slaveholders were the "fiends and wicked spirits" for abolitionists, and they hoped that ringing liberty bells would rid their country of the ghost of bondage and usher in modernity. Bells represented freedom and stood also "as mighty living monuments, of the industry, enterprise, and intelligence of the white man."[20]

Northerners argued that slaveholders had misappropriated bells. Masters, they maintained, used bells not to illustrate freedom but, rather, to regulate slave work and, more poignantly, to indicate slave sales. On both points they were right.[21] Darius Lyman explained clearly what abolitionists considered the principal differences between slaveholders' and democrats' understandings of the bell. In "The Cracked Liberty Bell," in 1857 Lyman continued an abolitionist practice begun in the late 1830s and appropriated the national icon of freedom and reconstituted it as distinctly northern. He depicted a slaveholder examining the bell, eying the inscription in particular. "These words are rather fanatical," Lyman had his slaveholder exclaim, adding, "Liberty should not be proclaimed to all the land. . . . That would disorganize society. . . . The sentiments on the bell . . . proclaim too great a liberty, and the bell is cracked because the doctrine is false." Lyman had the bell itself speak: "He who inscribed it designed me to utter the glad tidings of freedom without regard to race or sex . . . expecting that I should have naught but gladness to dispense, whenever my voice should be heard." But God cracked the bell, so the tale went, "seeing that I should speak ever after to a nation of hypocrites." Only a "miserable clatter" could now be heard, eloquent testimony to the divisive, sectional cracking of slavery in the United States.[22]

Because bells were often God's, their sound had power—power enough to cut through the noise of antebellum urban hubbub ("Clear as the tones of a cathedral bell above the hacks and drays of a city"). Bells were critical to the abolitionist crusade because their tones and sounds were not, both figuratively and literally, delimited spatially. The sound of the bell crossed borders, entered houses, and could not but be heard by even callous ears.

Right onward speeds the thrilling sound,
The Carolinas hear the tone;

.   .   .   .   .

Above Virginia's reeking plains
Redoubled swell the echoing strains,

.   .   .   .   .   .

That even the very dead might hear—

The peals were irresistible. In evangelical style they awakened the guilty by forcing their tones into ears; they penetrated territory from a safe cupola of liberty; they were ecumenical, untamable, and quintessentially national and international. As a symbol for nineteenth-century nationalism, globalism, and freedom the bell had no serious rival.[23]

Non-American abolitionists agreed. Religion, nationalism, and globalism were the focus of the thinking of Dublin's James Houghton in 1842 on the nature of the international bell of liberty: "There is something delicious in the silvery notes of even the tiny bells," mused Houghton, elaborating, "and how sublime the grand and melodious swell of the large-mouthed bell, flinging its magnificent tones to the winds of heaven. Heaven seems around and within us, when such peaceful and sweet music rests upon our ears." *Dulscire nostris resonat tinnitibus aer* (the heavens resound sweetly to our jinglings). On the bell's tone as spatially limitless and transcendent, he remarked, "So it is when we wake the echoes of our mountains; the sound bursts forth in full and magnificent volume, wandering from hill to hill, momentarily getting fainter until it dies away, seeming never entirely to lose all trace of its sweetness." Houghton also contrasted the bell's tones with the noise of slavery: "We made lake and mountain ring again," which were "dreaded sounds in the ears of 'soul-drivers,' all the world over." Lastly, Houghton heard the bell of liberty as global: "The liberty Bell is fixed on the world's watch tower—devoted watchmen have its cord."[24]

Literally and figuratively, U.S. freedom would be communicated through the *Anti-Slavery Bugle* of New Lisbon, Ohio, and the annual *Liberty Bell*. Indeed, in 1839 the first issue of the *Liberty Bell* began with a sonnet that, the authors claimed, had been inspired by the inscription on the Philadelphia Liberty Bell. It is worth quoting at length not least because it suggests how abolitionists heard as much as saw slavery, injustice, resistance, and victory:

It is no tocsin of affright we sound,
Summoning nations to the conflict dire; —
No fearful peal from cities wrapped in fire
Echoes, at our behest, the land around: —
Yet would we rouse our country's utmost bound
With joyous clangor from each tower and spire,
Till yon dark forms of mother and sire,
Lifting their sullen glances from the ground,
Shall stand erect exultingly, while near
LIBERTY passes by, with lofty greeting! —
The hills are shaken by the shout of cheer
From slaves made free, and friends long parted meeting.
Join, thou true hearted one, —oppression shaming!
LIBERTY through the land, to *all* its sons proclaiming.[25]

Writing for the *Liberty Bell* in 1842, John Pierpont added a nationalist flourish: "Our Liberty Bell! Let its startling tone / Abroad o'er a slavish land be thrown." The bell would smother slavery's awful register:

Let the Liberty Bell ring out—ring out!
And let freemen reply with a thundering shout,
That the gory scourges and clanking chains,
That blast the beauty of Southern plains,
Shall be stamped in the dust; — [26]

Combining the ancient power of dismissing evil with the modern sound of democratic freedom, the bell proved the ideal symbol for antislavery-ites.

All of this amounted to an abolitionist project both to convert other northerners to their cause and to make slaveholders hear freedom and the evils of bondage. Whatever separated various antislavery groups, they shared the conviction that slavery was an aural assault on U.S. freedom whose remedy was to make people listen to what they so clearly heard. William Lloyd Garrison captured the essence of this mentality in the first issue of the *Liberator* in 1831: "I will not equivocate—I will not excuse—I will not retreat a single inch—AND I WILL BE HEARD." The evangelical, revivalist, reform-minded impulse of many abolitionists and, more broadly, some northern elites undoubtedly helped give intellectual and religious justification as well as shape the style of the abolitionist belief in the efficacy of the shouted word. William B. Richards of Massachusetts, a man not unknown to consume "a liberal dose of antislavery reading," suggested the political efficacy of heard proselytization rather than silent prayer when he noted in his diary in 1846, "One hour of holy effort is worth a thousand of quiet church slumbering—or rigid penance." Abolitionists reveled in these aural combats, and the idea that the sins of slavery—the moans as well as the silences of the institution—had to be shouted to deaf ears united a fairly disparate group of men and women.[27]

This insistently aural combativeness, the need to make themselves heard, explains why abolitionists were so hard on northern men who kept quiet. Northern silence on the antislavery cause, abolitionists believed, was a threat to the hum of freedom and democracy and future peace and prosperity. African American antislavery activist James Forten Jr. made this point in an address in 1836. "I can tell you, my hearers, if the North once sinks into profound silence on this momentous subject, you may then bid farewell to peace, order, and reform." Abolitionists lambasted brethren who kept mum on slavery. "Father Mathew's decision to be silent, while in this country, on the subject of slavery" caused a furor, wrote Wendell Phillips from Boston in 1849. The Irish teetotaler Mathew was castigated for voluntarily "consent[ing] to be gagged" on the matter. "We deny the right of any man, especially one of commanding influence, to be silent on one question, that he may enhance his weight on others." His "silence on slavery to advance teetotalism" was undemocratic. Abolitionists were under a moral imperative to speak out.[28]

Just as published travel and slave narratives taught antislavery ears to hear black subjectivity, so abolitionists now attempted to ensure that others heard not only black agency but also the horrors of slavery. Escaped slaves such as Frederick Douglass helped set in motion the politicization of the sectional conflict by appealing to abolitionists' ears and using their aural sensitivities to heighten, refine, and sharpen their humanitarian sensibilities. Abolitionists happily obliged not simply by drawing on slave songs but by attempting to communicate—through the spoken and written word—the entire orchestral range of slavery's horrors. Theodore Dwight Weld, trained as a revivalist preacher by Charles G. Finney, proved masterful at making socially deaf audiences hear. For Weld, slaveholders were guilty of ultimate sins, "and human nature, with her million echoes, has rung it round the world in every language under heaven." Humanity "has uttered her testimony against slavery with a shriek ever since the monster was begotten." "Whoever denies this," maintained Weld, needed to have his ears opened. "Clank the chain in his ears," he blustered, "and tell him they are for *him*." Beriah Green wondered how abolitionists would combat the evil of slavery: "How shall we obtain the right answer? By shutting up our eyes? And closing our ears? And holding our tongues?" No, it was maintained, abolitionists had better listen then speak, and quickly: "Let us therefore speak at once, and emphatically . . . while our voices can be heard."[29]

If abolitionists sounded liberty, the enslaved had to learn to hear its call. Do not "bow down in silence to what tyrant's bid," the *Colored American* advised bondpeople in 1841. Instead, they should listen carefully to the North:

Come! Rouse ye, brothers, rouse! A peal now breaks
From lowest island to our gallant lakes:
'Tis summoning you, who long in bonds have lain,

. . . .

Hark! How each breeze that blows o'er Hudson's tide,

. . . .

The spirit tones are mounting: louder still
From out the din, where noble cities rise
On Mohawk's banks, the peals ascend the skies.
Responding sweet with morning's opening praise,
The sounds commingle . . .
In chiming music with the village bells.

Then came the warning to the master class:

The captive in his hut, with watchful ear,
Awaits, the sweet, triumphant songs to hear,
That shall proclaim the glorious jubilee,
When crippled thousands shall in truth be free.

Slaves, fancied abolitionists, were waiting to hear the North speak.[30]

An important aural signature of the abolitionist movement was volume, in contrast to the dishonest mutterings of slaveholders. "Faithful watchmen, they do not whisper while Jonathan's house is on fire," readers of the *Liberty Bell* heard in 1843. "May their lungs hold out until Brother Jonathan is thoroughly awakened to his guilt and his danger."[31] The fifteenth annual meeting of the American Anti-Slavery Society repeatedly condemned slaves' "clanking chains" and pointedly asked members to "show me a body calling itself by the name of Christ which stops its ears to the cries of the poor, and needy, and oppressed." Fearing God meant hearing oppression to these men and women. And once the shame had been heard, they had to ejaculate: "I do not see how it is possible for a person to be virtuous and not cry out against slaveholding as a damning sin," thundered William Lloyd Garrison in 1849. Make slaveholders hear bells of liberty, he argued, and "there will be no more clanking of chains, there will be no more shrieks and groans of those under the lash. . . . The day will have come for the jubilee song to be sung." First, men had to find the courage and spirit from within to speak out. It was, reported the *Colored American* in 1839, "the duty of every abolitionist first to abolitionize his own heart then those of his family, neighbors, town and state, before he overleaped its boundary and sounded the tocsin of liberty in the South." For Garrison, reluctance to protest was unfathomable: "Why then is it that the dweller on the banks of the Penobscot trembles and dares not whisper a word against" the foul institution? He gave a partial answer when he described the life-threatening activities of American "mobs . . . howling against their persons and their houses" during the brouhahas of 1836. Be that as it may, good men could no longer ignore the "groans of heart-broken millions [that] come up on every breeze." Partial or gradual emancipation would mean nothing. Should only southern border states abolish slavery, they would merely "send their slaves to clank their chains on Southern plantations." Complete silencing of the institution was the only answer.[32]

Free your slaves, abolitionists shouted to masters, and you will see that they share your appreciation of quietude. Free men of color in the North, maintained antislaveryites, had developed refined aural sensibilities—better, in fact, than some of the white lower orders. One, after all, was found reading in the countryside in a grotto of "unbroken quiet" on Independence Day 1828. "His zeal in the acquisition of knowledge, and his taste in this retiring from the noise and bustle of the city, on a day peculiarly attractive to persons of his class" demanded applause. Others, alas, were "indulging in revelry and riot!" But the point was clear: African Americans were not noisy by nature. If southerners truly wanted the quiet life, suggested James Russell Lowell in 1845, they should allow the slave freedom, for under bondage chattels "can never attain that quiet unconsciousness so necessary to a full and harmonious development." Nor could slaveholders themselves. Abolitionists well knew that what frightened masters the most was the "stealthy tread" of their bondpeople. One of the more radical and abrasive abolitionists (and there was fierce competition for that title), Stephen Symonds Foster, recognized that "the South never sleeps" because of the threat of slave insurrection. He also knew that only by recruiting listeners better than they could slaveholders hope to hear the silent slave walker. That, believed Foster, was why masters kept "a troop of blood-hounds standing sentry at every door!" Immediate abolitionism as announced by the Garrisonians in 1833 hit on one benefit that planters must have found appealing: the immediate freeing of slaves would cure jitters, and planters, for once, "may dismiss their fears, and sleep soundly." But masters had already made the calculation: fitful sleep and nervous twitches at every noise were the price to pay for retention of their organic society. Whatever its shortcomings and inconsistencies, abolitionism had the intellectual wherewithal to identify a central irony of southern slavery: insist on quietude at your peril.[33]

Moreover, some antislaveryites maintained that if slaves were noisy, as masters argued, they had only themselves to blame because they had created a context where the sounds of immorality were inevitable. According to Kentuckian James A. Thome, slavery's professed prohibition of "intercourse between the families and servants, after the work of the day is over" deprived slaves of appropriate moral interaction, leaving them to the mercy of an immoral, aural world: "The slaves, thus cut off from all community of feeling with their master, roam over the village streets, shocking the ear with their vulgar jestings, and voluptuous songs." "This pollution," believed Thome, "is the offspring of slavery; it springs not from the *character* of the *negro*, but from the *condition* of the *slave*."[34]

Northern free labor gave rise to a social order whose register often soothed and reassured abolitionists and capitalists alike. Slaveholders heard things differently. Yes, railroads and the market were sound progress; yes, masters consciously promoted and financed modest industrialization in their society; yes, slaveholders partook of a humanitarian sentiment; yes, they believed in the desirability of sober republican democracy. But the mobocratic tendencies of wage labor were a discordant, dangerous aural bazaar to southern planters. For masters, noise was any activity that threatened the harmony of their social order, and they proved unforgiving of those who challenged its day-to-day and abstract functioning. The tenacity with which elites in each section refused to habituate themselves to one another's soundscapes because of the competing ideologies each represented is testimony to the importance of the heard world in helping shape their sectional consciousness. They heard their own soundmarks as unique to their region and refused to let foreign noises mask them.[35] Southern masters would not listen to, much less be persuaded by, abolitionist or free wage labor arguments, no matter how much Garrison and his mob shouted. Should they shout too loud, should they try to worm dangerous sentiments into the ears of slaves and nonslaveholders, should they attempt to inscribe plantation serenity with the timbre of free labor, then they would be silenced.

What, precisely, could the master class do in response to northern criticisms? The Old South's masters had few choices. They did not, for example, make a distinction between retaining slavery and acquiring a qualitatively different quiet life of the sort offered them by abolitionists. To be sure, early efforts to colonize slaves focused on the benefit of "getting rid of troublesome free Negroes, whose presence in the South was a constant threat to social tranquility." In this way awkward binaries sometimes lost their clumsiness. Southern moderates, Virginia's Colonization Society in particular, used language designed to make their arguments at once southern to the master class but palatably modern to northern modernizers. On one hand, they depicted abolitionists as "fanatical" and accused them of arousing "religious zeal in a crusade against peace and order and union." Colonization, by contrast, "hushes discord." Just as white men "subdued the wilderness, and made those vast solitudes, hitherto unbroken save by the war-whoop of the Indian and the scream of the eagle, vocal with the hum of industry and the songs of christian praise," African Americans could learn to do the same when returned to Liberia. Given sufficient "engineers, carpenters, masons, blacksmiths, and other tradesmen," African states, populated by "the civilized negro" would resound with "the hum of industry" and the "sound of the church-going bell." Most masters rejected such thinking, arguing that slavery's retention was the key to preserving their quiet, orderly, organic society and all the white and republican independence that it seemed to guarantee.[36]

The aural dimensions of southern culture meant that slaveholders were extremely sensitive to whatever they heard migrating from north of the Mason-Dixon line. At the national capitol southerners encountered not quiet serenity but, as William J. Grayson described it in the 1830s, "tumult and disorder," and southerners believed that antislavery societies were "doing all they can to destroy our domestic harmony."[37] If they could not accommodate these sounds, their only choice, it seemed to many, was to silence them.

Slaveholders thought gag acts legitimate because they considered them an attempt to resist what has been called "sound imperialism." In the most literal sense, this means that "a man with a loud-speaker is more imperialistic than one without because he can dominate more acoustic space." Abolitionists' use of the image of the sound of the bell represented, to southern ears, an unapologetic desire to dominate southern space with the sound of freedom and thereby impose their cultural authority on U.S. space and time. Abolitionists did not deny this tendency; rather, they applauded it.[38]

Little wonder that slaveholders worried. What would lower orders hear in abolitionist rantings? Masters fretted even more as the transportation and communication revolution made it easier to hear such sedition in print. In 1847 the Mississippi Supreme Court declared that from "the pulpit many . . . hear . . . doctrines announced and enforced, and embrace them as articles of faith." As long as these doctrines were sound, then all to the good. But when these doctrines were considered dangerous, the master class went to great lengths to deafen the hearers to certain matters, which meant silencing the talkers. Whatever else they were wrong about, abolitionists were correct when they argued that slaveholders "fear the PEOPLE; they are alarmed at the very idea of power and influence being possessed by any portion of the community not directly interested in slave property. Visions of emancipation, of agrarianism and of popular resistance to their authority, are ever floating in their distempered and excited imaginations."[39]

In emotionally potent public speeches whose sentiments resounded when captured in print, the master class made clear to themselves and their audiences that it was the noise of isms that northern foes were attempting to spread not just westward but also south. Through voice and in print, North Carolina representative Thomas L. Clingman claimed that northern agitation over the expansion of slavery "remind[ed] one of the struggles of Chance and Tumult in the reign of ancient Night, when Chaos sat as umpire." Abolitionists were "common disturbers of the public peace" whose noise served as a distinct aural counterpoint to the "great body of the Southern people" who were "quiet and silent" folk. Were he correct in his characterization of southern folk, Clingman and the master class would have been in trouble. After all, it was principally through, as Hinton Helper put it in 1857, "noisy discussions of village and State politics," centered on "the 'manifest destiny' theory," the "stealing of all territories contiguous to our own," that many illiterate southerners were introduced to the aurality of sectionalism.[40]

Southern newspaper editors demonstrated a knack for capturing the aural essence of the sectional conflict, choosing aural representations that upped sectional tension and consciousness. The editor of Jackson's *Mississippian* heard the characters and social relations of the North and the South at the height of the debate on the Kansas-Nebraska bill. Southerners, he maintained, had demonstrated "calm, deliberate judgement" during the debate, "which characterize the action of the people" generally. "Look to the North, and what do we realize?" Harsh words emanated from the *New York Tribune:* "the insane ranting of Fessenden . . . the sickly cant of Sumner, . . . the horrid screeching of Lucy Stone, and her unsexed compatriots . . . and the disgraceful orgies of tumultuous assemblages of all ages, colors, conditions, who make night hideous with their frantic howlings." Northerners as a group screeched as much as women—hence the reference to Lucy Stone's unsexed compatriots. But what of the South? "In the South, scarce a ripple seems to agitate the surface of society. All is calmness. . . . We hear of no burnings in effigy. . . . We listen to no furious declamation. . . . We show that we are controlled by *reason*—not by *passion*. . . . We prefer leaving such weapons to the blusterers, fools, and fanatics, to whom they appropriately belong."[41]

Uncharacteristic northern silence posed threats, too. Alabama's William F. Samford, a staunch southern rights advocate and agricultural reformer, heard the Compromise of 1850 in conspiratorial terms because he believed that "despotism was always a sly, cunning, cowardly, quietly speaking bloody hypocrite. It never avows its designs. It is by stealthy encroachments that it labors to compromit and then destroy the rights of the people." There were, to be sure, decent northern men, but as Alabama governor Arthur P. Bagby argued in 1851, they "have become powerless," caught as they were "amidst the roar of tempest created by howling fanaticism" of abolitionists.[42] Such men could no longer be heard above the noise of the North.

Whatever subtle distinctions southerners made between sound and unsound northerners, when push came to shove, they lapsed readily into blanket denunciations of all northerners. "A Citizen of Alabama" in 1860 began a *De Bow's Review* article by lambasting New England antislavery poet James Russell Lowell. The Alabamian considered Lowell and his ilk filth, "literary mosquitoes, whose buzzing is more dreadful than their stings. . . . They belch forth their venom against the South from the pulpit—from the bench—on the hustings—in the halls of Congress—in the social circle, and in the wilderness. Even the worn artisan, half-fed as he is, and the wearied girl at the factory loom, ignorantly, heedlessly, hopelessly, join the same cry." These "chants against a system of servitude" were simply too much.[43]

Rather than knitting the Union, the transportation and communication revolution succeeded "only in bringing the opposing views of the sections into more concentrated action" and enabled sections to hear one another. Even the quiet telegraph "disturb[ed] the harmony of the sections" and hardly encouraged "men of the North to cease their frantic clamor." Reports and actual visits did not lessen consciousness of differing sectional soundscapes. Of the North, southerners said, "Already in her thousand towns and villages are heard the never ending din of the loom, the laborious groans of the forge, and the shrill shrieks, which proclaim that the Cyclopean powers of steam are being spurred to their tasks of mechanical toil." Unlike the South, where "labor and capital are still in harmony," northern modernity had given rise to isms that would rip the social order and erupt in "Chaos." Nor was there much hope for a reversal. Urbanization and industrialization—and all they represented— were not being counterbalanced by a return to the countryside. It was "this influx into towns, and depletion of the country, [that] must be regarded as highly prejudicial to the prospects of conservatism." Unlike in England, "where the honor attached to the possession of the land" and a social conservatism prevailed among capitalists, the northern addiction to the "excitement and bustle" of city life threatened to perpetuate the excesses of modernity. But when "we turn to the South, we behold a state directly opposite." "The traveller finds her cities comparatively small; increasing in no disproportionate ratio with the country at large. He sees and hears in them little of the restless turmoil and ceaseless hum of business that characterize the great northern cities and manufacturing towns." This did not mean that southern towns were impoverished—far from it. Southern towns were industrious places. But the point was clear: it was not urbanization per se that was the most pressing worry, it was urbanization northern style that was deeply troubling. The growth of northern towns had upset the delicate balance of town and country, a balance that, when correct and even, elevated the "tone of society" and, when out of kilter, threatened to unravel the centuries-old conservatism woven by organic social relations first established in the serenity of the countryside and slavery.[44]

How, then, to combat the aural threat posed by northern society? One tactic was to play on northern fears of economic silence and its implications. One of Henry Field James's literary characters warned the North in 1856 that continued agitation would cause northern industrialists to miss southern cotton. Without cotton, northern society would collapse into disconcerting silence punctuated by the noise of the impoverished mob: "The factories are all still—in place of the busy hum of industry, we hear wailing and lamentation. The operatives are deprived of employment—their wages cease, and they know not what to do."[45] Slaveholders clearly played on the Yankee fear of the stillness of economic recession and the cacophony of social protest that accompanied it.

Northern noise could also be fought with the sacred sound of southern thunder. "Re-elect me," Preston Brooks asked South Carolinians of the Fourth Congressional District after his caning of Charles Sumner, "with an unanimity which will thunder into the ears of fanaticism the terrors of the storm that is coming up on them." Northern threats were not always audible, for their manipulation of "the minds of our slaves" could not be heard. Such mind shapers were "working silently." This silence preceded the storm, for if abolitionism were successful, the South would eventually sound like a north whose "vagaries in moral and political philosophy are hastening her to anarchy [and] tumultuous chaos." Conversely, "No 'isms and schisms rankle in our hearts"; tumult "disturbs not our harmony." But the South had to speak out. "Silence and apathy with us may be deemed pusillanimity," and protest had to be made if the "Southern republic," that "self-adjusting machine[,] . . . would vibrate in harmonious unity."[46]

Another solution lay in the use of silence to disquiet northerners, a power masters understood well, courtesy of their bondpeople. Make no mistake, argued the *Charleston Mercury* during the secession crisis, northern noise in protest of the state's actions was just that—"frogs in chorus." South Carolina, by contrast, exhibited "a deep, calm feeling, very different from the excitement of a mob." Northerners should fear Dixie's quietude. The region's actions and protests were "not the uproar of school boys splashing in mischief" but, rather, "the quiet tread of Caesar's forces crossing the Rubicon." John C. Calhoun's 1847 speeches generated intense discussion among northern Democrats, which in turn was heard in a particular way by southerners. The *Tallahassee Patriot* said in 1847, "That is the main consideration at the bottom of all the noise which Democrats make about his speeches." When northerners declaimed against slavery, they borrowed a leaf from the abolitionists' book by upping the aural ante. As Calhoun put it in 1850, "The agitation and excitement in relation to [slavery] in every portion of our country forbids us to be silent."[47]

A final tactic was found in a critique of northern society and a simultaneous defense of southern social and economic relations. Southern proslavery thinkers were particularly sensitive to the aural world and the threats posed by northern isms to southern quietude. Abolitionism, commented Thomas Roderick Dew in his classic 1832 essay, was "wild and intemperate" and "subversive of the rights of property and the order and tranquility of society." Abolish slavery, he maintained, and "the Old Dominion will be a 'waste howling wilderness,'" a place where the comforting tones of slavery would be muted by the din of reckless democracy and immiserating capitalism. From the 1830s on, the master class could not afford to turn deaf ears to northern developments. If anything, hard listening was called for because the attack of wage labor and mobocracy could be silent, as indeed it had been elsewhere: "In reference to the west of Europe, it was the rise of the towns, the springing up of a middle class, and a change in agriculture, which gradually and silently effected the emancipation of the slaves."[48]

Upcountry South Carolinians echoed Dew's pronouncements of southern tranquility on the eve of war in 1860. Among the purposes of the constitutional compact, they maintained, was the intention to "insure domestic tranquility." The South had done its part for the North— their "large cities, their immense manufacturing establishments . . . would never have existed but for their connection with the South." On the brink of war, though, the loud northern attack would be stealthy and silent, the age-old calm before the storm. "The tiger," it was argued, "has no 'overt acts' before the final spring. Only let him get his springing place, and he crouches still and quiet as innocence itself. . . . Such is the fate, fellow-citizens, which our Abolition enemies . . . design for us." Rhetorical questions offered answers: "What of the domestic tranquility which Abolition secures? Look at the secret emissaries prowling about Southern homes and plantations. Look at the arms placed in the hands of slaves to destroy their masters. . . . These are parts of the 'domestic tranquility' you owe to the North." Abolitionists, then, aimed "to lay that tranquility at last in absolute ruins."[49]

Good masters, though, could parry such attacks. Henry Hughes's "Warrantors" were worthy of slaveholding if they paid attention to the heard world of plantation slavery and southern society. Warrantors' ears "must hear diligently" when judging "all cases over which they have lawful determination." In social affairs they had the power "to enforce cleanliness and quiet" in slave quarters. How would "the fulfilment of Warranteeism," "this progress, which is now a conception and a hope of all" manifest itself? Freed from northern influences, Hughes heard the southern future in the unmistakable register of the pastoral and organic: "Then, in the plump flush of full-feeding health, the happy warrantees shall banquet in PLANTATION-REFECTORIES; worship in PLANTATION-CHAPELS; learn in PLANTATION-SCHOOLS; or, in PLANTATION-SALOONS, at the cool of evening, . . . chant old songs, tell tales; or, to the metred rattle of chattering castanets, or flutes, or rumbling tamborines, dance . . . and after slumbers in PLANTATION-DORMITORIES, . . . rise at the music-crowing of the morning-conchs."[50] Arguing the positive goodness of slavery in 1837, John C. Calhoun maintained that "there is and always has been in an advanced stage of wealth and civilization, a conflict between labor and capital." The conflict—and its obverse—was heard as much as seen: "The condition of society in the South exempts us from the disorders and dangers resulting from this conflict; and which explains why it is that the political condition of the slaveholding States has been so much more stable and quiet than that of the North."[51]

Sectional tensions heightened sectional volume. Political ears were pricked, and abolitionism took on the noise of insane ranting to southerners. Abolitionism, claimed Richmond's *Enquirer* in 1857, "always clamored loudly for liberty and equality. . . . Its designing demagogues have been wont to shriek and wail . . . as they impose upon the popular mind of the North . . . their perverted portraiture of Southern slavery." The worry was that the movement would engulf all northern men so that "all the insane asylums in Yankeedom will be inadequate for the accommodation of its victims." The "abolitionists are leaping and weeping, kneeling and swearing . . . fuming, yelling and gesticulating, like so many bedlamites."[52]

In some respects the antebellum political crises had the ironic effect of silencing what abolitionists thought was an already dangerously quiet South. Political discussion, especially within the plantation household, developed a muted, taciturn aspect that was necessary in the presence of slaves and children. Following the publication of *Uncle Tom's Cabin*, for example, a political hush fell over the Eppes plantation in Florida. According to the plantation's precocious eight-year-old diarist in December 1855, "There is trouble in the air but I cannot find out just what it is, the grown folks keep very quiet when we children are around and if they are talking when we come into the room they stop talking right away." As the international sound of the Liberty Bell began its toll in the 1830s, some proslavery ideologues felt that the North's imperial tocsin was drowning out their voice to the point of aural isolation. William Harper's 1837 oration "Memoir on Slavery" reflects a growing sense of enforced provincialism: "If any voice is raised up among ourselves to extenuate or to vindicate, it is unheard. The judgment is made up. We can have no hearing before the tribunal of the civilized world." By contrast, northern whisperings of freedom inevitably reached slaves' sensitive ears, and these, combined with the folly of overly cruel masters, agitated slaves to the point where "society is kept in an unquiet and restless state."[53]

How else might slaveholders have reacted to the aural attacks by abolitionists and Republicans? To a master class accustomed to hearing even the slightest rustle as imminent danger, to a people who invested their identity in a constructed quietude and all the security, order, and harmony they heard in those sounds, the powerful and emotional aural descriptions and charges used by northern free wage labor advocates combined with southerners' own (selective) hearing of northern isms produced a profoundly defensive reaction. Not only could they hear the subtle strains of industrialism and urbanism beginning to crescendo in their own society and not only had their own insistence on quietude and their slaves' appropriation of it rendered them a highly skittish and aurally sensitive ruling class, but they now encountered a body of men and women whose own ears had been sensitized to slavery's terrible sounds and who insisted that masters hear clearly the evil and sloth of slavery and the sounds of modernity. Slaveholders were as worried about the aural shaping of their future in the West as they were about the growing dissonance within their own society that the abolitionist-Republican critique of slavery was in danger of unleashing. Among the many other reasons slaveholders fought the Civil War was a conviction that their soundscape—and all that it meant—was the most appropriate for the future of the United States and that their heard world was under an attack that would only get louder were it not silenced. The "incessant and terrible noises made all through the North" were sufficiently aggravating to reconcile southern leaders to the position that the "South certainly has become generally convinced that it is by hard blows, and not by loud blustering . . . that the sectional quarrel is to be settled."[54]

How fitting, then, that on Tuesday, January 8, 1861, Columbia, South Carolina, "was waked by the triumph-tones of our new bell, named so opportunely 'Secession'" to mark Florida's withdrawal from the Union. Days later the bell pealed again, telling "of Alabama's action." But it would take more than a secession bell to silence the Liberty Bell. Soon enough, the emboldening tones of the secession and other southern bells would have to be recast as cannon to silence the slaveholders' shouting enemy.[55]

(((PART IV

Noises Hideous,
Silences Profound,
Sounds Ironic

Listening to the Civil War
and Reconstruction

The American Civil War was a contest between sections concerning matters of monumental importance. Hundreds of thousands of lives were lost on the battlefield and at home over questions of bondage and freedom, states' rights and nationalism, and liberalism and conservatism. But because antebellum elites in both the North and the South had become accustomed to hearing as much as seeing these fundamental issues, they listened hard to the war and the reconstruction that followed.

Listening and hearing occurred in a context defined by the gravest constitutional issues and monumental slaughter, which were themselves inscribed with aural meanings that helped shape how and which events were heard. Listening to the heard war is helpful for several reasons. First, it muddies the tidy distinction sometimes made about the separateness of home and battlefield and contributes to a more recent emphasis stressing the conceptual necessity of blending the two, especially when trying to understand the Confederacy.[1] Second, listening shows how particular constituencies constructed the soundscapes of the Civil War differently. There were, in effect, multiple acoustic battlefields and home fronts during the war. Third, listening to actual and perceived soundscapes in the

Civil War South suggests how, in addition to all the other well-known forces that wilted the southern will to prosecute the war, the introduction of new noises and the muting of old sounds probably enervated white southerners and so helped erode their ability to resist their noisy northern enemy.[2]

While the unprecedented scale of the war, the industrial might needed to prosecute it on both sides, and the introduction of new technologies (such as rifled muskets) meant that the Civil War generally created different and louder sounds than any war that had preceded it, Confederates experienced new, jarring noises and silences on the home front to a far greater extent than northerners, who found many of the fewer qualitatively new noises compatible with their imagined and preferred future.[3] While mobilization in the North was, for the most part, in literal and metaphoric harmony with Federals' idealized and actual industrial, free labor soundscape, gearing up for war in the South became too northern for many white southern ears. Because the actual sounds of war were far fainter in the North, there was less adjustment to make than in the South, where noises of battle and strife increasingly encroached on the tranquil, idealized home front. For other southern constituencies, slaves in particular, the sounds of war were the welcomed melody of freedom.

The Civil War was fought with the aural idioms of antebellum sectionalism in participants' ears and minds. Those fighting the war on the battlefield and at home commented on the tumult of battle, the noise of military loss, the sound of victory, and the silence of defeat. Listening to how they heard the war allows us to appreciate the complexity of contemporaries' understandings of the conflict. Sounds of war and noises of military encounter were not spatially delimited: noises, shrieks, cannonading, and a host of other sounds echoed, literally and figuratively, on the home front, thus joining southern home and battlefield acoustically and metaphorically. In other respects, the Confederate home front especially had its own soundscapes. The combined effects of the war created not only what planters heard as the noisescape of freedom but also the silence of economic, social, and political ruin, a grave quietude that gained literal and metaphorical power through the South's substantial loss of bells. Slaves, too, heard the noises and silences of the war but used and understood these sounds differently, impregnating them with alternative instrumental and metaphorical meanings. For many ex-slaves interviewed in the 1930s, memories of the war were aural, and what they heard at the time proved useful. Slaves during the Civil War listened hard to the blending of military and plantation soundscapes in order to ascertain the location of Union forces and freedom.

Ex-slaves marked the end of the war by shaping the new southern soundscape in unprecedented ways. Freedpeople made aural their celebrations of freedom, reconfiguring the hated plantation bell into the sound of emancipation and the silencing of slavery, fulfilling, in effect, the abolitionist prophecy. While southern whites lamented the silence of defeat, the loss of many of their soundmarks, and what they perceived as the impending cacophony of black independence, ex-slaves celebrated the metaphorical and actual sound of freedom's bell. Planters responded by attempting to reimpose plantation quietude, not without success. One reason for that success is to be found in northern doubts about their own soundscape. Postbellum northern elites not only began to legislate more vigorously against excessive noise produced by a growing mobocratic voice in their own society but actively sought new avenues of quietude. To consume serenity, they happily listened and ventured southward.

The air was filled with the missiles of death, and the
earth trembled under the confused noise of battle.
—"The Repulse at Corinth," *Charleston Mercury,*
   October 15, 1862

(((8. Noises of War

We know more about the role played by acoustic shadows
during the Civil War than we do about the perception of sounds and their
phenomenological roles. While we know a little about the effects of wind,
altitudinal changes in air temperature, and wind speeds in shaping the
soundscapes and outcomes of several significant Civil War battles—particularly those at Chancellorsville and Gettysburg—we know much less
about how Civil War sounds affected and were manipulated by soldiers
on the fields and by civilians at home, and how each side measured their
progress, wins, victories, and losses by listening.[1]

Sounds of war had a long pedigree. In addition to clashing metal, the
use of noise by soldiers was an ancient military strategy, originating probably with the Greeks. Armies were sent into battle bellowing because their
"shouts and the clash of arms confound the hearts of the enemy." Because
of the disruptive tendencies of war, it was always the "din of Mars' alarm."
In preparation for an engagement with the British during the Revolutionary War, for example, a Lexington, Virginia, regiment paraded, "the

drums beat—the cannons roared and peal after peal swelled the general sound." These were distinctly "martial sounds."[2]

Such memories helped shape predictions of how future U.S. wars would sound. "The entire land," augured the *New York Herald* in April 1861, "will resound with the din of arms."[3] Henry Wadsworth Longfellow unwittingly captured something of the essence of the sound of America's first modern war in 1849 in "The Arsenal at Springfield." "This is the Arsenal," he began,

> Ah! What a sound will rise, how wild and dreary,
>
> .   .   .   .
>
> I hear e'en now the infinite fierce chorus,
> The cries of agony, the endless groan,
>
> .   .   .   .
>
> The tumult of each sacked and burning village;
> The shout that every prayer for mercy drowns;
> The soldier's revels in the midst of pillage;
> The wail of famine in beleaguered towns;
>
> .   .   .   .
>
> Is it, O man, with such discordant noises,
> With such accursed instruments as these,
> Thou drownest Nature's sweet and kindly voices,
> And jarrest the celestial harmonies?[4]

But the volume of this war and its unprecedented scale sent participants scrambling for metaphors. The Civil War was so loud that for many it could be compared only to nature's greatest sounds. Even northerners, who were more likely to have heard the decibels of factories, found themselves relying on natural and symphonic comparisons. At Camp Liberty, near Richmond, a reporter for the *New York Herald* heard thunder and war simultaneously. "The guns were large and the explosions loud," but they "were fully rivaled by the artillery of nature" in the form of a "violent thunderstorm." Confusion reigned: "There were also sharp cracks of musketry and sharp crashing thunder simultaneously with it, so that it was sometimes difficult to discern which was cannon and which was thunder."[5]

The viscerality of combat inscribed the war with noises of human pain, offering testimony not to America's progress but, rather, to the country's retreat to barbarism. "Cries of pain, and curses, and groans" were common on battlefields, and they affected both Confederate and Union troops. Injured rebels were not men enough to take their pain in silence. They "exhibit far less manly fortitude than our troops," noted one Union correspondent, because they "murmur at every pain." The perversities of human nature, especially in bloodthirsty contexts where winning becomes utterly meaningful, meant that some men found qualified grandeur in the sound of near-misses. "There is something grand in the whizzing and bounding of shot," wrote a Union officer in 1864, provided "no one is hurt."[6] These were the sounds of premodernity in a war that was characterized then, and has been since, as modern.

The war was premodern in other ways, too. Civil War soundscapes, particularly with respect to how some information was communicated, represented a throwback to an earlier age where intelligence was not carried through the telegraph's quiet wires. When lines were shredded or not nearby and when lulls in the sheer tumult of war made actual hearing possible, "the cheers come down the lines from the camps above and from far away in the distance. We doubt the news but we cheer—cheer as loud as any and the sound rolls along the camps away to the North." The Union army became a human telegraph whose audible throb stamped the sound of northern military progress onto a fighting South replete with its silent/howling wilderness: "Tonight we have built houses or shelter of boughs as we have no tents. So we are a village of 1000 men where yesterday at this time was the stillness of the woodland—or rather we are a city of thousands stretching over hill and valley for miles and the cheers roll along among the ghostly shadows of the trees from the many camp fires like the cry of fire from street to street." So John M. Bancroft of Detroit, Michigan, heard the shuddering march of the Union army on the quiet Virginia countryside in 1861.[7]

That much said, a good deal of aural mapping and acoustic intimidation took place during the war, stamping the conflict as civilized and efficient. Union "axes are detailed to cut down the woods where we are lying" in 1861 near Richmond, Virginia: "At the trees again; they are falling every which way axes ringing—merry voices calling—then comes the crash." Such sounds sharpened memories of the past, the homeland, northern efficiency, and the ways in which industrious northerners had paved the way for westward expansion: "Michigan boys know woodcraft to perfection. Never was land cleared quicker save by a hurricane—click, click,—and crash, crash." Silent balloons allowed reporters to hear as well as see camp life generally. One reporter for the *New York Herald* tagged along on a ballooning reconnaissance mission near Fredericksburg in early 1863.

"What a confused din of noises comes from the myriad of camps," he wrote. But the reassuring sound of activity could be discerned: "There is a group of soldiers hewing down a tree. . . . Loud sounds the axe, redoubling stroke on stroke."[8]

Northern aural descriptions of southern otherness continued into the war. Some northern correspondents listened to the natural soundscape of the Deep South and disliked what they heard. One reporter complained of the "gross and hideous frogs perched near my head and cracking in my ears" in 1862. Such a sound etched in his ears "the sacred soil of Mississippi." But others were struck by the quietness of the South's pastoral soundscape and gained an appreciation for the region. On the "beautiful Sabbath morning" of February 23, 1862, Lieutenant James H. Linsley of the 10th Connecticut Volunteers wrote from his ship anchored off St. Augustine, the "beauty and quiet of the scene seems to impress the same on the minds of all." By 1863 he waxed lyrical about southern soundscapes, remarking on the "still quiet of the atmosphere." As reported in "Opposite Fredericksburg" in December 1862, "Not a breath of air stirred in the frost-withered leaves." War invariably intruded: "The echoing roar of the cannon contrasted strangely with the serenity and stillness about us, only interrupted by the tread of passing cavalry and shrill whistles of passing trains." With delicious irony, however, Union troops fighting in the South seem to have used the serenity of the southern countryside to soothe ragged nerves. One soldier said simply of the Chesapeake region in 1862, "Scene: a fine quiet plantation, a very picturesque windmill for grinding corn . . . the water is quiet and calm." Such memories and accounts proved important after the war to a northern bourgeoisie anxious for quietude away from the worryingly discordant pulse of northern workers. The business of war, though, meant that aesthetic appreciations of the southern soundscape were ephemeral. From Kennesaw Mountain in the summer of 1864, General William Sherman looked and listened to the beautiful South: "The scene was enchanting; too beautiful to be disturbed by the harsh clamor of war." He then turned to the task at hand: the campaign against Atlanta.[9]

Battle sounds and silences took on important meanings to Federals. "Groans and oaths of the wounded were heard on every hand," followed by "an ominous quiet. No sound broke the stillness." Then "mutterings of the awful strife . . . began to be heard. Soon the din began. The voices of an hundred big mouthed guns began to vomit," wrote one Michigan infantryman years after the war. Lulls that foretold battle heightened tension. In 1861 "8-9-10-11-oclock all quiet," translated for one Union soldier into "Hours of suspense." Moments of quiet serenity and bites of inactivity often followed bloody carnage. Unaware of the irony, Edward Everett began his "Gettysburg Oration" with the words, "It is with hesitation that I raise my poor voice to break the eloquent silence of God and nature." From the South's perspective, his was a society that had already destroyed that quietude. The point remains: death warranted silence, especially during horrific war.[10]

War reaffirmed some old definitions of noise for northerners of all social ranks. "Petersburg is ours," exalted the *New York Herald* in April 1865, but not without an "infernal din" disturbing "the placid stillness of" sacred night. Federals found fighting hellish when it disrupted the traditional quietude of the night. The "infernal din of artillery . . . rings on the night air making it hideous with their discordant shrieks." But there was concord of sorts: "As if to keep harmony with such a scene of ruin, shells flew around hustling and whizzing on all sides."[11]

The noise of war bordered on the sacrilegious, for even Sundays were not spared. From John Pope's headquarters in Warrenton, Virginia, in July 1862, a *New York Herald* correspondent "aurated" [*sic*] "A Sabbath Morning in Camp." It had few of the keynotes of the sounds of Sundays past: "This is the Sabbath, and a more gloomy and dismal one I have never known. . . . We have no churchgoing bells, no pastoral voices proclaiming the Word, no choral anthems, no songs of the Sabbath school. . . . The air is vocal, not with the songs of birds and god-fearing worshipers, but with the clatter of feet, the rumbling of wheels, the shrieking of fifes, the rattling of drums." An engineer and balloonist in the 4th Michigan Infantry noted the disruption of Sunday quiet at First Bull Run in July 1861: "Today the gun's are firing . . . in the direction of Manassas. . . . All day long we have listened to the guns. One of the fairest of Sabbath days." Noises of war could sound like hell because the essential fabric of antebellum soundscapes was being ruptured. "The thunder of cannon was awful," remembered a Union soldier of Malvern Hill, "clash of arms, shouts of combatants, was deafening. Such a seething hell will never again be enacted on this continent." To win this conflict, men and women had to hear death. "The red flags of the hospitals" at Fredericksburg, "flap[ping] in the night wind with a mournful sound," joined "the babbling brooks" with their "solemn song."[12]

Sounds of Federal camp life were similarly discordant, principally because what was heard on the field echoed resoundingly in camp. Union dispatches from near Kennesaw Mountain, Georgia, in late June 1864 give something of the flavor. June 29 "passed quietly. There were occasional shots heard at the enemy from our batteries, but more for moral than material effect." Aural jitters were a two-way street. While soldiers enjoyed "noiseless slumber," shots rang out, and "every officer pricked his ears and listened." Tensions were high because "a stray shot perhaps being heard occasionally at intervals for fifteen minutes" kept Federals on guard. Camp registers themselves, though, sometimes threatened the stirring spirit of field sounds. A northern correspondent in 1862 heard field "drums and the accompanying fife play the inspiring air which has lost so much of its spirit by its too common use in camp."[13]

For the serenity-starved wounded, camp was no escape from noise. Soldiers riddled with cholera, surgeons advised, should seek "perfect quietude of body" to recover. "Oh to be here!" lamented John M. Bancroft in a camp outside Bull Run: "What pain—How weak one is; and all the boys are yelling and swearing &C as usual and one hears everything." Such tension was likely since relaxation in camp for the fit often translated into the freedom to use "every conceivable thing that any noise could be got out of," including musical instruments and drums. The psychological relief of making a racket in an environment that ordinarily required the most precise discipline of personal noise should not be underestimated, which helps explain why some recreated a bit too enthusiastically. "Some telegraph wires were laying along" tree limbs, recalled Michigan infantryman O. S. Barrett, and a "mischievous fellow in the rear drummed with his gun on them, causing a vibration of sound similar to a noise caused by a charging mule team running away. . . . If I could have known the fellow who caused the stampede, I think I could have mauled him." These were skittish men, highly strung by the aural tocsin of war.[14]

Sounds past anchored anomic men in the cacophony of the present: "I had heard [bells] in the old church near which my childhood had been spent, and in many a scene of later years, but never did they strike so clearly and melodiously in my ear as on that night of our bivouac," sighed one observer.[15] But tension in Union camps was never far away. Conflict stemmed in part from the elite's assessments of what they considered appropriate sounds. Patrician sensibilities were transferred to the field, where soldiers reacted accordingly: "Capt Chase was not in the fight" on Fort Wagner, South Carolina, seethed one Union soldier to his sister in 1863, explaining, "he was back in Camp but could be there if he wanted to he is a lazy fool after the fight we Boys through down our guns and equipment haver sack and lanterns and went to carrying the wounded into Camp he stuck his head out of the tent and wanted the wounded Boys not to make so much noise not to moan so for he wanted to go to sleep. how mad that made the other Boys." Then again, the soldier harbored his own prejudices: "There is a score of Nigger Boys out here drumming that belongs to the different Black Regiments they make an awfull noise and it sounds like Remptons old [illegible] wheel than any thing else."[16]

Away from camp, listening and hearing joined seeing and looking in shaping Union tactics and strategy. Take, for example, Federal responses to the fabled rebel yell. While hardly unique in warfare (Union troops also shouted, as did Native Americans), the yell was considered an important acoustic weapon in the Confederate arsenal because it not only buoyed morale but also, they hoped, instilled fear in Union troops. It seemed to work on several occasions and became central to Confederate military mystique, echoing long after the Civil War. Federals described rebel brigades as dashing "down towards us on a run, yelling like demons." Confederates "poured into our column, and, with a yell that sounded above the din, the enemy in solid column dashed upon us." "Our men were about to surrender," wrote Robert Dickinson, captain in the 21st South Carolina Regiment, "but the Louisiana Battallion came up and turned the Tide of Battle and saved the day; they came yelling at the tops of their voices Butlers & Tigers and remember New Orleans, which give the Yankees a scare and they took to their heels and fled in every direction."[17] But the yell did not always work. When fire was fought with fire, Union throats could sound equally fierce. When companies "followed with a shout that sounded above the roar and din of the artillery," Confederates knew their foes meant business and "fled in dismay."[18] "On came the rebels, yelling like a pack of devils, towards our brave boys, who were in line of battle," wrote a reporter of a skirmish near Fair Oaks in 1862. Union boys stood firm despite the yell, and the secessionists retreated. "The battle commenced with a rebel yell and ended with a Union cheer," and all was right again with the world. Moreover, Federals realized that Confederate sounds by themselves posed little threat unless there was substance behind them. "This afternoon the rebels gave a cannonading exhibition," noted a *New York Herald* reporter from Yorktown in 1862, "which, in a pyrotechnic and acoustic point of view might be set down as a brilliant performance." But because the Confederates missed their targets, "in every other regard [it] was a failure."[19]

Union military advances were sometimes gauged spatially, since sound was a good proxy for distance. General Philip Sheridan, for example, understood that he had gained so much ground near Richmond in May 1864 that his victories were "within the sound of the church bells" of the Confederate capital. Night's stillness was helpful in detecting Confederate action: "last night the whistle of their locomotives could be plainly heard at several points on our lines, indicating a rapid transportation either to the front or the rear—the latter very probably." War also trained reporters' ears. Outside Corinth in 1862, for example, a *New York Herald* correspondent commented, "By intent listening, we could plainly distinguish the rapid firing of artillery in the far distance." Intelligence was gleaned by cocked ears: "Though just discernible, each separate report was distinctly heard, and we came to the conclusion that Farragut and Foote were effecting a juncture, and that Fort Pillow was suffering in consequence." Military clues were everywhere if listened for. "It was evident from the quick rattle of the small arms, as the sounds approached nearer and nearer," remarked another reporter at Fair Oaks in June 1862, "that Casey and Couch were being driven by a superior force of the rebels."[20]

Relying on sound alone to read space and activity was risky. Sound shadows and windswept noises often confused troops listening for the enemy, and mishearing contributed to Union and Confederate losses in several important battles. Ascertaining the direction of military activity aurally, however, was probably quite accurate when sound shadows were not involved (most people can detect changes in angle of one degree when the audible source is straight ahead, although less accurately when the source moves to the side.) Thus Valentine Chamberlain of Connecticut, for example, gave the following account of a skirmish in Jacksonville, Florida, in 1862: "All the roads and avenues were strange to me. I did not know the force of the enemy. I could not see him only hear the clatter of his horse's hoofs."[21]

If military echolocation was sometimes helpful, so was the ability to control volume. Surprise attacks necessitated not just keeping "men out of sight" but also ensuring that "troops make as little noise as practicable," with responsibility "for all unnecessary noise" residing with commanding officers. The most successful retreats were necessarily noiseless. Joseph Hooker's withdrawal across the Rappahannock succeeded because it was unheard by Confederates: "With such noiseless caution was the retreat conducted, that the falling back of our skirmishers was unknown to the enemy." His return to Falmouth in May 1863 worked because his engineers knew how to lessen noise: "Pine boughs were spread upon the pontoons to prevent the noise of crossing, and about mid-night the troops commenced falling back." Similarly, the success of Union supply operations on Morris Island, South Carolina, in the late summer of 1863 was contingent on "the noiseless advance of a long and seemingly interminable wagon train, wending its way to the front." Silence was critical because these Union forces were under heavy night bombardment from Fort Wagner. Should Confederates get lucky, the teamsters "hurry their teams from one point of shelter to another." "Sometimes," though, "they accept the doctrines of chances, and take the screaming missile and its humming, whistling fragments, with no attempt to dodge them." Hearing the missiles was no guarantee of avoiding them because "faster than sound these missiles travel." Such were the perils of a modern war in which human hearing could not always match technology.[22]

Training guns and cannon on advancing Confederates during the siege of Yorktown in April 1862 was made easier if one understood something of the physical environment. Union officers knew that such maneuvers could be "performed in the most noiseless manner; for in the dead, dry sands of this region the rattle of a gun carriage wheel is hardly audible at the distance of thirty rods." Troops recognized that night marches demanded particularly sharp ears. Many a trudging Union soldier "could not see my team no more than a stack of black cats" but "could hear the other waggons," and this was sometimes enough to keep troops on the right tracks. The same rules applied to the enemy, though. "The Army of the Tennessee broke camp on our extreme left," reported the *New York Herald* near Atlanta in July 1864, "and noiselessly and quietly marched around our lines to the extreme right."[23]

Troops' use of sound and silence on the battlefield was important. Failure to navigate the Confederate soundscape could mean many things, including capture. Capture meant prison, and southern prisons meant hearing the Old South all too clearly. Union prisoners of war fretted about their incarceration. Libby Prison in Richmond, "made so famous to Northern ears," was a hellish cacophony. Here could be heard "the groans of the suffering" and, more upsetting from a northern perspective, the noise of "ball and chain" and "clanking" that could be described only as "a perpetual vulcanian din" befitting the sounds of slavery. Slaves, though, were on hand to help, and Union troops imprisoned with bondpeople learned from the best listeners and users of quiet. John L. Walker, who was incarcerated in Cahaba Prison, first noticed the baying of "the pack of hounds around the stockade." Through clenched teeth and "in an undertone" "an old negro" told him how to elude the dogs should he attempt escape.[24]

Avoiding Confederate prisons was no guarantee of immunity from the deadly aural effects of the conflict. War noises created tense moments. A "quiet" night could be followed by a deafening day when "the discharge of artillery drowned every other sound." At Perryville, noted a Union reporter in October 1862, "every sound was noticed and commented upon." The "braying of mules came across to our ears," followed by "a continuous roar of cannon, intermingled with the screeching of shells and the whistling of balls." Little wonder that "discord and terror" followed. Even "occasional lulls in the din," for the war-weary, assumed "a terribly ominous character." Pauses in sound, moments of comparative silence, "appeared like that dull, dead calm that always precedes the storm." This was dead sound with frightening implications. Hope was heard: "As I write my ears are continually assailed with the shouts of fresh troops arriving upon the field." Charles Henry Snedeker of the 124th Illinois Infantry commented on how the sounds of battle near Vicksburg alerted troops to the spatial quarter of military engagement, but he seemed far more concerned with the noises of war at night: "Was aroused at 2 O. Clock this morning by our Artillery opening fire on the Rebs. It was kept up for an hour a perfect roar. . . . Artillery firing every half hour through the night." Sleep was fitful, and sleeplessness could affect soldiers' acuity. Short of resounding victories, only national holidays could disrupt the cacophony of bombardment: "Independence day. The most quiet fourth of July that I ever knew. Not a gun was fired all through the night."[25]

Contradictions abounded. While one tactical aim was to silence the enemy, success in this regard produced new worries. Even though supporters crowed that the "din of war in the East appears to have been hushed ever since the battle of Fredericksburg," the enemy's silence could be disconcerting to soldiers on the field because meaning was difficult to attach to military quietude. "Our artillery is thundering on the flanks," wrote a *New York Herald* correspondent of Gettysburg on July 3, "but the enemy is strangely silent." He continued, "Down upon the meadows we hear the sharp crack of rifles, and the Minie bullets occasionally whistle unpleasantly near our heads; but the change since last evening is so great that one feels impressed with the comparative silence."[26]

Ultimately Federals heard military victory, and it could be silent or loud. "The New Ironsides," for example, bombarded Fort Wagner so fiercely in 1863 that Confederate guns were "completely silenced." Victory was heard when "the din seems to be lessening." When Union forces captured Fairfax Court House in July 1861, they marked victory aurally by sounding "the Court House bell, and all the tavern bells in the village." If Confederates could not be silenced, Federals could always revel in the sounds of their destruction. "Our mortar boats were throwing shell into Petersburg . . . with so much destructiveness," cooed the *New York Herald* in July 1864, "that several houses were set on fire, and the alarm bells were distinctly heard."[27]

Although some soldiers and officers had heard war before, particularly in the Mexican conflict, the sheer scale of the Civil War combined with the fact that most combatants were novices meant that sounds were often as new to Confederates as they were to Federals. First Bull Run introduced unwary reporters and green soldiers to war's fiendish medley. Federal cannonballs "pass over us with a sound that makes our flesh crawl," confessed one southern reporter. William L. Jones of the 48th North Carolina wrote in June 1862 from Petersburg: "We got orders to retreat, which we did amidst . . . cannon balls whizzing & bursting over our heads and all around us cutting the limbs and the trees down by our side." In an effort to make sense of this new soundscape, southern soldiers, like their Federal counterparts, drew on their pasts to understand what they now heard. Aural analogies quickly gained wide currency as men tried to convey the sheer volume and unprecedented sound of the war. Southern listeners, particularly early in the conflict, relied on natural comparisons and similes to convey the impossibly loud fighting. A reporter for the *Charleston Mercury* at First Bull Run compared the Federal "roar of musketry" to "the roar of distant and protracted thunder."[28]

Confederate troops understood their war in aural terms, and some particular battles had aural signatures and rhythms. The 1863 battle at Chickamauga Creek, for example, started off with the "roar of small arms . . . saying too plainly that the work of death had become a thing of certainty." Musketry added its own tempo: "Time and again it swelled and died away as line after line passed into the maelstrom of destruction." Crescendo followed with the "din of musketry and cannon." Then came the pause, peppered with echoes of death: "The moon looked down on the field through the dense smoke which settled everywhere in a silence only broken by the shrieks of the wounded." On the next day was heard "the terrific thunder" of battle renewed, a "pause in the din followed" by a "breathless silence." Finally, "Our cannon . . . shook the world to its very centre."[29]

As within Union lines, the heard and listened-for world was critically important to Confederate soldiers. Joseph Dill Allison, holed up in Vicksburg for much of 1861–62, wrote on May 12, 1861, that although there was "no duty today," officers demanded that "all [were] to be within hearing distance of the camp." Clues to Federal activities carried on the air. "We can hear the enemy's drums distinctly," wrote Allison in April 1862; "We must fight tomorrow." Night pricked ears, although sometimes to little effect. Allison acknowledged as much: "The Gunboat Indianola passed down last night. It was so dark and cloudy, we had to fire at the sound of her paddle, so she was probably not hit." A year later Allison heard impending defeat. "Sick lists very large, the men having lain in the trenches for three weeks with no protection from the sun," he wrote a month or so before the city's surrender. "They can't stand up," he continued, "without having a dozen bullets whistling around their heads, and to attempt to walk around is certain death." The noise seemed constant and unremitting. "Towards night," he wrote in May 1863, "the Mortars slackened off and we had a few moments quiet for the first time in four days and nights." A month later voluntary deafness had set in: "It would surprise anyone not accustomed to shelling to se[e] how coolly we take it." On the day of surrender he put the extent of the noise of war in graphic relief: "Strange to say all is quiet this morning. . . . This morning is the only quiet we have had in 47 days."[30]

Confederate soldiers listened so closely to the sounds and silences of battles because their registers contained important, life-saving clues. But care had to be taken. More than merely loud, the sounds of war were distinctly disorienting. As they did with Federals, acoustic distortions and maskings led Confederate troops astray. As one Confederate soldier recalled before the fall of Atlanta, "Have not heard the Artillery yet, as the noise and confusion would prevent, were the enemy at the very gates of the city."[31]

Strategies of silence were as common among Confederates as among Federals. The Confederate navy employed stealth technology of a premechanical sort. In October 1863 the "battering ram *Manassas*" caused havoc among Union ships off New Orleans. The "ram glided noiselessly down the current," punctured "the stricken *Preble*," and "judging from their piercing, agonizing shrieks," did the vessel serious damage. Union ships were confused by this silent attack: "Not a single shot had been fired at them yet, and such was the confusion on board all of the ships that they commenced ringing their fog bells, which proved that even yet they were so confused and ignorant of the nature of our attack that some of them supposed it to be a accidental collision, caused by the extreme darkness, and rang their fog bells so as to show the *Manassas* their position and enable her to avoid them." Other Confederate military tactics relied on silence, and Union officials recognized as much. Blockade runners in Charleston, Wilmington, Savannah, and Mobile in 1863 "noiselessly emerge from or creep into their sheltering harbors at a slow, stealthy pace, which is attended by no sound from screw of paddle."[32]

Confederates used silence well, not least because they had encountered some good teachers. Like the slaves some of them had left behind and the Indians a few of them had fought, Confederate soldiers cultivated the art of moving silently because they had learned that the price to pay for unnecessary noise could be high. "We lay perfectly quiet," recalled one soldier of a skirmish near Murfreesboro, Tennessee, "until the Yankees got within a hundred yards of us when we fired on them, and hardly left enough to go back to tell the tale."[33]

Like Federals, Confederates gauged victory and defeat aurally. Near Chattanooga in September 1863, "the trumpet note" of Confederate victory sounded, as did "the knell of defeat and death to the enemy." Triumph was heard, even above the clang of war. A "shout rises that roars loud as the artillery from our men," noted a southern observer at First Bull Run following a particularly successful advance. Even small victories were heard as much as seen. "No doubt," wrote one Confederate soldier to his brother in 1862, "we came off better than the Yankees did [because we] heard a powerful screaming & groaning among them." Men listened for the enemy's anguished caterwauls to confirm military victory.[34]

The noises of war made concentration difficult. "Your correspondent is put to his wits' end," learned readers of the *Charleston Mercury* in 1861, "by the necessity of writing at railroad speed in the midst of a confusion surpassing any thing he ever witnessed; for, through the open windows (the day is very warm) comes a din of a thousand commingled noises, not the least of which is the steady tramp of soldiers marching out to the great encampment." The slightest sounds that disrupted precious sleep during campaigns were grating. Mosquitoes "thundered" in fatigued men's ears, and "quiet slumbers" were disrupted "by the shrill whistle of the locomotive."[35]

Federals suffered likewise. By 1863 Union troops stationed on Morris Island, South Carolina, were so accustomed to the sounds of war that they found once-coveted silence deafening. Ordinarily, "from day light until dark there is one continued roar of heavy artillery, which slackens during the hot noon hours, but is never entirely silent." As irritating as this sound was, a one-night cease-fire proved even more disturbing: "The silence was so unusual that many in camp, accustomed to hear the incessant reports of heavy guns and mortars, the hawl [*sic*] of shot and the bursting of shell, felt that there was something needed to make it possible to sleep soundly." These men were aurally scarred, and memories of past and anticipation of future sounds warped with each decibel of war:

> If a big bass drum had been struck at intervals during the night the relief would have been grateful. Nothing but the deep monotone of the surf on the beach was heard to disturb the quiet of the night, and many were nervous and restless because the regular nightly cannonade had been suspended. It had become almost a necessity of their existence. What will the soldier do when he returns to his quiet country home, and hears only the shot of a fowling piece or the annual discharge of the immense brass six plunder on Fourth of July? He will sigh for the pomp and circumstance of war again. . . . Will not this rebellion radically change the spirit of the nation? It seems so to me.[36]

Screams of bullets kept Confederate soldiers in a state of nervous anticipation and fear because after whiz came likely death. In woods near Richmond, Alexander Beck of Georgia learned the significance of the whiz: "I am still near Richmond in the woods, near the banks of the Chickahomina River, where the Yankees are as thick as fleas on a sheep skin. And they [are] not what mite be called harmless creatures, (either) for every few minutes, whiz comes a mine ball, close to your head, cutting the limbs and leaves in every direction." With "minnie balls singing close to my ears," recalled Benjamin F. McPherson of Texas, "I thought every moment I would be killed." Sleep was as fitful for Confederate soldiers as it was for Federals. Similarly, "the stillness of night" could be "suddenly broken by the rattling of bells, blowing of horns, beating of pans" by men worn by the monotony of war and after a change in soundscape. At other times, men in camp were wisely quiet. Jokes were shared, but of necessity, they "go round without noise or tumult."[37]

Raw nerves and bruised cochlea were among the prices soldiers willingly paid for heightened listening. Like their Union counterparts, Confederate soldiers well understood that sensitivity to the heard world was critical if they were to identify the activities and direction of the enemy. Even before the fighting started, soldiers recited doggerel to this effect. In 1860 one wrote,

> Sometimes before a battle
> The Gen. hears a row
> He says "the Yanks are coming—
> I hear their rifles now."[38]

Confederates ignored acoustic clues to their peril. Mishearing could mean death or, at best, prison. Although prominent captured Confederates expected decorum in their treatment, they were often bitterly disappointed, as was Captain John Morgan, incarcerated in the Columbus, Ohio, penitentiary in 1863. Would he be treated as "courteously" as captured Union officers were in the South? asked the *Charleston Mercury*. Hardly. Instead Morgan was subjected to the "silent system, as it is called," which "almost destroys the intellect of a ten years prisoner." The treatment was gruesome, hardly becoming a man of Morgan's standing. This sounded too much like black slavery to white southern ears: "On the day of his discharge he hears his name for the first time, perhaps, during years. . . . At daylight the bell signals—turn out. . . . At another signal his number is called. . . . Breakfast signal sounds, and he falls in line." This was simply too much, smacking of "the barbarisms of the inquisitorial period." Other Confederate prisoners doubtless envied Morgan because many found their customary embrace of quietude utterly shattered by the terrible noise of Federal military prisons that did not use the silent system.[39]

Echoes of a progressive past, a martial present, and a modern future converged on the home front as much as on the battlefield during the war, especially in the North, which experienced more continuity and amplification of antebellum sounds than the Confederacy. Sounds of war and sounds of home mingled in cities where Union forces mobilized. Home to the army's transportation bureau, Washington, D.C., hummed with industrial life preparing for others' military death. By 1863 "the streets resound[ed] with the deafening rumble of heavy wagons, and where the children frolicked on the grass enormous stables and long gloomy workshops taint[ed] the air." The "Horseshoeing Establishment" was loud, its throb reassuring to those thirsty for victory. Its "immense scale" was home to "two hundred men, blackened with soot . . . [who] are beating out an anvil chorus that rings in a deafening peal upon the ear, and dies away over the housetops and across the placid river in a soft, melodious chime, like the music of the bells. . . . The roar of the bellows sounds like the rush of a tempest sweeping upon a forest." "Washington is rapidly assuming the aspect of a vast military camp," noted the *New York Herald* in April 1861, listening to mobilization: "The streets are crowded with regulars and volunteers, and warlike din resounds in every direction." In Brooklyn's naval yard "every factory and workshop resounds with the din of preparation." Such were the sounds of northern industrial capitalism gearing up to fight its first great offensive. New mingled with old, though, not least because preparations for the coming conflict relied on traditional northern civil signals. In January 1861 Brooklyn's militia units were ordered to "turn out and proceed to the armory on hearing ten strokes from the City Hall bell." In New York City the signal was twelve strokes. "Accordingly, every ear was strained to hear the alarm."[40]

War entailed sacrifice, which in part meant upsetting the sounds of the northern week. "War knows no Sabbaths, as well as no laws," grumped the *New York Herald* in April 1861. Mars demanded preparations that disrupted periods of customary quietude: "The streets, usually quiet and half deserted upon the Holy Day, crowded with people; . . . resonant with cheers and martial music; full of bustle, life and animation. The sound of the church bells, calling the multitude to worship the Prince of Peace was drowned by the roll of drums, calling soldiers to march to the wars."[41]

On the few occasions when the war came home to roost, northerners were hostage to the din of conflict and southern noise. A Cincinnati correspondent heard Confederates moving against Kentucky in 1862: "The excitement in Cincinnati is again at fever heat, and nothing is heard but the sound of the drum and heavy tread of armed columns proceeding to the front. All business is again suspended." Southern silence replaced the hum of commerce when economic sounds were dragooned for military use: "The city bells this morning . . . gave notice for the reserve force to assemble, and it was but five minutes after that the sharp rattle of the tenor drum gave us notice that detachments were on their way to headquarters." Tension heightened: "I am listening every moment for the sound of musketry and artillery," twitched the reporter.[42]

When rebel advances on Columbia, Pennsylvania, in June 1863 introduced the sounds of war to the city, the registers were initially injected into civic life. "It is noticeable here," observed one reporter, "that while

the cannonading was going on the church bells of Columbia were ringing to assemble the citizens to the ordinary divine worship of the Sabbath
day." Enter war's wretched strains: "The crackling noise produced by the
burning bridge, and the shouts and confusion of the people, all made up
a scene such as is seldom witnessed" by eye and ear.[43]

Union cities in imminent danger from invasion instituted fairly elabo-
rate warning systems. When it seemed that Robert E. Lee might invade
Baltimore in July 1863, the "signal men stationed on the different roads
had discharged rockets announcing the approach of a rebel force. . . .
Alarm bells were accordingly rung as was agreed upon, by order of General Schenck, as a signal for the members of the Union League to meet
at the different rendezvous and prepare for the defence of the barricades."
Such a military system, though, was bad for civilians' nerves: "That the
people of this city are very badly scared no one . . . can doubt. A general
nervousness and looking for some sudden and dreadful disaster pervades
the community. A man shouting in the streets, the sound of drum, or, a
horse galloping rapidly along will cause a general rush of the people to
doors and windows."[44]

Noises of war sometimes spilled onto the northern home front, disrupting civilian soundscapes. In this sense the war was total, enveloping
even those not directly fighting. In June 1863 S. M. Carpenter of Frederick City, Maryland, claimed that the "agony" of the Confederate incursion into the city was over, but not before the city's soundscape had been
jolted. "I must confess that the tumult exceeded anything of the kind ever
witnessed in my history." He shivered at memories of the "clatter of hoofs
[that] resounded through the town" with the arrival of the rebel cavalry.
"During Sunday morning everything remained quiet in Frederick. The
rebels roamed about the city and the people gaped and wondered. The
church bells were rung; but who could attend Divine service when gray-
backs were running loose about town?"[45]

Even when northern civilians did not hear the war literally, they were
introduced to the dreadful sounds of the battlefield figuratively, and print
again gave voice to distant action. Union soldiers wrote letters home
laden with aural descriptions in an effort to convey something of the
nature of war. As one New Hampshire soldier described the fighting
around Fort Wagner, South Carolina, in 1863 to his sister, "I now sit down
to answer your kind letter I received while in the trenches yesterday just
as I were reading the last line a Mortar Shell burst over head and flew in
all directions. . . . When it is dark the enemy begins to throw the Shells

faster they throw the Shells so high we have time to get out of the range of them we can hear the report long enough before the Shell reaches us but when they throw Chrousich [?] Shot from there Whitworth guns they reach the Parapet before the report."[46]

War's noise did nothing to silence criticisms of the South. Unionists and abolitionists particularly continued to stress the noisiness of slavery. Unionists spoke of how northern men had been wrenched "from their quiet homes" by the warring South. Federals behaved with much greater aural dignity, the argument went: "These brave battalions march to no impudent '*Dixie*,' but plant their measured tread to 'the music of the Union.'" Their loyalty was heard as full and sincere, while the Confederacy's secession sounded visceral, wild, and not unlike a red-skinned enemy northerners fancied they had muted. Although Union soldiers were arrayed against a foe "whose sole watchword is 'THE SOUTH,' no war cry of '*The North*,' '*The East*,' or '*The West*' fills their throats." They merely "roll out in mighty chorus '*God Save America!*' Against the 'insurgents[']" clamor for 'the right of self-government'" and "Slavery[,] . . . [the] sole key-note of the Southern traitors," northern Unionists attempted to give their cause a grave and dignified soundmark. Confederates would be shown Federal superiority. J. N. McElroy, the Union army's acting assistant adjutant general in July 1861, explained that were Union troops to enter Charleston, South Carolina, they would be met by a people who "have been told that we come as robbers and murderers of women and children." To counter the image, "we will march through in soldierly order, no man leaving the ranks or shouting or making any unnecessary noise."[47]

Whatever changes the war introduced to the northern soundscape, it also left much of it intact. The noises heard in mobilization and occasional Confederate advances notwithstanding, northern registers experienced continuity and change that proved harmonious to most. Many usual sounds remained. Even the reallocation of farm labor to women in the Midwest and the mechanization of northern agriculture during the war fed into a long-held northern respect for the sound of industrialization. Reapers worked by women made "a clicking sound that was pleasant to hear." That the war was far less disruptive of the Union's soundscape than the Confederacy's is also evident in the North's ability to mark civil and religious times with bells. Because the Union had less need to melt bell metal for cannon and most of their churches remained unscathed, their celebrations of victory were heard through both bell and cannon.

Traditional northern punctuation of the national year, in fact, augmented the war. Having enthused about July Fourth celebrations in 1863, one reporter commented, "After the signing of the Declaration of Independence, Adams prophesied that the Fourth of July would always be celebrated by the ringing of bells, the blazing of bonfires and the firing of cannon. But," he added, "we shall have a new cause for celebrating the day if it is made the anniversary of the beginning of the regeneration as well as of the first establishment of the nation." He spoke of Gettysburg.[48] When Vicksburg finally fell, the "voices of the people as well as of bells and cannon testified to the delight experienced by the fall of the Western rebel Gibraltar." That the Thirteenth Amendment was rung in was entirely fitting. Massachusetts governor John A. Andrew asked the president to alert him by telegraph "as nearly as possible the moment you will sign the resolution for amending the Constitution" because "I desire to echo it immediately by a national salute on Boston Common, with a chorus of all church bells of Massachusetts." The governor was obliged, and "the people manifested their enthusiasm by the ringing of bells and firing of salutes throughout the State."[49]

More mundane but equally important soundmarks also persisted. Christmas in the North was still marked by bells because there were still bells to ring. Northern snow permitted northern sounds during the war and allowed Yankees aural validation of who constituted the true Americans. Snow in February 1863 permitted New Yorkers to vent "long pent up enthusiasm of our sleighing population," and they sounded "jingling bells as was proper during such excellent sleighing." "Such is life," after all, "and such are the Americans."[50]

The aural marking of the somber was important for the North's sense of decorum and of continuity with its past during this period of conflict and social anomie. The end of the war in the North was marked by temporary quietude punctuated by bell tolls. Abraham Lincoln's assassination was an event that demanded aural scoring. On the day Lincoln died, bells tolled in the capitol and, within a week Lincoln's death echoed throughout the North. "In nearly every city," reported the *New York Herald* on April 20, 1865, "the church bells tolled and minute guns were fired." The progress and direction of the funeral procession was a matter for eyes and ears. Northern factory villages rang bells when the convoy passed through. Towns and villages along the route to New York City marked the arrival and departure of Lincoln's corpse with reverent silence and mournful tolling. As the procession "passed through Trenton the bells

Noise of the northern mob, New York City, 1863
(from "Sacking a Drug Store in Second Avenue," Harper's Weekly, *Aug. 1, 1863)*

of the city tolled"; New Brunswick residents fired guns and rang bells, as did denizens of Elizabeth and Newark. In New York City, "church and fire bells" sounded. Silence also marked the gravity of the event. Clocks were stopped in the city, and the northern crowd for once was taciturn. "It was not an ordinary crowd—pushing jostling shouting." No, the *New York Herald* reported, "a reverential stillness prevailed, unbroken and undisturbed." Nothing jarred the "monotonous stillness" as "slowly and silently the corpse was borne through the silent streets." The "booming of minute guns and the tinkling of distant bells" effected social silence. Contrast the sound of death, said the *Herald,* with "the shouts of the populace, the loud huzzah of enthusiasm" when Lincoln had visited four years earlier. Now all was "silent as the grave." "This has been a solemn and mournful day," it was said in Washington, D.C., on April 17, 1865: "the streets have been quiet." Solemnity was enforced, although how, precisely, is unclear: "No person, except upon official business . . . are [*sic*] allowed to desecrate the spot or disturb the solemnity of the place even by the sound of a footstep." Likewise in the White House, where the "officers and domestics . . . moved about with noiseless tread."[51]

The North paid other prices for victory. Most importantly, the war accelerated the productive capacity of northern labor and the social unrest that went with it. Throughout the Northeast and Midwest, workers protested the draft law of March 1863 and rioted against free black workers who might lower already measly pay rates. The disturbances that followed struck a chord in the ears of elites who now heard with painful clarity what southern slaveholders had always feared: the noise of the mob. Antislavery women, horrified at the attacks on northern blacks, cast the rioters in terms southern masters well understood. They spoke of "the menacing cries and groans of the multitude," and they heard "the rabble" who indiscriminately attacked "quiet" and peaceful African Americans. Following New York City's draft riots of 1863, the *Herald* ran a piece titled "The Latest Tumult in the City." The source of bourgeois angst was clear: "The draft tumult which opened in blood still continues, creating a partial suspension of business. Yesterday was marked by further depredations on life and property. . . . The general commerce appeared to stand still," much to "the chagrin of traders." Rowdy mobs scored the calm of business with discord. The press attempted to capture the noise of social dislocation through drawings conveying the smashing of that quintessential middle-class barrier to plebeian public noise, glass. The *Herald*'s editors understandably expressed some relief that on July 19, the "Sabbath stillness of the city yesterday was unbroken."[52]

Increasingly, northern elites understood George Fitzhugh's 1862 commentary on the war. "The mob rules despotically among our enemies," he chortled. As for the Northwest, "Its population is a spiritless rabble." He went on, "To avoid civil discord, by keeping the people engaged in foreign war, has been the common policy and practice of statesmen in all ages and all countries. It is thus with the North. She fears the unemployed, destitute, agrarian mob of her large cities. . . . She has to choose between domestic war and war with us."[53] Northern elites were beginning to sense he was not far wrong.

In rich Northern cities could be heard the boom of guns,
the scream of whistles, the shouts of surging hosts
greeting the crowned victors.

—Sarah E. J. Boyles Dilworth, "My Reminiscences of
the Sixties (1861–65)"

(((9. Confederate Soundscapes

Confederates heard the war and listened intently for victory,
hoping that abolitionists would be silenced in the noise of conflict and
the shouts of a southern victory." They were horribly disappointed. Confederates never heard the sweet chimes of ultimate victory, and the process of defeat altered their soundscape profoundly. Unlike most northern civilians, many southerners on the home front heard the war literally
as well as metaphorically. For them the conflict was an acoustic hemor-
rhaging, a disruption of a way of life built on social quietude and tranquility. The master class found it increasingly difficult to control and contain the soundscape of the Confederacy, especially as the war went on and
invading Union forces brought sounds of freedom and industrial might,
noises the master class had long loathed.

Confederate soundscapes depended on how sounds from the past were
remembered and conceptualized and, more immediately, on the extent
and nature of Union advances. Sounds of war came quickly to Virginia in
particular. Early in the war, "the cool western breeze which [rippled] the

tasselled corn into endless waves, [came] laden with the hum of war from the distant Alleghanies," and the "dull rumble of the ambulances, bringing in the sick and wounded from the front, [echoed] on the hillside."[1] The onrushing sounds of war reaffirmed the Confederates' commitment to their heard pasts, while counterpointing how new, different, and disruptive the present sounded or, in some instances, did not sound.

For most Confederates the initial phases of the war were quiet, more reminiscent of the Old South than of the noisy future they feared. Witness the early wartime correspondence of Savannah River rice planter Louis Manigault. In late March 1861, usual sounds prevailed. "Mr Capers tells me all things are going on well on the plantation," wrote Manigault to his father; "I find thus far all very quiet, and we will occupy ourselves in Threshing the Crop." Manigault's soundscape sounded utterly peaceful. "Macon," he commented, "is quite a pretty Town, of about Ten Thousand inhabitants, and it is quiet there as if we were still a part of the United States." By November strains of conflict remained distant, and the noises of war did not muffle the happy throb of Manigault's industrious plantation or disrupt the tranquility of slave labor. "Our people are working well and Cheerfully," smiled Manigault, elaborating, "I yesterday rode over the entire plantation with Mr Capers, and I have no reason to complain of any thing. In every direction the steam of the various Threshers indicated that the usual plantation work was going on. . . . It is difficult to realize, in the midst of our quietude, that *nearly* within sight, on Daufuskie Island, the Yankees have entered Mr Stoddard's House."[2]

Little had changed by December. In fact Manigault was more preoccupied with the customary disruptions to plantation life than he was with the possibility that new forces would interrupt his serenity. At times Manigault could hardly believe his eyes or ears: "Every thing on the plantation, and as far as the eye can reach, is going on in our usual quiet manner," he told his father on December 5, 1861, "and the News Paper . . . alone indicates that we are having troubles in our midst." Be that as it may, Manigault still felt on guard at this time of the year. He explained: "Christmas is always a very bad time for Negroes, and it is always a God-Send (any year, but far more so this) when that Holyday is over, and we all resume our quiet plantation work." He listened to customary plantation tranquility, anticipating the social noise of war and impending aural disruption, and took extra precautions: "Mr Capers wishes me to mention that the three Negroes now Confined should by no means be allowed at any moment to Converse or see each other."[3]

Even by the end of the war some places still sounded traditionally southern. Savannah's soundscapes were among the least affected. In October 1864 John Ransom of the 9th Michigan Cavalry "hear[d] bells ring about the city of Savannah. Very different from the city of Richmond; there it was all noise and bustle and clatter . . . while here it is quiet and pleasant and nice. . . . Don't seem as if they could both be cities of the Confederacy. Savannah has probably seen as little of real war or the consequence of war, as any city in the South."[4]

Even when strains of war began to reach their home fronts, Confederates essayed to preserve the heard world of their homeland and past. Slaves, most obviously, continued to have their behavior regulated and checked, and so, too, did suspect whites. A rightly nervous northerner in Columbia, South Carolina, made both points in early 1861. Johnson M. Mundy, who helped design Columbia's State House, explained to his sister, "It may . . . prove interesting to you, to have some account of what is said and done in the 'Palmetto State', from one who is an eye and ear witness." "Everything about me indicates Revolution," Mundy fretted. "The people cry 'oppression', and 'give me liberty or give me death' is upon every tongue, excepting the *slaves,* who have been taught to fear, and submit to their rightful masters." Mundy felt peculiarly conscious of his own enforced silence: "You may think that a person from the North would not find himself very much at home surrounded by such elements. And you think rightly. He is under great restraint, as his words and actions are closely watched, lest [they] conflict too strongly with Southern principles." Confederates attempted to control the political soundscape of the wartime South, hoping that the noise of southern Unionists would be "hushed" with victory.[5]

The nature of this particular war meant that southern soundscapes would change regardless of the immediate military outcome. Victory required gearing up the South's military might through rapid industrialization, a process that rendered the Confederate soundscape more northern and more modern. Social quietude, many hoped, would return with Confederate victory. Loss, though, would alter the southern soundscape permanently. Surely a defeated South would be remade in the image of a society that coveted the hum of industrialism and the throb of liberal democracy.[6]

In some respects the traditional southern endorsement of the sound of industriousness proved sufficiently flexible to accommodate the upped volume that accompanied mobilization. Louis Manigault suggested as much when he alluded to the comforting throb of his steam threshers in 1861. Likewise, Susan Bradford Eppes recalled that the beginning of the war on her Florida plantation sounded initially like a protracted Sunday. So that the plantation's "white men . . . might enlist, Saturday the mills shut down—the mighty throb of the huge engine ceased—but then it always stopped for Sunday. Monday morning silence still reigned." In the longer term, the Union's naval blockade and the demands of war encouraged planters to introduce "new industries." As a result, "the whirr of the spinning-wheel, the clump, clump of the loom, and the shoe-maker hammering . . . grew to be familiar sounds" on some Civil War plantations. In this way, at home and on the battlefield, the Civil War increased in volume. Eppes heard the sound of gearing up for war on neighboring plantations: "Carpenters were called in and new loom patterned. . . . Soon . . . the whirr of the wheel and the noise of the baton could be heard in almost every home."[7] These were comforting melodies scored with patriotism and industriousness and laden with hopes of victory.

War's sounds were incorporated relatively quickly into the evolving Confederate soundscape, even if this incorporation was sometimes grudging and temporary. How long, precisely, Confederates were willing to put up with these new noises for the cause of the South, no one quite knew. But put up with them they did, at least for some of the time. In 1864 the *Charleston Mercury* described the aural novelty of the city's first shelling and illustrated how quickly Charlestonians had become accustomed to the noises of war. At a "time when the bombardment was still a novelty to our people," Charlestonians sat awake in the early hours "listening to the monotonous sound of the cannonade kept up on the enemy position from the batteries on James Island." A hotel guest "was startled by the noise that . . . resembled the whirr of a phantom brigade of cavalry, galloping in mid air. My first feeling was that of utter astonishment," and the "deafening explosion" of the shell hardly lessened his anxiety. He compared the "awful rush and whirr" of the shell to a meteor. Traditional urban cues were then used in the context of the unusual: "A watchman was running frantically down the street and, when he reached the corner just below me, commenced striking with his staff against the curb—a signal of alarm practiced among the Charleston police."[8]

Southern comfort with the sound of industrious mobilization was fractured by the sheer scale of preparation necessary for effective prosecution of the war. If the Union intended to make the South modern, then it in part succeeded not least because preparing for the conflict altered the region's soundscape, enfranchising new sounds and upping the volume of old ones. Even at the beginning of the war in Richmond, correspondents noted how plans for the conflict increased noise levels and altered the social timbre by accelerating industrialization. Richmond was "amid all the tumult and bustle of war. . . . The clatter of military trappings, the rattle of heavy freight wagons and the measured tap of the drum constitute the music of our times," reported one observer. Even the nocturnal southern soundscape, so long the idealized refuge of quietude, was disrupted: "At night most of our citizens retire into their dwellings. . . . The shrill whistle of the locomotive during the still hours announces the arrival of more troops." Government sounds replaced civil ones, and martial preparation silenced business registers. Late in the war in Richmond the "rumbling of government wagons monopolized the privilege of making a noise during the day, nearly all other vehicles having been withdrawn from the thoroughfares of business and mercantile operations in a great measure suspended."[9]

Social noise also increased dramatically. George Fitzhugh heard "the senseless clamor of home-keeping people" in the Confederacy. Voices of protest grew louder, especially from traditionally quieter southern constituencies such as women, who began to attack the state in protest against extortion and inflation, signaling to Confederate elites that a predominantly northern form of mobbing was making its way south. Confederate leaders worked hard to reassure themselves that southern gentility and quietude would prevail. "All things considered," readers of South Carolina's *Yorkville Enquirer* learned in 1863, "there has been very little complaining or croaking in the South." Indeed, the "croaking in the South is but a gentle whisper compared with the thunder tones in the North uttered by the mobs in New York and elsewhere." But protest in the Confederacy was becoming too loud to ignore. Southern wartime mobs, like their northern counterparts, were depicted as glass-breaking, property-damaging, grim-faced rioters whose noise suggested social disintegration. Northerners in the South at the beginning of the war confirmed the increase in volume and the shift in timbre. On a trip from Richmond to Nashville in November 1861 a *New York Herald* reporter noted Confederate noisiness: "The contrast between the South and the North . . . is most remarkable. When I come into the Northern States I find the people pursuing their usual peaceful avocations, without any apparent disturbance or interruption, as if the blast of war had never sounded in their ears," which, of course, it had not, literally, for most. Go south,

Southern cacophony, Richmond, Virginia, 1863
(*from* Leslie's Illustrated Newspaper, *May 23, 1863*)

though, and "war is on every tongue. . . . Day and night you hear nothing but war shouts."[10] In some regions of the Confederacy the pressures of war, even by late 1861, began to alter a southern soundscape invested in social and political quietude.

Southerners on the home front did not always have to see the war to know that it was happening. Many heard it and resented the nervousness that booms and blasts generated. Charlestonians, for example, condemned Confederate officers testing the alertness of their troops. When Colonel Ripley at Fort Moultrie started "a heavy cannonading, which continued for several hours [and] was heard in the harbor," in June 1861, "to see whether our garrisons [at Forts Sumter and Beauregard] were mindful of their duties in the night as well as in the day time," Charlestonians were upset at what they considered unnecessary, nerve-wracking noise. As the *Charleston Mercury* commented, "If Col. RIPLEY should continue the discipline, does he not bid fair to rival the fire alarm telegraph, with its ringing of bells, in alarming the community?" Military sounds threatened civil ones.[11]

Southern constituencies on the home front interpreted the sounds of war in myriad ways. Early in the war Sarah Morgan Dawson of Baton Rouge sought protection from a Union relative in New Orleans. Caught in the middle of a skirmish en route, Dawson found comfort and strength in being a part of the aural world of war. Male relatives had told her about "the singular sensation produced by the rifled balls spinning around one's head; and here I heard the same peculiar sound, ran the same risk, and was equal to the rest of the boys. . . . I think I was rather proud of it." Once in New Orleans she disdained what she considered the rudeness of Confederate women who spat at Union soldiers and who breached southern gender conventions by raising their voices. "'Loud' women, what a contempt I have for you!" she declared, excoriating those who "bellow[ed] like any Billingsgate nymph!" for transgressing southern aural etiquette. Elite women could still offer respite, however. Well-bred nurses ("two ladies," no less) in Charlottesville at the beginning of the war were seen and heard moving "about the room, noiseless, active, efficient."[12]

The war added aural layers to the southern soundscape and in the process generated additional meanings for broadly understood and recognized social cues. Charlestonians complained of the "strange commingling of sounds," namely, "the soft, sweet tones of the church bells inviting the people to the house of prayer, and the boom of cannon and crash of shells summoning the unfortunate soldier to the dark abodes of eternity. The one was the gentle messenger of peace and life, the other the dread summoner of suffering and death!"[13] Similarly with the sound and meaning of cannon. Most Americans understood cannon firing as celebratory. War, however, supplemented the meaning of this sound. Captain Hugh Black of the 6th Florida Infantry noted the overlap and conjoining on October 4, 1862, "the day appointed for the inauguration of the rebel Military Governor Haws" in Frankfort, Kentucky. "The booming of cannon," quite typically, marked the inauguration. Sounds of war quickly encroached, however, changing the meaning of the scene. "Scarcely" had the sounds of the inauguration cannon "passed away before the firing of hostile cannon commenced nine miles from Frankfort on the Louisville Road."[14]

Some wartime measures helped quiet southern urban soundscapes. Fear of being shelled or shot sometimes dampened the ambient volume of New Orleans. "This town, with its ten thousand soldiers, is more quiet than it was with the old population of seven thousand citizens," Sarah Dawson commented, adding, "With this tremendous addition, it is like a graveyard in its quiet, at times." But peace was short lived as Confederates stepped up efforts to retake the city. "Ever since I commenced to write," lamented Dawson on August 5, 1862, "the sound of a furious bombardment has been ringing in my ears."[15]

Fancying they could hear delicious bites of victory, civilians listened actively for clues of military progress. "Yesterday evening the attack commenced at Port Royal and during the day there were numbers of persons gathered together on the battery listening to the booming of the cannon which tho' faint could be distinctly heard," wrote one South Carolina listener in 1861. "West, the fierce struggle for the Mississippi Valley is at last begun," reported the *Charleston Mercury* in 1863, "and, amidst the din and tumult of the unequal contest, we think we can distinguish the shouts of victory. Doubt and discord reign in the councils of our enemies. With us, all is confidence." Successful battles and winning the war would be loud, reckoned a Georgia volunteer in 1862. "Victory," he maintained, "shall be resounded to all nations of the earth."[16]

As the war grew protracted, though, anticipated sounds of victory were muffled by noises of successive defeats. By 1864 impending defeat resounded "in the din of the bombardment." "Let us hope," cheered the *Charleston Mercury* in July 1863, "that our enemies will never light bonfires, ring their bells, fire cannon and serenade President LINCOLN, at the fall of Charleston."[17] Such hopes went begging. Loss meant noise for Confederate troops and civilians alike after the surrender. They encountered jubilant "drunk Yankees, and they were so noisy," and "drunken soldiers and Negroes carousing, shouting, embracing, and preaching." What they found after the war confirmed long-held fears: restless blacks "cracking whips, throwing stones, and yelling at the tops of their lungs." Captain Benjamin Wesley Justice of Lee's Army of Northern Virginia captured the essence of why, by 1864, Confederates had come to loathe the sound of war and why, in aural terms, they had fought it in the first place. As he wrote to his wife in May 1864, "I am heartily sick of blood & the sound of artillery & small arms & the ghastly, pale face of death and all the horrible sights & sounds of war. I long more intensely & earnestly for the sweet rest & quiet of home than ever before."[18]

Southern soundscapes were altered permanently by the cacophony of defeat, providing sure testimony to the death of social tranquility. Fleeing Atlantans in 1864 left what sounded like a Yankee Sodom: "Mingling in the air are sounds of the Sabbath bells, the shrieks of a score of locomotives, the rumbling of cars and carts, the shouts of drivers and unceasing hum of a modern Babel. The spectacle is more melancholy than interesting." Southern soundscapes past were drowned by Confederate noisescapes present. Charleston's Christmas sounds in 1862 were masked by war's noise: "How the gladsome pealing of our ancient chimes, which were wont to usher in the morn of Nativity, are drowned today in the din of a ruthless war."[19]

Everything Confederates feared came true, though not all at once. Even though northerners must have wondered if their impending victory would stimulate the South to accept Yankee isms, reports from 1865 suggested it would not. From Savannah, Georgia, little was heard: "In the city all is quiet and as calm as a Sunday morning in a country village," wrote a correspondent from the *New York Herald*. Plainly, the "quiet streets" and the "talk in low tones" by "Men in all colors" indicated an Old South marked by "idleness and a singular lethargy." Only "the resuscitating influences of a band of keen, sharp, wide-awake business men from the North . . . stir them up for a moment." A *New York Herald* correspondent captured the aural dimension of Union victory in October 1863, should it come to pass. "In the future," he offered whimsically, "we may make further progress into the heart of the sacred soil, till our cars run into the streets of Richmond, and the dome of the rebel Capitol itself re-echoes the shrill whistles of our proud engines."[20]

Beyond the noise of Union victory, military loss also meant that a blanket of disconcerting silence enveloped the southern home front. This reminded some of past tranquilities. Charleston's residents had become "accustomed to the howl of rushing shell. . . . Now that quiet and safety are insured they propose to repair and live comfortably once more." But a return to the quiet life—and all the meaning that Confederates attached to that term—was not easily achieved. Silence there was, to be sure— too much so, perhaps. War destroyed Charleston's commerce so that "the silent streets" no longer hummed with the industriousness and economic vibrancy grounded in slavery. A *New York Herald* reporter stationed in Port Royal listened to Beaufort at the end of 1861 and heard only ruin. "The streets of the village were silent," he noted, concluding, "The silence of death was over the deserted streets." "The whole country on both sides of the James, the Rappahannock, and the Potomac" by August 1862 "had been reduced to a desert. . . . The houses along the road are as silent as death. The inhabitants have fled; the inclosures are all broken down; the windows and doors staved in; the very desolation sits brooding in silence over the land." In 1865 a Union sailor described his "first setting affoot in Charleston" and offered the following assessment of the health of the city: "The lower portion of the city was entirely deserted long before our occupation and its warehouses, mansions, churches hotels &c abandoned to utter solitude and dilapidation. Shells from our Morris Island lodgment dropped almost daily into this part of the town committing havoc at pleasure with masonry & roofs. We see traces of them everywhere." But there were signs of life and confirmation of victory to northern ears: "Further up, the hum of industry is now started." [21]

Other southern constituencies projected their own meanings onto— and derived their own advantages from—the sounds of the Civil War. The war brought new sounds to the plantation, and slaves noticed. "I never had heard such noises in my life," recalled Easter Reed: "I hadn't never heard a fife or a drum." "I remember what a roar and din the guns made," recalled Susie King Taylor following the firing on Fort Pulaski. Sometimes the strangeness and volume of cannon booms frightened them, just as they did whites. When she heard the war, Dinah Watson, a young slave in New Orleans, struggled for comparisons and opted for the natural: "The cannons was roaring like thunder." Slaves also remembered the meaning mistresses attached to the sounds of war. Ex-slave Sarah Debro heard her first cannon with "Miss Polly," who said, "Listen, Sarah, hear them canons? They's killing our mens." [22]

Novelty aside, slaves knew to listen and look during the Civil War. Georgia's Amanda McDaniel "didn't see any of the fighting but I did hear the firing of cannons." Ella Johnson saw and heard Sherman's crash through Georgia: "We could see the sky all red and hear the distant roar of guns, but that was all." That, apparently, was enough to enable a few bondmen to locate the rough direction of Union lines. "A few of the Negro slave men joined Sherman's army," Johnson noted. Old tactics of hard listening and cultivating silence proved handy for other slaves: "I . . . was awakened late in the night by the sound of a wagon, and hushed noises. I knew that the ex-slaves (as many as possible) were escaping."[23]

The wartime aural world of slaves was one of continuity and change. On one hand, they recruited hard listening, as they had always done, to ascertain intelligence; on the other, news of impending victory loosened tight tongues, relaxed taut vocal cords, and allowed the slave community to influence more fully than ever the soundscapes of the southern plantation. As he did with many things, Booker T. Washington captured the essence of the process: "The news and mutterings of great events [during the war] were swiftly carried from one plantation to another. . . . As the great day drew nearer, there was more singing in the slave quarters than usual. It was bolder, had more ring, and lasted longer into the night."[24] As they had before the onset of hostilities, slaves during the war communicated important information to one another through song. They sang of God when they meant Union troops; they sang of slavery when they meant freedom. But veiled meanings did not escape sound-sensitive slaveholders, who sometimes jailed slaves for singing because they understood the communicative function of their songs. Patrols during the war clamped down on slaves' religious gatherings, trying to quiet an increasing volume of freedom.[25]

Slaves, though, had cultivated the art of aurality. Old skills, especially slaves' ability to control volume, still proved indispensable. "I also remember the whispering among the slaves—their talking of the possibility of freedom," recalled Mary Gladdy of wartime Georgia. Mary Barbour remembered the coming of Union forces and her father's admonitions for silence in an effort to reach freedom's lines. Her father woke her at night, "all the time telling me to keep quiet. One of the twins hollered some, and Pappy put his hand over its mouth to keep it quiet." The strategy worked; they reached Union lines in New Bern, North Carolina.[26]

Louis Hughes escaped from his master during the war by remembering the importance of silence from antebellum days and by recognizing that the new sounds of war gave him some acoustic room to maneuver. With two companions Hughes determined to escape from Tennessee, and he was guided by the wise head and ears of "Uncle Alfred," who "cautioned us not to speak above a whisper" so as not to alert "the rebel troops camped on both sides of us." Alfred scouted, returned, and reported, "I can not hear a sound . . . so let us go on." The small group ventured onward with cultivated stealth: "There was no talking above a whisper, for fear of being heard by the soldiers." Vigilance reigned: "Alfred and I had made a turn around the place, listening to see if we could hear any noise, or see any trace of soldiers." Luck proved elusive: "Suddenly I heard the yelp of blood hounds in the distance. . . . The sound came nearer and nearer, and then we heard men yelling." Their defense was olfactory and aural. Alfred told Hughes, "Let me oil your feet." The reasoning? "He had with him a bottle of ointment made of turpentine and onions, a prepa-ration used to throw hounds off a trail." Feet anointed, they ran, while Alfred's advice on the art of successful escape echoed quietly in their ears: "Don't let the brushes touch you." Had Alfred followed his own advice, escape would have been possible: "[Alfred] ran through the bushes with such a rattling noise one could have heard him a great distance. He wore one of those old fashioned oil coats made in Virginia; and, as he ran, the bushes, striking against the coat, made a noise like the beating of a tine board with sticks." Their pursuers soon caught up with them.[27]

Returned to his plantation, Hughes still kept an ear out for the chance to escape. The plantation was, by his account, a quiet place during the war, at least among the slave community. Even tepid news of freedom made slaves "unspeakably happy." That was precisely the point: thoughts of freedom usually remained silent unless context (or mistake) dictated otherwise. "They were afraid to let the master know that they ever thought of such a thing," continued Hughes, "and they never dreamed of speaking about it except among themselves." Only distance from mas-ters' ears and the acoustic absorbency of the cabin provided safety: "They would laugh and chat about freedom in their cabins." Hughes kept his ears pricked for clues of freedom's coming. "One winter night," he recalled, "I was awakened by a rumbling noise like that of heavy wagons, which continued steadily and so long a time that I finally concluded it must be an army passing." It was the Yankees; Hughes followed them to Union lines and continued all the way to Canada.[28]

Slaves listened hard during the war. Muttered news by careless whites could mean the difference between enduring slavery and enjoying free-dom. Because whites knew as much, their lips pursed tighter as war pro-gressed. Susan Bradford Eppes recalled that during the war her slaves "took to listening under the windows at night and 'Totin' news.'" Such behavior put whites on greater aural alert. "We never were allowed to listen to the conversations of the white folks," remembered Mary Jane Simmons of Georgia. Masters and mistresses were at once quieter—a delicious irony given their own admonitions to slaves—and even more sensitive to what they heard. Eppes conveyed some sense of skittishness. "Screams and groans . . . were heard in the back-yard," remembered Eppes, adding, "it is needless to say we were a nervous set in those days so we rushed in the direction of the out-cry."[29]

But the war introduced sounds that were beyond white control. Ike Derricotte recalled that everyone on his Georgia plantation, slave and free, listened for the whistle of "dat old Georgia train." He explained:

> Not many folks was able to take de papers den, and de news was in 'em was from one to two weeks old when dey got here. All de men dat was able to fight was off at de front and de folks at home was anxious for news. De way dat old train brought 'em news was lak dis: if de southern troops was in de front, den dat old whistle jus' blowed continuously, but if it was bad news, den it was jus' one short, sharp blast. In dat way from de time it got in hearin', evvybody could tell by de whistle if de news was good or bad and, believe me evvybody sho' did listen to day train.

Slave and free listened to the same sounds but undoubtedly heard different meanings in the whistles.[30]

If Confederates chafed at having to keep mum in the presence of slaves during the war, they also experienced another, equally upsetting form of silence. Whatever small victories and large losses Confederates experienced during and at the close of the war, they were less able than Federals to use bells as aural signatures to mark such events. "The Ordnance Bureau of the Confederate States," reported the *Charleston Mercury* on April 3, 1862, "solicits the use of such bells as can be spared during the war, for the purpose of providing light artillery for the public defense." "While copper is abundant, the supply of tin is deficient to convert the copper into bronze. Bells contain so much tin that two thousand four hundred weight of bell metal, mixed with the proper quantity of copper, will suffice for a field battery of six pieces." Donating bells was patriotic, argued Confederate officials: "Those who are willing to devote their bells for this patriotic purpose will receive receipts for them, and the bells will be replaced, if required, at the close of the war, or they will be purchased at fair prices." Those heeding the call could send their bells to one of ten depots in the Confederacy. For proper reimbursement, "persons and congregations placing their bells at the service of the Government are requested to send a statement of the fact, with a description and weight of the bell, to the Chief of the Bureau of Ordnance, at Richmond." So anxious were Confederate forces for bell metal that they sent submarines to recover "two large brass church bells" from "the bottom of the Pamunkey," which had been sunk by retreating Union forces in late 1862.[31]

God's men were sometimes reluctant to part with the Lord's sounds. When lobbied for his church's two bells in March 1862, Catholic bishop William Henry of Natchez, Mississippi, hesitated to deploy God's bells for Mars's cannon. Confederate leaders, including P. G. T. Beauregard, who issued the original call, countered by pointing to the loyalty of planters and stressing that the temporary disruption of southern soundmarks through the appropriation of church bells would preserve the integrity of future southern soundscapes. As he explained to a recalcitrant Father Mullon of Saint Patrick's in New Orleans in March 1862, "The call which I made on the planters of the Mississippi Valley to contribute their bells from their plantations to be cast into cannon is being so promptly met, that I am in hopes of being spared the necessity of depriving our churches of any of their appendages." While Beauregard agreed with Mullon that "our wives and children have been accustomed to the call, and would miss the tones of 'the church-going bell,'" he argued, "if there is no alternative we must make the sacrifice, and should I need it I will avail myself of your offer to contribute the bell of Saint Patrick's Church, that it may rebuke with a tongue of fire the vandals who in this war have polluted God's alter. . . . I can only hope that the day is not far distant when peace will once more bless our country and I shall visit again a quiet home." Father Mullon had a point. When church bells remained unmolested, the familiarity of their sound—their reassertion that the South was still the South—provided solace. For the most part, though, planters and congregations responded to the call.[32]

Not all bells were melted down. Some remained in Petersburg by the end of the war, and Atlanta retained some, too, with its evacuation on September 18, 1864, marked by "the Sabbath bells . . . tolling their solemn chimes." Some bells were too important to the coordination of Confederate military and civilian life to melt. In addition to "boatswains' whistles; the fife and rolling drum," "the musical bells of . . . ships striking the passing hour—bell answering to bell and echo[ing] back again on the passing breeze" were critical for shipping, martial and merchant.[33] Bells remained important for alerting Richmond's denizens of immediate danger. The New York Herald reported in 1864 that "the bells of the city were rung and men were rushing through the streets, crying, 'To arms! To arms! The

Yankees are coming! The Yankees are coming!'" Some bells were kept in case of emergencies. Illinois infantryman Charles Henry Snedeker testified to this during the siege of Vicksburg in June 1863: "There was a large fire in Vicksburg about midnight. I could distinctly hear the fire bells ringing in the city."[34]

When church bells remained unmolested, the familiarity of their sound and their reassertion that the South was still southern offered consolation. Near Atlanta at the end of the war an "old village bell" still called "the devout to public worship." So familiar was the noise of war that parishioners tuned out the guns and heard only the sound of worship: "Their songs were being responded to by Sherman's guns, with which they had grown too familiar to be interrupted by them in their devotions."[35]

Confederates went to some lengths to protect their campanological soundmarks. William J. Grayson wrote in his diary on June 12, 1862, "The bells, today, were taken down from the steeple of St. Michael's Church to be sent to Columbia" because Charlestonians feared Union forces would steal them.[36] Confederates were also loath to hear Federals using southern bells to celebrate northern victory. "Two contrabands came in today from Charleston," wrote Percival Drayton to Gideon Welles in June 1862: "Like the rest I have seen, they represent the harbour as completely blocked up. . . . Another piece of information from our runaways was that the banks had been removed to Columbia as well as the Church bells, as he said to prevent our ringing them if we got into the city." If Confederates could not celebrate either precious wins or toll increasing losses with their ever fewer bells, they were damned if Yankees would appropriate southern sounds for their celebrations.[37]

The Union took all of this as good news: melting sacred cannon denoted desperation, they reasoned. The blockade was having desired effects, reported the *New York Herald* in April 1862, for "it is plain that the rebels are deficient both in cannon and small arms." Indeed, the "recent appeal, asking for the church bells and the bells used by the planters to be sent to the rebel foundries, to be cast into field pieces, is a palpable proof of their deep need of artillery." Surely, mused Federals, this was, at long last, the gasp of a dying cause, the sigh of a grubby, backward country unable to fight a modern war. Their antebellum ears had not betrayed them: the South was an anachronism. "This call upon the churches for their bells to be cast into cannon is not a new expedient of war," noted the *Herald;* it "is an old device of the Mexicans." This was good news, for "we cannot perceive that they have derived any advantages from it in the way of inde-

The silence of ruin, Charleston, South Carolina, April 1865—view of ruined buildings through porch of the Circular Church, 150 Meeting Street (Library of Congress, Prints and Photographic Division, reproduction no. LC-B8171-3448 DLC)

pendence, after forty years of such experiments as this of Jeff. Davis and his confederates." Anyway, "we are strong in the belief that within a few days we shall have such glorious tidings from the Southwest and from Virginia as will satisfy the Southern people of the folly of converting . . . their church-going bells into the horrid music of field artillery."[38]

Northerners were even more pleased when such bells were captured. *Harper's Weekly* crooned in 1862, "Four hundred and eighteen rebel church bells, which had been sent to New Orleans in response to the call of General Beauregard, and captured in that city, were sold in Boston on the 30th ult." This was some loss for the Confederacy and no little gain for the Union: "They weighed together upward of one hundred thou-

The silence of defeat, Charleston, South Carolina, April 1865—view from roof of the Mills House, looking up Meeting Street, with ruins of the Circular Church in center (Library of Congress, Prints and Photographic Division, reproduction no. LC-B8171-3440 DLC)

sand pounds, and brought about twenty-four thousand dollars," about 100 small, 100-pound cannon lost by the South.[39]

Fewer bells meant many things. At times Confederates on the home front managed to do without, sometimes to the benefit of their nerves and improved civic coordination. So common were bell alarms during the war that southerners became insensible to their foreboding. Although in Petersburg in 1864 the ringing of "the Courthouse and engine bells" at the common sight of advancing Union troops prompted citizens to respond "with their usual alacrity," many citizens eventually became inured to alarms. Richmond officials observed and tried to overcome the problem. In May 1864, "the reserve forces of militia were called out—not by a repetition of the unearthly clangor of bells but by the more quiet method of printed notices." It seemed to work: "The difference in the effect of this mode of assembling the people was manifest in the calmer and less bustling, but equally prompt manner in which it was responded to." This new device—born partly of necessity and the melting of bells but also a product of fatigue from nerve-wracking tocsins—worked when immediate alarm and action were not required: "The idea of alarm is so inseparably associated with the din of fire bells, that whenever they sound they invariably produce an undue excitement, and on such occasions especially produce unnecessary and painful agitation among those whose peace should be most particularly consulted."[40]

By the spring of 1863, Union reports revealed the "very deplorable" appearance of Charleston, noting, "There are but three bells of any size or consequence, the remainder having all been taken down and melted up for cannon." Melting bells for cannon and the attendant disappearance, even temporarily, of familiar sounds is a good example of contemporaries hearing in the absence of the heard. Sounds past and future, as from belfries, invested people with faith in what they heard, and without familiar sounds, Confederates felt dislocated. St. Phillip's of Charleston donated its bells for the Confederate cause, and "almost 115 years of silence from our bell tower" followed until new ones were installed in the 1970s. So, too, in Columbia, South Carolina. Emma LeConte lamented the silence of "the old town clock [she meant bell] whose familiar stroke we miss so much" following Sherman's razing of the city. "I miss greatly," mused Kate Cumming at the end of the war, "'the sound of the churchgoing bell.'" "On making inquiries," Cumming learned "that all the bells had been taken" from her church "to make cannon." The decrease in bells meant that Confederates found it difficult to mark their wretched losses. When early in the war southern cities still had bells in numbers, Charleston honored the dead at Manassas by suspending business for a day and issuing aural marks for the deceased: "The tolling bells, the heavy beat of the muffled drum, and the melancholy dirge for the departed soldier, resounded mournfully over the house tops and through the streets of our quiet city."[41] But the Confederacy's greatest defeat was greeted by relative silence.

In addition to the devastating loss of loved ones, the war caused profound changes in the South's soundscapes, at times making it sound northern to their ears. The deprivation of familiar sounds also took its toll on southern civilians and soldiers who were impotent to toll even their defeats. Conversely, the war was an acoustic victory for the enslaved, albeit temporarily. Not only did they use the sounds of conflict to help locate freedom's lines, but they also marked freedom by appropriating the sounds of slavery and injecting their own sounds into the Confeder-

ate and postemancipation South more vigorously than enslavement allowed. Ultimately the war not only inaugurated profound debates about
the nature of freedom and citizenship but also changed the soundscape of
a region. The predominant soundmarks of slavery were gradually replaced
by the haltingly slow keynotes of free wage labor in a postwar South that
proved reluctant to relinquish its Old South echoes.

St. Louis now re-echoes the voice of New York,
and San Francisco that of St. Louis.
—Benj. T. Tanner, *An Apology for African Methodism,* 1867

(((10. Sounds of Emancipation, Reconstruction, and Reunion

If the Civil War ruptured the Old South soundscape, the volume and timbre of emancipation shattered it. Freedpeople now helped tune the soundmarks of a society that had regulated and consumed their sounds. Yet ex-slaves' aural contributions to the postbellum South were muted within a few years of the beginning of Reconstruction. Whatever political influence the planters lost as a result of the war, they retained enough economic and social power to reimpose an aural world that, while less comforting than their soniferous gardens of old, was nonetheless vaguely reassuring, albeit in the context of something that was beginning to sound like a free wage labor society. Postbellum southern elites were not alone in hearing new worlds and attempting to reinscribe acceptable sounds because the Civil War amplified voices in northern society that clamored in the ears of an increasingly reactionary bourgeoisie. The deafening register of the northern working class, particularly after the wartime riots and the panic of 1873, caused the northern bourgeoisie to wonder whether they had gone too far in their great experiment. To ponder the matter and recuperate from the growing din, they turned, of all places, to a quiet South whose register sounded more like the days of their antebellum past and less like their increasingly cacophonous present.

Freedpeople conceptualized, measured, and understood the coming of freedom and the immediate aftermath aurally and visually. For one North Carolina ex-slave, seasonal sound helped her situate and remember the timing of freedom. Union forces came to her plantation just before Christmas, when the ground was frozen. "De sojers come gallopin' up to de house," she recalled, "dey horses feets clickin' on de ice." Sam Polite of Beaufort, South Carolina, remembered, "When gun shoot on Bay Point for Freedom, I been seventeen-year-old working slave." Sometimes the novelty of the sound of Union forces proved difficult to interpret. A former Alabama slave recalled, "Us hear dat trompin' soun'. It didn't soun' lak no ever'day marchin'. It soun' lak Judgement Day." [1] It was.

News of nearby Union forces often inspired celebrations. "When word came dat dey was comin'," recalled Nettie Henry, "it soun' lak a moanin' win' in de quarter. Ever'body . . . march' 'roun' de room an' sorter sing-lak." Sarah Louise Augustus remembered that slaves on her plantation "began to shout and pray" upon hearing of their freedom. "On de day dat de slaves wuz sot free in Texas," remembered Andy Williams, slaves marked the date aurally, as had millions of antebellum white Americans: "Den Mamy broke loose er shoutin' an' she sang an' hollered all day." "Maw," told Harriette Benton, ex-slave from Georgia, "she says dat de Yankees done come pass our plantation and dat when dey come, de whole Jasper County started screaming and hollerin' like dey's all crazy." Freedom meant that "de Lord has heard de groans of de people," and as a result, the people asked, "haven't we a right to rejoice?" [2]

Slaves greeted freedom with shouts of joy, and the announcement of freedom itself was sometimes loud. "At every station from Lake City to Tallahassee," recalled James M. Dancy in 1933, "the train was stopped and an announcement made through a megaphone to the negroes living on the plantations which lined the road that President Lincoln had declared them free and equal citizens." [3] Emancipation upped the volume in the South.

Freedpeople gauged their freedom by testing previously forbidden aural worlds and quashing sounds of old. They were free now to shout with louder voice and free enough never to have to hear the sounds of slavery again. As one freedom song put it,

Dey'll be no hoe ter handle,
No plow ter hold de eye,
We'll jis' turn loose a shoutin'
In de Sweet Bye an' Bye!
No moans, no groans, no weepin'
On wings us all will fly,
Dere's a happy time a comon'.[4]

As for most Americans, the sound of the bell for southern slaves and freedpeople held multiple meanings. For many it meant work; for this reason plantation bells were sometimes silenced by African American federal troops. When a company arrived on the Ball family's plantation during the war, they found the work bell and smashed it. Given the centrality of sound in regulating slave bodies and behavior, it is hardly surprising to find that emancipation from bondage was often understood in terms of the sounds of clock and bell. As Mattie Williams recalled, emancipation was simply "de day when freedom rung out." For others the horn replaced the bell: "[My master] blow the horn and everybody come to the house, he said to us, 'Well you is all free, you don't belong to me no more.'" So central was aural work time to slavery that freedom was measured in terms of freedom from its sound. As ex-slave Fannie Berry of Petersburg, Virginia, put it, "Child an' here's another one we use to sing. 'Member de war done bin when we would sing dese songs. Listen now":

Ain't no more blowin' of dat fo' day horn
I will sing, brethern, I will sing.
A col' frosty mornin' de nigger's mighty good
Take your ax upon your shoulder.
Nigger talk to de woods,
Ain't no mor' blowin' of dat fo' day horn.
I will sing brethern, I will sing.[5]

Freedom's bell sounded sweet because it ended old time and started new.

Thus, the sound of freedom took on new meanings even as it was announced with old sounds of enslavement. "I will never forget the day of freedom," said Lewis Williams: "All the hands was in field when old Marster come from Milam with the news. He blow the horn and everybody come to the house." James Bolton heard the doubleness of the plantation horn: "One mawnin' Marster blowed the bugle his own self, an' called us all up to the big 'ouse yahd. He tole us, 'You all jes' as free as Ah is.[']"6

The coming of freedom was more than a moment of amplified sound, however. Like slave culture, freedom's complexity was too profound to be captured and expressed in one register alone. Rather, freedom was punctuated by silence and sounds of celebration. For Mary Anderson emancipation was an unforgettable crescendo. At the outbreak of the war, "I heard something that sounded like thunder, and Missus and Marster began to walk around and act queer." Then there was silence: "The grown slaves were whispering to each other." Thunder followed again: "Next day I heard it again, boom, boom, boom." Then came the news: "At nine o'clock, all the slaves gathered at the great house, and Marster and Missus come out on the porch and stood side by side. You could hear a pin drop, everything was so quiet." Only then was freedom greeted with "whooping and laughing."7

Freedom was also often a quiet affair, met with secrecy and the gravity of silence that the occasion demanded. This was especially the case with the first whispers of Union presence: "We heard it whispered around that we were free," remembered Sampson Willis, "but it seemed to be a secret." Other slaves used sharp ears to glean news and then passed along messages quietly. "One day, I hears Marse Eden a tellin' he wife dat he guess he 'ud hab to read dem 'free papers' to de niggers in de mo'nin," said one ex-slave from Texas; "I slipped out jes' lak a mouse, an' hunt my Mammy an' whisper an' tell what I done heard." Even the emancipators were sometimes perceived as new, but somehow familiar, enforcers of quietude. Said Gillam Holman of South Carolina of a Yankee teacher who came south to teach freedpeople during Reconstruction, "She have a long switch and she use it, too, if we break de rules. I got a few lashes one day for whisperin'." Within a few hours of learning of their freedom, many newly freed slaves realized that the ties between master and ex-slave could not be severed utterly, at least for the time being. Joyous song gave way to gloomy silence, and "one by one, stealthily at first, the older slaves began to wander from the slave quarters back to the 'big house' to have a whispered conversation with their former owners as to the future." To exploit freedom fully, freedpeople turned to some of the aural strategies they had mastered under bondage. Booker T. Washington, for example, tried to overhear conversations about education while working in a coal mine: "In the darkness of the mine I noiselessly crept as close as I could to the two men who were talking. I heard one tell the other [about] . . . the school established for the members of my race." For most slaves, though, freedom entailed their right and ability to shape the southern soundscape in ways that had been forbidden in the antebellum period. "After the war was over," remembered Milton Marshall of South Carolina, freedpeople "built brush arbors for to hold meetings in." Gone were the old hush arbors. Emancipation meant the freedom to "sing and pray and shout."[8]

Abhorring the new soundscape, elite southern whites did their best to restore their version of a southern soundscape. The end of the war and slavery brought changes to the region's keynotes, changes that spoke directly to white southerners' loss of control over black labor. Susan Bradford Eppes recalled that many hands left her Florida plantation after the war and that their absence introduced new sounds. Unmilked cows now emitted a "piteous lowing," while before they had been comforted into quietude by black hands. For white southerners who wanted to end Reconstruction before it had hardly begun, the debate over freedpeople's rights was, like antebellum discourse over the right to expand slavery, one of "din & turmoil." George Fitzhugh heard in Reconstruction nothing but "indecision, confusion, chaos."[9]

The departure of labor from plantations left whites with fewer familiar aural bearings. Without slaves to ring house bells, plantation mistresses and masters were strangely lost. Freedom for slaves meant freedom from the sound and command of the breakfast bell for whites; it also meant no breakfast. Eppes found the absence of familiar bells and the ensuing silence eerie and confusing. She testified unwittingly to her dependence on the summons of the slave's bell after the war: "The New Year dawned clear and cold. Snuggling down in our blankets we waited to hear the rising bell—it was getting late—could it be we had not heard the bell? Alack and alas, no bell had rung. The white folks were all alone on the hill—not a negro was there—one and all they had gone—stealing away in the night." In an act as pitiful as it is staggering, the family lay in bed waiting for the bell to ring.[10]

The departure of ex-slaves from plantations also meant the reallocation and amplification of sounds for many southern whites because it removed freedpeople from the mechanisms that had governed slaves' aural production. Freedpeople's protracted meetings after slavery lasted days, the participants "making noise enough to really distract every body in the vicinity." Emancipation made freedpeople noisy because it freed them to express their religious convictions with unprecedented volume. Hush arbors gave way to full-throated religious expression, and the pent-up sound of generations of slavery finally found release. Susan Bradford Eppes thought so, especially with freedpeople who lived in the countryside. Urban freedpeople's religion was more self-disciplined and quieter. "We have given you the plantation side of [black] religion," wrote Eppes, "in the towns and cities it is somewhat different." There, black congregations were "quieter, neater"; voices were "well modulated," and the music was tasteful and restrained. Noise and race often blurred, and some freedpeople themselves sometimes found the religious exuberance of neighbors and friends intolerably noisy. Ambrose B. Hart of Florida related the story of a black man who allowed religious meetings in his house: "They had honey figuzzled him with their dancing and hollowing and stamping and jumping up and down and keeping dat ole nigger awake at night just as long as he would put up with it and now they must hunt up some other house to tear up as they had done ruined his already." [11] On the whole, emancipation for elite southern whites marked the end of their quiet, disciplined, orderly way of life. In its stead came a noisy black class and "political meetings of a very excitable character." "The turmoil and confusion of the times must have resembled, to a marked degree," wrote Eppes in 1925, "the Union-labor spirit of today." Finally the noise of the mob had come south. Secession's attempt to further "reduce the Negro to unconditional quiet and submission" had failed. [12]

Sound was as contested under freedom as it had been under slavery. Whites were no longer as confident in their ability to shape the soundscape simply because they could no longer command with certainty the social and economic relations that undergirded it. For their part, freedpeople were determined to test the boundaries of their liberty, and since their aural and oral expression was central to that endeavor, conflict was likely. At times postbellum whites simply surrendered in the face of the new soundscape. When thousands of freedpeople celebrated July Fourth at the South Carolina College Chapel in 1865, whites could hear only "strange Negro songs . . . a terrible noise" and "shut themselves within doors" to mute the sounds of black freedom. [13]

Such apparent capitulation was short lived. Accustomed to years of mastery, southern whites did not relinquish their idealized soundscape and what it represented without a fight. Slaveless masters readily used extralegal force to smother freedpeople's noise. In July 1868, for example, John Schofield approached four black Virginia women singing in their home. Schofield asked them to stop, and when they refused, he beat them.

Aural battles between slave and patrol were resurrected in the postbellum South, but now they were between freedperson and Klansman. Secrecy and attendant silence—social and strategic—were essential to Klan success. The organization's constitutions and bylaws were replete with rules governing silent communication. Those who joined promised "never [to] reveal to any one not a member of the Order of the ***, by any intimation, sign, symbol, word or act, or in any other manner whatever, any of the secrets, signs, grips, pass-words, or mysteries of the Order." Members were admonished to keep quiet: "Hush! Thou art not to utter what I am; bethink thee! It was our covenant!" Just as the Klan's social invisibility relied on silent maneuvers, so its intimidation of blacks and Republicans was premised on its militaristic use of sound. Night raids on Republicans and soft Democrats were coordinated by gesture and audible whistle. Such intimidation "had a most quieting effect on the Negroes," according to one observer.[14]

Market realities also proved conducive to planters' efforts to reimpose social quietude. Labor contracts with freedpeople included provisions that had little to do with work but more to do with the reimposition of an acoustic order that echoed the old master-slave relationship. The first article of agreement between workers and Samuel G. Page of South Carolina signed in 1866 required laborers "in all cases to avoid curseing and swearing in his presence." George Wise understandably had few immediate takers on his plantation after the war. His proposed contract stated: "Impudence, swearing, or indecent and unseemly language to; or in the presence of the employer, or his Family, or agent; or quarelling or fighting; so as to disturb the peace; will [be] fined one dollar for the first offence; and if repeated; will be followed by dismissal." Article 3 was similarly revealing: "No general conversation will [be] allowed during working hours." If ex-slaves refused to sign, others would. On Louis Manigault's Savannah River rice plantation in 1875, Irish laborers proved more than satisfactory: "These men occupy any ordinary negro-house, and are quiet and orderly in demeanor." Francis Butler Leigh's solution to the problem of freedpeople's presence was complete sensory isolation during nonwork hours. She aimed to "improve" her husband's Georgia plantation "by removing the Negro houses away from where they are now, close to this house, to where I can neither see, hear, nor smell them."[15]

After the war old concerns about social order in southern cities were similarly resurrected, not least because they had never gone away. After slavery there was a more dangerous and rapidly growing urban rabble

whose potential for disturbing the peace was high, and southern municipal ordinances during the 1870s targeted lower orders. Atlanta's antinoise
legislation of 1873, for example, was enumerated under a municipal section titled "Peace, Good Order and Morals." Noisemakers included persons who "sing to a public audience any sacrilegious, indecent, vulgar or
lewd song." Anyone "who shall be found drunk, hooting, hallooing, or
making any other unnecessary or unusual noise, to the disturbance of any
citizen" was fined and jailed. Those who worked all day had to be careful how they relaxed at night: "Any person who shall make any noise at
night, calculated to disturb the public peace, or to annoy any of the citizens, shall (if done in the presence of the Marshall) be arrested by him."[16]
Police in Mobile, Alabama, were empowered to deal with "tumult, riot,
insurrections" and to "preserve order, peace and quiet." Imitating "any of
the signs, signals, devices" used by police was still a crime, and counterfeit
sounds were still not tolerated. The use of bells and taps to direct officers
to fires and riots was maintained; the only difference was that city growth
necessitated increasing the signal to eight taps. Precisely the same noises
were prohibited and exactly the same soundmarks for various functions
echoed in Mobile after the war.[17]

Modest changes were afoot, though. A few white southerners reconciled themselves to postwar realities by listening to freedpeople's progress. The editors of the *Savannah Republican* thought that they could hear
the progress of northern-run Bethlehem High School in 1866, and what
they heard pleased them. The school and students took studies seriously,
and a modern South was within reach: "This is the right road to prosperity, peace and reconciliation of both races, that will again sound the
hum of industry, and set the wheel of progress in motion." Similarly,
southern boosters and proponents of the New South began to laud the
sound of industrialism. By the 1880s southern courts were in the process of embracing industrial noise as necessary. An 1883 Alabama injunction against the erection of a steam gin, granted initially because the gin's
"noise . . . disturb[ed] the quiet enjoyment of the adjoining proprietor"
and inflicted "unseasonable noises at unseasonable hours," was later re-
pealed. The courts reasoned that the noise was merely "a trifling discomfort or inconvenience suffered by the party complaining." Besides (and
this was the key), steam-powered cotton ginning "is admitted to be a
useful business, which is common to the country, and one which should
not be discouraged."[18]

For the most part, however, white southern soundscapes remained intact in mind and, sometimes, place. Whatever else the war had done, it had not changed fundamental beliefs among the region's ruling class about the nature of southern society. From Camp Lee, Virginia, in 1866, George Fitzhugh found satisfaction in being in the countryside, "where the vulgar hum of industry is never heard, where intrusive visitors seldom disturb, and where silence, peace and quietude reign supreme." The southern response to northern wartime gestures at restoration continued. As one southern newspaper editor blustered during the war, "The Southern ear is inexorably shut against all overtures which look to reconstruction. The Southern man who listens to such a whisper is a traitor." Such sentiments echoed for years.[19]

After the Civil War, northern elites differentiated more sharply between the aurality of democracy and that of capitalism. For most elite antebellum ears, the two remained indistinguishable, which is why abolitionists and Republicans could agree on the necessity of liberal capitalism generally. But northern labor unrest during and after the war encouraged closer listening to different registers. Conservative and moderate Republicans during Reconstruction still applauded the sound of industrialism but became increasingly critical of the political and social noise of labor. The working class's increasingly loud critique of industrial capitalism and its willingness to use organized labor and full-throated democracy to challenge it encouraged even some liberals to make the shift. The beginnings of pro-industry/antilabor aural distinctions were partly antebellum in origin but principally a product of developments during the Civil War and Reconstruction. Certainly, wartime protest in 1863 pricked already wary ears, but what really consolidated this new aural construction was the panic of 1873 and attendant labor unrest, whose dissonant echoes were in some ways louder than the registers of victory of 1865.[20]

For many postbellum boosters the turmoil of war had been worthwhile. After all, the West now echoed with the sound of capitalism. In San Francisco in 1883, "every street, lane, and back lane [was] a ceaseless buzz of industry." The immigrant workforce had learned the discipline of capitalism, too: "Twelve o'clock comes. At each stroke of the bell pour out squads of docile workmen." Neither had the war interrupted the essential sounds of northern progress. "Wars may come, the elements may be unpropitious, crops may fail," gushed one promoter of Indiana in 1867, concluding, "No matter! The prosperity of the city is not retarded. The hum of industry goes on. Labor reaps its full reward." This was the sound of ineluctable uplift. On Thanksgiving 1876 Rev. Benjamin Arnett of the Ohio African Methodist Episcopal Church applauded "the busy hum of industry" he heard undergirding U.S. material progress. In 1869 J. W. Meader similarly gauged New England's progress over two centuries when he listened to manufacturing towns in New Hampshire. Of Manchester he said, "Pavements resound with the tumultuous rush and rattle of busy strife in place of the stealthy trail of the lurking savage; where monster mills send forth the pleasant hum of industry to take the place of the war-whoop." It was only a matter of time until these soundscapes worked their way southward. If Native American sounds of premodernity had been muted, why not those of a defeated South? That West Virginia in 1870 "should lack inhabitants, or the hum of industry, or the show of wealth" was now "an absurdity." Just as the "inquisitive Yankee" would inscribe the woods of a sparsely settled Wisconsin in 1875 with keynotes of liberal capitalism and ensure that "the wooded hills resound with the hum of industry," so, too, would he rescue the prostrate South from quiet oblivion.[21]

Some old northern standbys of aural definition persisted into the postbellum period. The depression of 1873, for example, silenced northern economic productivity, which in turn caused observers to lobby for a system of government-backed internal improvements. Only then "will the busy hum of industry and the noisy murmur of prosperity be heard again in the land." In Nebraska, where the hum of industrialism was supposed to resound, a slew of misfortunes had reduced the Nebraska City Distillery to silence in the early 1880s. The "whole place" should "be sounding with the pleasant hum of industry, instead of lying an inert ruined mass, a tacit reproach to the town."[22]

The mechanization associated with the Age of Capital, "a period of unprecedented economic expansion presided over by a triumphant industrial bourgeoisie," was entirely appropriate because it seemed to promote the comforting sounds of economic expansion while quieting an increasingly boisterous laboring class. "No corn raiser of the North plants by hand," maintained Edwin Q. Bell just after the war, explaining that "even the ungainly potato has been reconciled and is now quietly and regularly inserted in its bed by the unthinking machine."[23] While elite northerners heard much to reassure them by 1873 (the registers echoing the nation's 75 percent increase in industrial production since 1865 must have sounded musical), they also heard strains that worried them enormously. Unprecedented levels of immigration and urbanization and an increasing maldistribution of wealth accompanied rapid industrialization. Wage earning was becoming a permanent condition of the laboring classes, and with that status went elite fears of social dislocation and unrest. Worsening economic conditions in the postwar period were not lost on skilled and unskilled workers, who increasingly organized to combat wage slavery with what some called a "moral economy." "The masses," declared a union leader, "will never be completely free" until they had "thrown off the system of working for hire." Echoes of wartime protests rang loudly in 1871, with rioting in New York and bloodbaths in Paris. The comforting twang of liberal capitalism, it seemed to northern elites of all persuasions, was beginning to take on the distinct cacophony of class warfare. Demands by workers for government-sponsored economic and social legislation to protect them from the vagaries of a volatile market encouraged even liberal reformers to insulate politics from working-class influence. Even before the tumult of 1873, many reformers "retreated from democratic principles altogether, advocating educational and property qualifications for voting, especially in the large cities, and an increase in the number of officials appointed rather than elected."[24]

The panic of 1873—the first crisis of industrial capitalism—heightened anxiety among northern elites, hardened their reservations about free labor, and undermined their faith in inexorable progress. By the end of the year about half of the country's iron furnaces were silent, and the disquiet of economic recession grew in volume. Handmaiden to the silence of economic recession was the growing voice of labor and, more disconcerting still, the march of thousands of onomatopoeic tramps who roamed the country. The unemployed demonstrated in the North and Midwest. In 1874–75 textile workers, railroad laborers, and miners went on strike and used their collective voice to protest worsening conditions. The antagonism between labor and capital reached a head, and northern elites heard with clarity what proslavery ideologues had augured: the terrible throat of a property-threatening mob.[25]

Sounds once considered vaguely distasteful from the antebellum period echoed loudly in the context of this crisis. Tramps and beggars came in for particular criticism. Shiftless, indolent, and dangerous, this class of roaming scum, intimated even philanthropists, "whine at your houses," assume a "hoarse voice" to deceive charitable souls into giving alms, and were "clamorous for bread." Divorced from the disciplining tendencies of the wage labor market, beggars represented a serious threat to order. Perhaps the democratic, capitalist experiment had gone too far, some suggested. Elite men found the comforting strains of their worlds attacked from all quarters, and even their customary custodians of social quietude seemed less than reliable. What had been a voice in the late antebellum period from northern feminists was now a deafening roar. When Elizabeth Cady Stanton and others insisted that the logic of equity under contract also apply to marriage, northern elites shuddered. Wendell Phillips called Stanton's critique "noisy [and] alien." Even whispers concerning the right to divorce created horrible echoes in patricians' ears. "Were this discussion admitted," commented one observer sympathetic to equality in marriage, "conventions would be a mob, and it would be years before the riot would cease sufficiently for those who demand nothing but equality, to be heard." To many postbellum northern patricians, social dislocation was conveyed through what they heard, and in all instances they heard the cracking of an older, more paternal, more disciplined form of capitalism that while tolerant of the shout of democracy had always been wary of the voice of the mob.[26]

Disconcerted though they were, northern elites had an aural architecture on which to rebuild and stabilize the postbellum soundscape. Most obviously, they returned to colonial and antebellum legal statutes. Further industrialization in the postbellum period amplified noise in northern cities, and just as they had in earlier periods, municipalities went to some lengths to quiet the urban soundscape and define noise. Civic authorities passed a flurry of antinoise legislation directed not against the pulse of economic modernity but the growing noise made by laboring people. This class dimension to aurality is illustrated clearly in the legislative underpinning of noise pollution in late-nineteenth-century New York. Antinoise regulations were construed along instrumentalist lines in favor of what industrial capitalists—and those they had in their pockets—considered useful sound and nonuseful or unnecessary noise. The rhythmic timbre of the factory represented progress, testifying to the desirability of U.S. industrial capitalism. The sounds generated by workers, conversely, were noise. Singing, frolicking workers' determined exploitation of precious leisure time—as delimited by the factory bell—and the sounds of such activities were discordant strains, unfortunate by-products of the industrial machine. Consequently, postbellum New York's antinoise legislation targeted not the greatest quantitative noisemakers, the factories but, rather, the working classes.[27] When citizens in Massachusetts complained in 1884 that local factory bells rung "at 5 o'clock in the morning" disturbed the peace, the court sided with the mill owners on the ground that "the benefit [of alerting factory operatives] was greater than the injury and annoyance."[28] Similarly, where the courts had sometimes enjoined railroads from disrupting church property and peace in the antebellum period, they now found that the "interruption of the worship of God by complainants in their accustomed places of public worship or in their own residences on the Sabbath day is not such a private nuisance as will be restrained by injunction." So, at least, ruled a Pennsylvania court in 1867.[29]

Antebellum ordinances echoed loudly. Statutes regulating and prohibiting the public playing of music and peddling were introduced, revised, or refined. Peddlers in particular came in for renewed regulation since they were considered on the margins of wage labor discipline. While auctioneers were encouraged to be "loud" in describing the quality of their goods, less respectable, common peddlers were noisy because they were of the lower orders and their shouting of wares disrupted legitimate commerce. But even an auctioneer "or other person, who shall . . . cry for sale . . . within the streets, alleys, or commons of the city of Chicago," according to an 1872 ordinance, "shall be deemed guilty of a nuisance" without "written permission of the mayor." Vagrants were similarly defined as those "who shall disturb" schools or churches. Even porters who made "any unusual noise or disturbance" during the execution of their business were similarly penalized, as were people "in or about any railroad passenger house" who made "any loud noise."[30]

Anyone "making any improper noise, riot, disturbance, breach of the peace, or diversion" was similarly frowned upon in Chicago and could be fined up to $100. The price of disturbing God's sound and silence was slightly cheaper. Those "who shall disquiet or disturb any congregation or assembly met for religious worship, by making a noise" that "disturb[ed] the order and solemnity of the meeting" were fined up to $50. Civil authorities did the listening for the urban bourgeoisie. Chicago augmented the powers of the police in the 1870s by empowering them to arrest anyone violating the "Order, peace and quiet" of the city. The definition of noise was left to the discretion of the police officer.[31]

Although it was doubtless difficult to enforce, some of this legislation was essential to burgeoning urban growth. In 1872 Chicago allowed essential municipal authorities, in this instance the fire brigade, to perform their duty free of howling spectators whose din might render important orders inaudible. The ordinance empowered fire marshals to arrest "any person hindering, resisting, conducting in a noisy and disorderly manner."[32] Lest the "discharge of any cannon, gun, fowling piece, pistol or fire arms of any description" also muddle the city's soundscape, Chicago prohibited such activity. Apparently the problem was serious because an additional ordinance was explicit in its intent to reduce "FALSE ALARM" and "STREET NOISES." In an echo of the Old South's urban ordinances, persons "imitat[ing] any of the signs, or signals or devices adopted and used by the police department" could be imprisoned for up to three months. Anyone who "shall willfully give or make a false alarm of fire or watch, or who shall employ any bellman, or use or cause to be used any bell, horn, or bugle, or other sounding instrument, or who shall employ any device, noise or performance tending in either case to the collection of persons in the streets, sidewalks or other public places, to the obstruction of the same," would be fined. An idea of how serious Chicago's authorities considered this particular offense may be gathered from the $25 fine, while selling deadly poison carried a penalty of only $5 for each offense. Functional alarms produced sound and necessary noise; unnecessary signals were dangerously misleading to civic authorities, could cause crowds to congregate, and contributed to already clamorous "street noises."[33]

Some sounds of northern progress were also subject to qualification, and older sounds were sometimes preferred to newer ones in the process of that redefinition. "No railroad company," for example, "shall cause or allow the whistle of any locomotive engine to be sounded within Chicago" except when the whistle "may be absolutely necessary to prevent injury to persons and to property other than their own." The locomotive's bell was given a freer hand, presumably because it was deemed less grating, historically legitimate, and equally useful in preserving life, limb, and property. In fact, Chicago railroad companies had to erect "a bell tower not less than fifteen feet in height" to warn citizens of oncoming trains at crossings over the din of city noise. Citizens willfully ignoring the aural cue by dallying on the crossing "after being notified by the ringing of the car bell" were fined if still alive. Ordinances tried to enforce the coordinating function of bells in postbellum cities by regulating the sound of industry and mandating the aural obedience of citizens.[34]

Ideally northern elites wanted respite from the new tumult and rasping of boisterous wage labor capitalism gone awry. They wanted to hear both quietude and older sounds of progress that resonated with the gentle hum of industrial activity and harmonized with the timbre of virtuous workers. For some, the place to rediscover such keynotes was not in their own society but to the south, which echoed their own antebellum past where labor supposedly worked rather than rioted, and where the early strains of industrialism offered a comforting soundscape free of the noise of class discord.

So disenchanted were northerners with the modernist excesses of their own society that at times they reveled not so much in the reassuring register of a slowly modernizing South but in what they perceived as the "noise" of "doing little work" on postbellum plantations. The South offered northern elites tired of class conflict, social dislocation, and general freneticism a "welcome retreat from northern modernity." Because northern elites were never entirely comfortable with surrendering their own work ethic and the society they had made and fought so hard to advance, they tended to listen for comforting echoes in the South, registers that offered an escape from modernism but assured them that the South was beginning to modernize along a trajectory they had chartered in the antebellum period. What they wanted to hear, in other words, was southern quietude combined with a reassuring keynote of gradual, sustained, and controlled modernization and sober freedom. Recuperating physically and mentally in a place that sounded more feminine because it was quiet was music to their bruised ears.[35]

Postbellum northern evaluations of black southern free labor were, compared with abolitionists' scathing critiques of slavery, far more benign and betrayed a certain comfort with the realities of the postbellum South. Northern listeners heard something to covet in southern fields. Instead of grisly slashings and clankings, they now reveled in freedpeople's singing and found soothing "an occasional snatch of some camp-meeting hymn." They thought Florida especially a soniferous garden and healthful retreat from the noise and din of northern industrialism. In some respects Florida had always been a sanctuary of quietude. At the height of the Mexican War, for example, St. Augustine was considered a quiet alternative to "the din of war," and in 1853 it was reckoned that Key West's mild climate lent itself to a natural "tranquility." But now Florida sounded even better because slavery had been silenced. Even the state's nonnative population sounded sweeter than the noisy rabble from Eastern Europe currently being dumped on northern shores. A northern writer in 1868 characterized St. Augustine's Minorcan population as a "very quiet, peaceable and orderly people."[36]

The physical desolation of the postwar South allowed promoters of the quiet South to ground some of their claims materially.[37] Boosters bent on making the South a favored destination for harried Yankees frequently used the South's bells as a metaphor for sectional rapprochement as well as a cure for nervous dispositions. In 1885, for example, Charleston's St. Michael's bells were made audible in a nifty little piece of southern holiday propaganda, *Winter Cities in Summer Lands*. The church's bells, the pamphlet noted, had previously acted as "war's rude alarums" when "they were molded into cannon, and spoke in tones of thunderous anger, rather than in peaceful call to worship, or were broken and hid away that the same precious metal might be made into bells again." "They ring out now the same sweet sounds they did in days of yore." Past grievances hardly echoed in this new world, and what had been noise to antebellum northerners was now repackaged as the sentimental sound of an aurally neutral past. Some promoters made even St. Augustine's old slave market antiseptic: "Whenever a *sale* was to take place the bell in the cupola would be rung to notify the public."[38] The sound of aged pedigree was peddled vigorously, and St. Augustine's cathedral bells were especially noted. These oldest of American "cathedral bells," it was declared, "have chimed a lullaby through all the hours."[39] Tourists were also lured south by quiet and pleasant sounds, which were invariably pastoral and often feminine. Mockingbirds "and many a companion songster," one 1885 pamphlet assured all visitors, "fill the days and even the nights with melody," and Virginia's "feminine grace" was confirmed by the presence of relaxing "silent pools" and springs.[40]

Florida's promoters were acutely aware of the significance and regional nature of soundscapes, and in crafting and polishing the state's aural attractions they did not make the mistake of denigrating northern soundscapes. Southerners had learned much about aural dichotomies in the antebellum period, and they now applied the lessons thoughtfully. Rather than luring northerners with the blunt binary of a noisy, boisterous North versus a quiet and sedate South, they projected Florida as having a soundscape that was attractive but not alien. "There is a merry jingle in the silvery sound of sleigh-bells which fascinates and makes us love to linger in the icy Northland," admitted Reau Cambpell in his 1885 *Winter Cities in Summer Lands*. The southern purl was warmer, though: "But there is a soft melody in the whistle of the mocking-bird to charm and make us forget it all in the land where every day is Summer." Anyway, Florida was simply more fun and provided a harbor from northern disturbances. In Jacksonville one could hear from the river "a melody of laughing voices, blended with the splash of oars." Hot climates demanded open windows at night, but only melody, not noise, would drift in: "From the shore, through open windows, there may be heard sounds of revelry." Even southern peddlers sold in euphony, and visitors were urged to listen to their "seductive voices."[41]

Promotional literature did not laud only quietude, however. Those responsible for targeting northern vacationers knew too much about northern ears to be so simplistic. Sponsored by railroads, real estate companies, and the fledgling tourist industry, Florida's postbellum promotional literature identified the preferred soundscapes of "Tourist, Settler and Investor." As a result, the literature featured the preferred quiet of those fatigued by northern noise but also tried to reassure investors that business was humming, if not booming.[42]

Not all was pastoral sotto voce. The sound of the modern braided with the trill of nature to produce comforting euphony, postbellum southern style. Pamphlets were littered with pictures of "the train in motion" running through whispering Floridian pines. Should the scenic beauty of the state distract attention from its progress, "the sounding gong recalls us from our contemplation" and reminded visitors to listen to the train whistles of Florida's developing economy. Anxious to promote the state as a dynamic place friendly to northern investors, promoters resurrected old sounds of progress. Concerning urban development on the outskirts of Jacksonville, for example, one booster in 1889 echoed sounds from the early part of the century by claiming that towns had "been laid off in lots, and the sound of the hammer and saw are heard in every direction." And after a hard day of investing in modernity, where better for northern moguls to relax than in one of Florida's fine hotels equipped with "Electric Bells"?[43]

Quietude, however, figured most prominently in promotions for the New South. Come to Jacksonville, a slew of late-nineteenth-century pamphlets cooed; come to a place that is "home-like and quiet, yet sociable and progressive." Come to a region that mixed the best of Old South serenity with New South progress. If Jacksonville were not to one's taste, then the "shaded nooks and quiet corners" of St. Augustine surely would be. Old sounds rippling from St. Augustine's "ancient belfry and chime of bells over two hundred years old" assured visitors that they would experience the sounds as well as the sights of the South's ancient past. But the sound of the past had to be made easy to hear, particularly for invalids. Consequently, travel in Florida was a quiet affair because one could simply glide on "roads . . . smooth and hard," and sandy lanes ensured that "the carriage rolls noiselessly along." Should visitors feel hardy and covetous of returning to nature, they could always stay in a seaside town, such as Rockledge, where "on its outer edge beats the Atlantic, the roar of the surf coming distinctly to the ear." Now that slavery was gone, Florida was a place of "quiet beauty," the ideal refuge from the noise and bustle of northern social and economic dislocation.[44]

A nervous northern population had much to gain from the "quiet and cleanliness" of St. Augustine's streets. The town was an acoustic paradise, home to "quiet warmths of sunlight," where even the instruments of noisy war, "the wheels of the cannon," became "trellises for peaceful vines." Florida's hotels were among the best for frazzled Yankees because they promised "freedom from noise."[45] The countryside was even better. Hinterlands had their own melody, sometimes described as "music which is wild and weird." Rivers whisked the adventuresome to eerie marshes whose polyphony would astound and stun them into awed silence: "The splash of the wheels making the waters ripple under the lilies; the cry of a frightened bird; or out in the darkness, the whip-poo-will's call, wooing its mate—all combine to produce a strange, wild melody."[46]

Although Florida was the acoustic dream, other places in the South were also lauded for their harmonics. Age gave Charleston its pedigree, which was echoed in ancient bells. Visitors were told "the chimes of St. Michael's bells will ring softly from the distant spire, a lullaby, which shall be your reveille as well, when another happy day is born." The South's dulcet tones generally, its "placid" and "balmy" sounds, were enough "even to forgive a carpet-bagger."[47] Sounds helped facilitate political reconciliation.

Northern travelers did not always realize promises of postbellum southern quietude. One correspondent for Pennsylvania's *Trinity Record,* for example, found Jacksonville's soundscape worryingly northern. In 1887 the correspondent began a series of southern sketches with the observation, "Those of my readers who have crossed the Jersey City Ferry at New York and have been assailed by the yelling mob of hack-drivers and hotel runners . . . can form some idea of the tumult which awaits one on alighting from the cars at Jacksonville. . . . The English language is hurled at one in stentorian tones in almost every form of accent and dialect conceivable." Jacksonville was no better. It was "a city out of place . . . a Yankee town on southern soil."[48] In a climate where the "thermometer stands at 70 and people sit with their doors and windows open," sound and noise disrupted domestic peace. Erastus Hill of Chicago found as much when he relocated to Florida in the late 1870s. "I was somewhat disturbed in my rest," he wrote from Gainesville, "by the niggers in the night and some other things. I realized more fully than ever before the force of the saying that Saturday night was nigger night. They kept up a shouting and holering till I went to sleep." In a pre–air conditioned, hot climate, freedpeople gathered outside at night, and their sounds drifted into houses whose windows had to remain open if their inhabitants were not to feel suffocated. Blame rested partly with transplanted Yankees themselves. "The hotels are all northern property," wrote L. D. Huston of Palatka, Florida, in 1874, which helped explain the attendant increase in southern urban noise: "O, the crowds, packs, caravans of people! Cough, cough everywhere." Advertising and promoting the quietude of the postbellum southern soundscape was one thing; trying to maintain it in the face of quiet-hungry, noisy northerners was another.[49]

Nor was the southern reception of northern visitors always as cordial as promotional literature suggested, especially in the early postbellum years. Northerners in the South in the immediate postbellum period held their tongues as tightly as they had during the war. Taciturnity was the prime defense against southern resentment. Accent and politics gave these visitors away, and for many silence was the preferred choice. So said George Franklin Thompson, a Freedmen's Bureau inspector who traveled to Florida in 1865. His first encounter was with a landlord in Jacksonville who, Thompson believed, had failed to wake him at the correct time. Fuming, Thompson "turned to a very quiet man seated on a bench on the porch and directed my conversation to him in particular when he very quietly observed, 'you will get used to such things after you have lived in this country awhile.' Never! I said, I would rather die than be so stupid." The remainder of the conversation is worth quoting precisely because it was never actually heard beyond the two men. Thompson continued:

> Then you are not a native of Florida are you? Oh, no. I was from New Hampshire and came to this country fourteen years ago. . . . I informed him that I was from Massachusetts which seemed to unloosen his tongue, and looking first to the right and then to the left to see if anyone was watching him or listening to the conversations, he said "come this way I don't like these people to see me talking of these things for the fact is, the Election is but just over and they are like a hive of bees or nest of hornets which has been stirred up with a sharp

stick and they are buzzing around trying to find victims upon whom they can vent their spleen."

When spiteful southerners buzzed like wasps, northerners fell silent.[50]

Then there were the echoes of slavery and northerners' own racial prejudice. Northern teachers who ventured south just after the war found a formerly enslaved population whom they deemed able to learn quietude but who had been rendered tumultuous by bondage. When Margaret Newbold Thorpe, a Quaker missionary who taught at Fort Magruder in Virginia, first encountered freedpeople, she expressed such sentiments precisely. "Their respect for 'white ladies['] awed them into quietness." While this suggested that freedpeople could learn the civility of silence, "with each other they were like snarling puppies, not so much from ill humor as the result of a training that had taught them neither self respect or respect for any one but those with white skins." From Thorpe's perspective, Old South training in quietude failed dismally because it was imposed from without rather than cultivated from within (as had been the intention of the silent system of antebellum northern penitentiaries). During 1866, when she became more acquainted with her "pupils," things seemed to improve a little. Thorpe delighted, just as had masters, in "listening to their dear and often beautiful voices singing the weird plantation hymns and songs." And she was grateful to "Griffen a powerful man with corresponding voice," who as "some of the people began to sound an alarm [of fire] . . . asked had they 'no more 'sideration for the ladies than to have them skeered by the cry of fire.'" But the sounds of the New South and the recently freed still hung heavily on Thorpe's ears. The "constant wailing and groaning" of spirituals grated on her, and she was horrified at the answer of a young boy when she asked him to explain his bruises: "Mammy done lick me 'cause I wouldn't shut up the racket." "Scratch the skin of a slave," she sighed, "and you will find a tyrant."[51]

Northern travelers to the postbellum South sometimes heard in freedpeople what they heard in their own laboring classes. They spoke of encountering "a crowd of rampant, bawling blacks" at the Charleston terminus in 1872. Ambrose B. Hart of New York, who bought a plantation near Lake City just after the war, considered southern freed blacks as noisy as northern free whites. His letters give the unmistakable impression that he considered his workforce noisy and lacking the self-discipline of quietude. Hart found even the tone of a bell aggravating when pulled by freedpeople. "The negroes are called together at their meetings by the ringing of a large cow bell," he wrote his brother in 1868, revealing the essential cause of his antipathy. "It has just now sounded for afternoon service and they will soon be in full blast." What Hart disliked most was the noise generated by African American religion. No longer required to scurry under hush arbors and murmur into upturned pots, the South's freedpeople released centuries of silence. Hart found it infuriating. He, after all, was a man who appreciated "that blessing and quiet satisfaction—church." His workers did not, as theirs was a religion of shouts and of "howwowing, screeching, dancing, stamping and jumping up and down." Even their secular sounds were noises that told of indolence and indifference to progress. "These negroes," he wrote in 1867, "are the most inordinate guffawers I have ever heard of. They are laughing, yelling and singing from morning until night. They have no care on their minds." African Americans, felt Hart, were, at best, noisy like children; at worst, "debased." "The nigs," he wrote in a little piece of nastiness he called the "Negro Character" in 1870, "sing, shout and go on worse than ever. It seems as if they are actually going back into barbarism." They were in a state of perpetual "perfect uproar," unable to quiet themselves and appreciate the solemnity of silence.[52]

The irony is that men like Hart did much to pollute the New South's soundscape. Hart himself built steam engines for ginning and grinding, which he sometimes kept going all weekend. For all of Florida's noise, though, Hart could not quite free himself from the romance of the southern soundscape. In 1868 he sounded like an antebellum planter who gauged industriousness and appropriate productivity with ears different from those of his northern counterpart. As he wrote to his sister, "Thirty one plows are busily at work in the fields. . . . The sounds are the singing of birds and the incessant crowing of cocks who send forth their notes in every direction as every negro family keeps more or less of chickens. The anvil is sounding in the shop only a hundred yards from the house while the neighing of a horse or the braying of an ass occasionally awakes the echos far and near." It was aural bliss.[53]

What are we to make of all this listening to what was heard? Obviously, a thousand sounds that were heard have not been listened to in this study, and much work remains to be done on rescuing historical aurality from the grip of ocularity as well as exploring the significance of the other senses in nineteenth-century America. That much said, this study invites several conclusions. Methodologically, listening to what was heard proves essential if we are to understand how ruling classes constructed their sectional consciousness and defined and exercised their power. The diffusion of printed words and texts in the first half of the nineteenth century did nothing to detract from the authenticity of the heard world. Endless spoken aural descriptions and images were recorded permanently in print and not only kept aurality alive but amplified its importance by enabling the mass communication of aural sectional identity and consciousness. These constructions and the articulation of sectionalism, heard so clearly in the context of political debate in the 1840s and 1850s, owed much—indeed, cannot be properly understood if read apart from—elites' everyday aural environment. How elites heard northern workers labor and resist, how they heard the forms of slave expression and the silences, sounds, and noises embedded in these and other mundane social soundscapes cannot and should not be divorced from the political, sectional debates that led to civil war. The heard worlds of antebellum America were at once social, cultural, economic, and political, and not listening to the past in these multiple, interrelated contexts impoverishes its texture and denies us access to new understandings and reaffirmations of earlier interpretations, particularly those that stress the importance of slavery to the creation of sectional consciousness and the coming of the Civil War. Listening to the heard past in its intimate and messy entirety is at once ambitious but necessary if we are to avoid becoming hostage to the deafness of intellectual and scholarly parochialism.

Interpretatively, listening to the heard worlds of nineteenth-century Americans shows that modernity; capitalism; freedom; constructions of gender, class, and otherness; and the content and expressive forms of southern and northern nationalism contained distinct and meaningful aural components. Workers especially understood the power of silence and the control of sound as tools of effective resistance to their enslavement and exploitation. None of this was lost on northern and southern elites, who gauged the successes and failures of their own societies aurally as much as visually, hearing threats to the integrity of their communities from within and, increasingly, from without. The aural representations, metaphors, and analogies sectional ruling classes employed to critique one another were clumsy and binary. For southern ears the North was noisy, and for northern hearers the South sounded at once worryingly silent and disturbingly cacophonous. The descriptions used to convey the various meanings contemporaries attached to sections could hardly communicate the entire scope and range of sounds in the region that they purported to represent. But by the closing decades of the antebellum period, contemporary ears heard the aural symbols as precise, authentic, and utterly real. Used to hearing "otherness," and familiar with constructing and manipulating core values and social and economic relations aurally within their own societies, sound-sensitive elites heard sectionalism as material and ideological differences between North and South became more pronounced. As such, notions of "the North" and of "the South" gained terrible credibility so that ruling classes heard one another in ways they did not like, even though in reality some of those differences were not always as great as they liked to hear. Images of the South and conceptions of the North crystallized because of contemporaries' hearing, and it is the nature of their hearing that helps explain how sectional identity became meaningful and antagonistic. Whatever modern registers were heard in the Old South, whatever tones of mild conservatism tinkled in the liberalizing North, and whatever aural similarities the two regions shared, they were far too faint to mask an increasingly loud, emotional, and resoundingly authentic sectionalism that rendered the soundmarks of each region at once intensely important to the socially powerful men and women who lived there and violently and emotionally repugnant to those who did not.

It was noise that moved the multitudes, held the public ear, and like magic, swayed the public heart.

—L. H. Holsey, *Autobiography, Sermons, Addresses, and Essays,* 1898

(((Sound Matters: An Essay on Method

Since this study argues for the importance of hearing and listening in a context that traditionally stresses seeing and looking, I hope I may be forgiven for using a mixed metaphor to profile the point. Writing about heard worlds of the past is not unlike groping for illumination in a dark room and finding all of the light switches at once. The room is dark because of the scarcity of readily adaptable conceptual frameworks that historians ordinarily use to help them shape evidentiary categories and arguments; the lights blind and dazzle because the evidence for the heard past is ubiquitous, powerful, and deliciously evident once listened for. The historian of aurality confronts a dilemma: turn on all lights at once and risk befuddlement and disorientation by the cacophony of sounds past, or flip a selected few switches, enough to light principal historical keynotes that contemporaries used to navigate and construct their environments, others, and their senses of self and place. At risk of emphasizing certain sounds, wittingly silencing a good deal, and not posing the heard and seen worlds as oppositional, this study turns on a few, key switches to shed light on some of the most important sounds and how they were heard by contemporaries in the United States in the nineteenth century. A few of these sounds are familiar, for they pertain to vocability and the much-privileged world of music (itself paraded as the art of sound in so much of Western culture). But voices and music are treated not so much discreetly (as has been the tendency in most historical writing) but, rather, as some sounds among many. I am most interested in the nature of the broader aural fabric and how it was constructed by elites; how it was contested by workers and slaves, and particularly how that contestation affected the worldviews of ruling classes in the antebellum North and South; and how soundscapes were heard along increasingly sectional lines.[1]

Think of what we miss if we do not listen to the heard past. Then, as now, what was seen and what was heard were related intimately. Mute a film or television program (one that is intended to be heard). The images are less potent because the visual is less meaningful without the acoustic. The timbre, register, and score of the film are what give it excitement, appeal to our emotions, and impart dimensionality and flavor to the visual. Now mute history so that the events of the past are recounted solely through the eyes of its participants. Missing and distorted are texture, meaning, and depth, not to mention storylines themselves. Indeed, new storylines—causal and explanatory—begin to emerge once we start to listen to hearing.

Historians are just beginning to understand the heuristic poverty of not listening to what was heard in the past and of continuing to practice ocularcentrism. Recent and forthcoming work notwithstanding, historians have been inclined to examine the past through the eyes rather than the ears of historical actors and have tended not to listen to (and, ergo, not hear) the aural worlds of the past. Thanks principally to some fine studies of deafness, we probably know more about the unheard past than we do about the heard, and so it seems entirely reasonable to begin to add acoustemology to visuality in our investigations of the past.[2]

There is no compelling reason why we should "read" the past solely through the visual in order to gain "perspective" or why we should fixate on the postmodern "gaze" of hegemonic authorities' world "views" (even our language betrays a fetish with the seen and comfort with ocular tropes). Several people have suggested reasons for this obsession with visuality. In the context of his fascinating discussion of colonial "soundways," Richard Rath notes, "What people did to construct and respond to their soundscapes is well-documented, but because of the focus on vision in the late modern academic sensorium, it has been with a few exceptions, ignored outside the realm of speechways."[3] Douglas Kahn explains further in his fine and, to some art historians no doubt audacious, study of the aurality of avant-garde twentieth-century art. "Sound inhabits its own time and dissipates quickly," observes Kahn in his effort to "hear past the historical insignificance assigned to sounds." The ephemerality of sound—even while contemporaries clearly lived in and recorded in writing the meaning of what they heard—points to another reason why historians have been slow to hear explicitly what Kahn calls the "historical register." The early modern revolution in print culture, the assumed and growing ascendancy of vision in the eighteenth and nineteenth centuries, and a modern tendency to privilege sight over sound has, quite wrongly, inclined us to expect historical actors to appreciate and wring meaning from their worlds visually rather than aurally. Nineteenth-century Americans did not doubt their ears and trust their eyes as much as we might think. Sight and sound were not always oppositional, and contemporaries attached authenticity to what they heard both literally and figuratively. Print culture, laden and infused with aural representations (ranging from forthright metaphor, to metonymic transfer, to simple symbolism), was not necessarily a silent medium, as nineteenth-century orators well understood. By listening to how contemporaries heard their ideological and physical environments and understanding how that aurality was communicated, we rescue the legitimacy of historical aurality and the ephemerality of sound and its social meaning as constituted in the past lives of various peoples.[4]

When soundscape theorist R. Murray Schafer asked, "What is the relationship between man and the sounds of his environment and what happens when those sounds change?" he posed a historical question. A history of how certain sounds made by particular peoples and objects have been constructed aurally may be tackled in several ways. One could examine sound in the objective: a proposition that posits history becoming louder as society modernizes and industrializes. Thanks to exhaustive work by the World Soundscape Project, it is theoretically possible to demonstrate that the United States became louder as it modernized by measuring increases in decibel levels (levels in sound pressure) over time. But such an emphasis serves to deafen us to the social and historical construction of the heard world. Instead, and again following the lead of acoustic theorists, I rely largely on subjective, cognitive understandings of sound and the heard world in an effort to historicize and contextualize the matter.[5]

Contextualization, as Hillel Schwartz has shown, is critical to understanding the meanings of historical sounds. Although we live in a world where the definitions of noise and sound parade as axiomatic and exquisitely neutral (as reflected in the specious objectivity of, for example, noise ordinances and decibel gauges), in reality we have inherited a bourgeois rubric of what is noise and what is sound, and to assume that the past was the same gives historical transcendence to that construction. If understandings of modern sound and noise are in part invented, what of past auralities? How were sound and noise defined when decibels had not been devised and could not be measured in any positivistic manner? What people heard was assessed subjectively within the context of specific power relations, which were shaped by the shifting peculiarities of time and place.[6]

The need to contextualize the heard worlds of the past is apparent if we attempt to treat sound and noise as objective, transhistorical phenomena. Consider the nominal objectivity of noise and sound. Scientifically, ambient noise levels are measured in decibels for a given period of time. The intensity of sounds is gauged in decibels on a logarithmic scale of power, and the range of sound intensities accommodated by the human ear is expressed on a scale of 120 dB (any intensity above about 120 dB, known as the threshold of hearing or the threshold of pain, can cause hearing loss).[7] While most modern societies define noise quantitatively as well as qualitatively, no definition has proven wholly satisfactory, not least because the subjective interpretation of what is noise intrudes on objective efforts to define it in terms of weighted frequencies and decibels.[8] By some definitions, noise is simply any aperiodic signal, but by this standard, transistor radios cannot be considered noise. Describing noise more qualitatively as an "imbalance" in the soundscape or as "unwanted sound" begs questions. By whose hearing is it an imbalance? By whose definition is it unwanted?[9] Even the claim that noise is invariably pejorative is misleading. We still speak of quiet noises, for example, and Psalm 95:2 counsels us to "make joyful noise unto him." Inevitably, then, we deal with generalizations, and we must define noise in the aggregate. Noise is unwanted sound, and the trick is to identify who considered it wanted and unwanted and why in a specific historical context.[10] None of this was lost on nineteenth-century Americans. Contemporaries recognized that sound was socially constructed and mediated. "Music," explained one writer for a New York newspaper in 1812 "is the science of sounds." "But," he queried, "what sort of sounds[?]" He answered his own question by arguing for the subjectivity of sound: "The sound that would be music to one ear, perhaps to another would be more grating than the creaking of a wagon wheel."[11]

Historical soundscape studies should not ignore the physical basis for the sounds that were heard and interpreted, and they should necessarily listen to the perception of sound. Subjective appreciations of sound and noise were, and are, necessarily related to the acoustic environment. The relationship is hardly direct or proportional, but it does exist. Metaphors, analogies, similes, and a host of other forms of representations are not invented in vacuums; earthquakes or tornadoes cannot sound like trains until trains are invented and embedded in listeners' social hearing, for example. Conversely, small actual differences in sound (decibels) or timbre (acoustic and social texture) may be amplified and distorted through the skeins of social and economic relations and modes of production. Because a soundscape may be both an actual environment and an abstract construction, it is important to treat it as both.[12]

Because historical sounds are so ephemeral, we must contextualize what was heard and resist the temptation to imagine that sounds past actors heard and listened for are the same as ours. Antebellum Americans operated within carefully delimited aural vectors that were in turn shaped by what they heard in their own societies. We must also extract some sounds from others and study them in isolation simply because some contemporaries did so. But we should not forget the larger context in which this extraction occurred. Contemporaries isolated regional sounds and imbued them with specific meanings while ignoring accompanying sounds and, in effect, practiced selective listening and partial hearing.[13]

Although it is difficult to grasp how utterly meaningful and potent the heard world was to antebellum Americans, a variety of work by literary theorists, psychoacousticians, and social scientists gives us some idea of the gravitas contemporaries attached to what they heard. Some historians of auditory culture have rightly found work by psychoacousticians helpful in alerting them to the situational (temporal and geographical) variables shaping how people in a variety of contexts heard and interpreted their worlds. This work is helpful not least because it suggests that how and what people hear demands careful consideration in order to understand the emotional and ideological value antebellum Americans invested in their heard environments.[14]

Work by psycoacousticians shows that what is heard is highly situated and utterly historical. "Sound perceiving," writes Stephen Handel, "must always be anticipating, hearing forward, as well as retrospecting, hearing backward." Hearing is an active sense, one that is fundamentally historical in cognition: "The 'present' gets its meaning from both preceding and succeeding sounds." As the sentinel of the senses, hearing is also passive because it is always on. People who have lost it "describe the loss of the dynamic, ongoing, time-varying, characteristics of the world." In other words, passive hearing and active listening root people to time and place and give those contexts meaning. As Steven Feld has remarked, "As place is sensed, senses are placed; as places make sense, senses make place." Hearing was important for delimiting, albeit in clumsy ways, what it meant to be "southern" and "northern" to many Americans, particularly in the last thirty or so years before the Civil War.[15]

K. D. Kryter has examined the situational or contextual basis for defining unwanted sound and explains that acoustic "unwantedness" falls into two basic categories. The first "is that in which the sound signifies or carries information about the source of the sound that the listener has learned to associate with some unpleasantness not due to the sound per se, but due to some other attribute of the source." Here sound and noise are reflected in metaphor or analogy, for it is "the information [conveyed] to the listener that is unwanted." "This information," maintains Kryter, is influenced by the past experiences of individuals as well as the anticipated future experiences of group-conscious individuals. Kryter suggests that communities with common backgrounds, memories, and anticipations "will fairly consistently judge among themselves the 'unwantedness,' 'unacceptableness,' 'objectionableness,' or 'noisiness' of sounds that vary in their spectral and temporal nature." Sounds that are sudden, unexpected, and/or deemed alien to the group are considered noise even when the decibel level is low. In such instances, although actual sound pressure might be small, the Perceived Noisiness Levels can be high.[16]

The question of acoustic evaluation, then, cannot be disassociated and decontextualized from social and historical meaning. L. N. Solomon has found that people can meaningfully evaluate sounds on a variety of axes, including magnitude (heavy-light sound); aesthetic (good-bad); clarity (clear-hazy); security (safe-dangerous); and familiarity (familiar-alien). Sounds carry meaning. Certain sounds connote activity or inactivity, and judgments about the nature of those sounds are embedded in the listener's context in which the sounds are heard.[17]

How people evaluate sound and its meaning(s) is due, in part, to "whether they consider the noise-producing activity to be important for their social and economic well-being and whether they believe that the noise is a necessary by product of the activity that produces it." Two other factors are also important in shaping the meaning of particular sounds. Listeners judge sounds negatively when they "believe that those persons responsible for the operation and regulation of the noise-producing activity are concerned about their (the exposed population's) welfare" and if they think that the sound will physically damage their property. All in all, there is ample evidence to suggest that while rooted in the physical production of sound, perceptions of noise and sound are precisely that—attitudinal, situational, and contextual. Although annoyance with particular sounds can sometimes decrease over time, "there is little evidence that annoyance due to community noise decreases with continued exposure."[18] To appreciate the importance of the acoustic environment historically, it is imperative to understand how meaning was attached to particular sounds, to realize the depth of that commitment, and to begin to listen to what those sounds or noises were in a particular historical context.[19]

R. Murray Schafer has termed the sounds of a given community "keynotes." Schafer maintains that the "keynote sounds of a landscape are those created by its geography and climate: water, wind, forests, plains, birds, insects, and animals." True enough, but keynote sounds—sounds that imprint "themselves so deeply on the people hearing them that life without them would be sensed as a distinct impoverishment"—may also be produced by specific configurations of social and economic relations and modes of production. Because these sounds are both assumed and cultivated deliberately, they may be termed either keynotes or soundmarks, and they serve as experiential mediums locating people in time and space. Slavery had a particular and meaningful keynote to antebellum slaveholders; for northern elites, sounds of freedom, capitalism, and industrialism had their own soundmarks that, combined, constituted part of what it meant to be northern. Critically, "once a soundmark has been identified, it deserves," argues Schafer, "to be protected, for soundmarks make the acoustic life of the community unique." Sounds, then, serve as anchors to regions, as acoustic identifiers of community, and as a result, if those soundmarks are threatened by alien strains and rhythms, communities interpret those sounds as threats to their identities and ways of life.[20]

The question remains as to how differences in keynotes are communicated and how the threat becomes palpable. So many slaveholders who had never witnessed northern democracy and wage labor firsthand and so many abolitionists and free wage labor advocates who had never seen slavery became so exercised about the "other," in part, because figurative representations, metaphors, and analogies of aural sectionalism (of the acoustic construction of otherness) exerted a powerful influence on their minds and acquired an authenticity by virtue of their insertion into local and national political discourse. Hearing one another abstractly and, at times, literally and not liking what they heard caused passions to flare and for imagined and substantive differences to become real even as they remained sometimes loosely metaphoric. Although metaphorical cognizance is based in "non-metaphorical preconceptual structures arising from everyday bodily experiences," metaphorical understanding is more than the simple transference of the material to the disembodied abstract because, as Christopher Tilley and others have shown, "we do not just employ and construct metaphors but we live through them." Grounded in materiality, "metaphorical concepts structure perception, action and social relationships" while simultaneously reflecting the texture and meaning of those actions and relationships. The everyday representation and metaphor becomes reality itself even as the abstraction, while grounded in structure, may not necessarily reflect the entire complexity of that same material reality. Even mild metaphors attain a distinct vividness and credibility precisely because they link subjective and objective experiences, provide "a way of mediating between concrete and abstract thoughts," and enable us to remember and project (spatially and temporally) a complex set of events in an easily communicable and seamless manner. Projections and metaphors, then, become part of the reality by which people navigate and conceptualize their worlds. They are inextricable from everyday life, culturally specific, and work to glue social groups who use the same representation to link personal experience with collective identity.[21] Modal metaphors, particularly those that play on the heard world, have, according to Bernard Hibbitts, a profound specificity and tangibility and in their familiarity have the power to "obscure and distort." The image becomes real and "so compelling" that it is no longer "recognized as a metaphor, it redefines truth on its own limited terms." Simon Schama says the same of visual landscapes: "Constructs of the imagination projected onto wood, water, and rock" are so powerful that once even a distorted, highly rarefied construction establishes itself, "it has a peculiar way of muddling categories, of making metaphors more real than their referents, of becoming, in fact, part of the scenery."[22]

Since aural imagery was both subjective and objective, since it united communities in space and time, since its meaning was so affective and profoundly emotional, and since the heard world carried such historically specific meaning and weight, it becomes understandable why contemporaries in antebellum America used modal, aural images not only to construct—with a basis in material reality—their preferred soundscape and society, but also to identify and subsequently critique challenges to their respective societies. Sounds—and the images used to convey them and the ways in which particular ears listened to them—carried enough weight to prompt people to destructive action.

NOTES

ABBREVIATIONS

Am. Sl.
George P. Rawick, ed., *The American Slave: A Composite Autobiography,* 1st and 2d series, 19 vols., continuously numbered (Westport, Conn.: Greenwood Press, 1972); supplement series 1, 12 vols. (Westport, Conn.: Greenwood Press, 1977); supplement series 2, 10 vols. (Westport, Conn.: Greenwood Press, 1979)

AU
Special Collections and Archives, Auburn University, Auburn, Ala.

CWF
Research Queries Files, John D. Rockefeller Jr. Library, Colonial Williamsburg Foundation, Williamsburg, Va.

DU
Manuscript Department, William R. Perkins Library, Duke University, Durham, N.C.

FEC
Nineteenth Century Promotional Literature, Florida Ephemera Collection, P. K. Yonge Library of Florida History, Department of Special Collections, Smathers Library, University of Florida, Gainesville

FSU
Special Collections Department, Robert Manning Strozier Library, Florida State University, Tallahassee

LC
Manuscripts Division, Library of Congress, Washington, D.C.

MHS
Massachusetts Historical Society, Boston

MMC
Miscellaneous Manuscripts Collection, Department of Special Collections, Smathers Library, University of Florida, Gainesville

NEMV
Gary Kulik, Roger Parks, and Theodore Z. Penn, eds., *The New England Mill Village, 1790–1860* (Cambridge, Mass.: MIT Press, 1982)

SCL
South Caroliniana Library, University of South Carolina, Columbia

SHC
Southern Historical Collection, Wilson Library, University of North Carolina, Chapel Hill

Swem
  Manuscripts and Rare Books Department, Earl Gregg Swem Library,
  College of William and Mary, Williamsburg, Va.

INTRODUCTION

1. Lerner, *Grimké Sisters*, 3–7.

2. Ibid., 279.

3. Bobo, *Glimpses of New-York City*, 10, 11, 32, 17, 21, 58–59, 25.

4. Ibid., 95, 96, 97, 98, 97, 144, 171. See also Franklin, *Southern Odyssey*, 13, 39, 149, 177, 178, 180, 187.

5. Fitzhugh, *Cannibals All!*, xiii, xvi, 149–50.

6. For a discussion of the theoretical, conceptual, and methodological under-pinning of this study, see "Sound Matters: An Essay on Method" at the end of this book. On ocularcentrism, see, for example, Jay, *Downcast Eyes;* Levin, *Modernity and the Hegemony of Vision.* On historical anthropology of the senses, see Corbin, *Time, Desire and Horror;* Corbin, *Foul and the Fragrant;* Classen, *Worlds of Sense;* and the very helpful discussion in Schmidt, *Hearing Things,* esp. 16–21. See also the fine study by Sterne, "Audible Past."

7. My thinking here has been influenced by Freehling, *Reintegration of American History,* esp. 253–74, and Hobsbawm, *On History,* esp. 71–93.

8. Bailey, *Popular Culture*, 205, 206. P. T. Barnum once remarked that when attempting to cheat customers (and avoid being cheated himself), "Our eyes and not our ears, had to be masters. We must believe little what we saw, and less that we heard" (quoted in Greenberg, "Nose, the Lie, and the Duel," 11). But we should be careful not to assume that the predominance of print and visual culture, consumed and produced so vigorously by nineteenth-century Western elites, necessarily precluded their ability and tendency to define power, class, and otherness by listening and hearing. Likewise we should not assume that pre-literate nineteenth-century cultures always expressed themselves exclusively in oral and aural terms. On these important matters, and for an interpretation different from the one offered here, see Hibbitts, "Making Sense of Metaphors," and Rath, "Sounding the Chesapeake"; Rath argues that the "sonic representation" of colonial bells and other sounds was "largely set aside in favor of the visible world over the next two centuries" (7). See also Bruce R. Smith, *Acoustic World,* 10.

9. On technological innovations in capturing and recording sound, see Kahn, "Introduction"; on the antebellum American orator as a "scholar-agitator" of the audience's intellect and emotion and the explicit recognition that printed versions of speeches inspired the reader emotionally, see Warren, *Culture of Eloquence,* 2, 18–19, 111–12, 142. Note, too, Grant, *North over South,* esp. 9–10; Lehuu, *Carnival on the Page,* esp. 7, 19, 23. Historians of antebellum rhetoric and oratory have recognized how public speakers did much to "crystallize commonly unarticulated notions into hardened opinions," although they have re-

mained insensitive to the role aural imagery played in that process. See Auer, *Antislavery and Disunion*, vi. For observations on how metaphors, representations, and political talk can be deliberately unclear in an effort to preserve order as well as the use of candid imagery to provoke conflict, see Myers and Brenneis, "Introduction," 14–16; Brenneis, "Straight Talk and Sweet Talk," 69–71; Weiner, "From Words to Objects to Magic," 175–76.

10. Rath, "Sounding the Chesapeake." Although it is probably true that the past was, measured in decibels, quieter, we must be careful not to sentimentalize historical soundscapes. Certainly some sounds could be heard over greater distances in the eighteenth century. In 1745 guns fired in Philadelphia were heard as far away as New York, and by 1891, "It is probable that no weight of artillery could now be heard from city to city!" Partial paving in the mid-eighteenth century meant that coaches and carriages echoed less than they did in later years. As Watson reported in 1891, "Old persons have told me that before the city was paved, and when fewer carriages were employed, they found it much easier than now to hear distant sounds" (John F. Watson, *Annals of Philadelphia*, 1:213, 2:492, 3:131). Colonial Williamsburg did not have paved or cobbled streets—as the contemporary depiction would have it—and so the roll of carriages and the sounds of horses and of walking did not echo as much. Instead the clay and sand helped muffle noise. See Goodwin, "Streets of Colonial Williamsburg," 14, and "Authenticity of Street Conditions," 27, "Streets," CWF.

11. Powers to Pappas, Mrs. Rutherfoord Goodwin to Kellems, and "Bruton Church Bells Found A-Pealing," "Bell Founders," CWF. On the use of town criers and post horns in New England and southern colonies by the middle of the eighteenth century, see Gibbs to Dwyer and Mary R. M. Goodwin to Graham, "Posthorns," CWF. Snow could disrupt soundscapes. A severe winter in Williamsburg in 1737 delayed the mail for six weeks and silenced the post horn. See Edward M. Riley, "Excerpts," and Mary R. M. Goodwin to Graham, "Posthorns," CWF. Even as imperial sounds were extended in a sort of acoustic sphere from London to the colonies, they necessarily underwent change. Colonial bells were either imported from London or cast in the New World by the early eighteenth century. Colonial America, precisely because it attracted immigrants from both Britain and the Continent, became distinctly something other than either English or Continental in its soundscape. English bells, for example, tend, even now, to be swung full circle, which generates a "notably fuller and more sonorous tone" than bells on the Continent, "which favors the hanging of bells 'dead'—that is, not swinging—as 'chimes,' or as 'carillons,'" which originated in the Low Countries. What distinguished colonial America was its new world of sound, which braided both English bells and Continental dead chimes. On the two systems, see *Whitechapel*, 5–6, 8, "Bell Founders," CWF. On Whitechapel Foundry bells exported to the colonies and the United States, 1752–1900, see Hughes to Tucker, "Bell Founders," CWF. From 1901 to 1986 Whitechapel Foundry exported another twenty-seven bells to the United States. On sound and empire, see Bruce R. Smith, *Acoustic World*, 288–90.

12. Garstin, *Samuel Kelly*, 104, 106, 108. Animals as well as humans shaped

Williamsburg's soundscape. In winter "the trumpeter [swan] . . . utters a sound like a trumpet," and the sounds of perennial ducks, cows, sheep, chickens, and turkeys added to the chorus; quoted in M. Stephenson to Frank, CWF. See also Edward M. Riley to Miller, CWF.

13. On regional variations in temperature and rainfall, similarly high levels of population density from 1660 to 1775, and modest colonial manufacturing generally, see Paullin, *Atlas of the Historical Geography of the United States,* plates 3E–5, 60A–E, 133. An obvious regional difference in sound was born of variations in snowfall. Travelers throughout the eighteenth and nineteenth centuries commented on the distinctive sound of sleigh bells, particularly in northern cities. As one French observer said of New York City in 1826, "Elegant city sleighs are drawn by well-harnessed horses wearing collars hung with bells. This precaution is necessary, since horses galloping on deep snow do not make enough noise to attract the attention of pedestrians." Plainly, such bells were largely redundant in the lower South but also less used in rural regions of the North. See Milbert, *Picturesque Itinerary,* 22–23. On the sounds of colonial Charleston, see Rogers, *Charleston,* 76; Coclanis, *Shadow of a Dream,* 4.

14. *Pennsylvania Gazette,* Sept. 29, 1779. In the 1750s, 1760s, and 1770s established settlers listened back to a presettlement American past and heard only "a howling Wilderness." See *Pennsylvania Gazette,* Aug. 25, 1757, Jan. 16, 1766, Apr. 19, 1770. See also Rath, "Sounding the Chesapeake," 4; Nash, *Wilderness and the American Mind,* 1–83.

15. John F. Watson, *Annals of Philadelphia,* 3:209, 201, 398. For comparative purposes, consult Corbin, *Village Bells.* See also Meade to Riley, Riley to Meade, Middleton to Alexander, and "Whitechapel Bell Foundry Data," "Bell Founders," all in CWF; Boland, *Ring in the Jubilee,* 40–41.

16. See Mrs. Rutherfoord Goodwin to Lay and to Kellems, CWF. The town bell was important enough that men were paid to ring it (Richard Sykes of Springfield, Massachusetts, was paid a shilling a week to do so). On the presence of a town bell and watch house in Springfield, Massachusetts, in 1645 and 1719–20, see Mason A. Green, *Springfield,* 75, 94, 214. On the religious function of bells in colonial Virginia, see Rath, "Sounding the Chesapeake," 5–8, 20. Like the call of the wolf "which defines the territorial claim of the pack to an acoustic space," the church bell delimited not simply the parish and denomination specifically but the sound of God generally. Those within earshot were both within the community of the parish and also reminded of God's continual presence. See Schafer, *Tuning,* 39. People—the elite in particular—wanted discipline through sound; they asked for church bells; they gave servants bells; in short, they needed the social function of sound. Without bells eighteenth-century Americans would have experienced difficulties in coordinating urban life, economic transactions, and the organization of labor.

17. Bruce R. Smith, *Acoustic World,* 330, 332; Rath, "Sounding the Chesapeake," 15–17. On the silence of Native Americans, see Bruce R. Smith, *Acoustic World,* 315–17. Immigrants from African and European countries speaking different languages populated colonial America. Heard together, especially in

cities, accents and languages conspired to produce a veritable Babel if not cacophony. They also served as regional identifiers. As Fischer has explained it in terms of speechways, the regional twangs, whines, and drawls of early America were related to English dialects and, it should be added, European and African ones, too. Plainly, the regional nature of immigration to the colonies gave particular areas in colonial America an aural distinctiveness through voice and accent. That much said, accent was insufficient to mark the southern, middle, or New England colonies as aurally distinctive or to imbue whatever aural regionalism did exist with profound, enduring, and consequential political meaning. See Fischer, *Albion's Seed*, 257–64; Joyner, *Shared Traditions*, 163; Rogers, *Charleston*, 76–79.

18. The ability of the church to impose silence was testimony to its power. See Schafer, *Tuning*, 201–2; Hull, "Some Passages of God's Providence about Myself," 154, 160.

19. John Edwards, "An Ordinance for Establishing a Guard and Watch in the City of Charleston, and for Repealing all Ordinances respecting the Guard in Charleston," Feb. 20, 1796, in Edwards, *Ordinances of the City Council of Charleston*, 144; John F. Watson, *Annals of Philadelphia*, 1:62. African American music performed during elections in the North produced "sounds of frightful dissonance" to elite white ears; see Berlin, *Many Thousands Gone*, 191.

20. Meranze, *Laboratories of Virtue*, 99.

21. Analysis based on CD-ROM word searches for "noise" and "sound" in the *Pennsylvania Gazette* for 1728–55. Specifically, see the following constructions of who and what produced negative sound or noise in the *Gazette:* on criminals, thieves, and "the ignorant sort," Jan. 27, 1730; on danger (snakes and earthquakes particularly), Oct. 28, 1731, Feb. 28, 1733; on runaway slaves, Sept. 12, 1745; on Native Americans, July 2, 1752, July 17, Oct. 16, 1755. For things and people considered sound or that produced sound, see June 3, 1736 (the logic of elite men); Jan. 7, 1742 (good Christians); Jan. 24, 1749, Feb. 11, 1752 (sound money); Nov. 20, 1746 (words and accents); May 18, 1749 (liberty). On colonial women's disorderly speech, see Kamensky, *Governing the Tongue*, esp. 19–22, 96–98.

22. See, for example, Persons, *Decline of American Gentility;* Kasson, *Rudeness and Civility;* Halttunen, "Confidence Men and Painted Women"; Farrell, *Elite Families;* Sheidley, *Sectional Nationalism*, esp. 188–200; Grant, *North over South*, esp. 34–35.

23. On the equation of modernity and noise, see Schwartz, "Noise and Silence," esp. 2–6.

24. Schwartz, "Indefensible Ear." For a helpful overview and argument concerning the varieties of northern critiques of slavery, see Ashworth, "Free Labor, Wage Labor, and the Slave Power." Although, as Ashworh maintains, northern Democrats harbored concerns about the threat of wage labor (as opposed to free labor) to republican virtue, there was a good deal of agreement by the 1850s among northern politicians of all stripes that the sounds of slavery and the silences they believed undergirded the system were deleterious to a broadly

understood northern way of life. For similar Democratic and Republican under-standings of wage labor and, especially, a common loathing of the Slave Power, see Gienapp, *Origins of the Republican Party;* Gienapp, "Republican Party and the Slave Power."

25. Quoted in Young, *Domesticating Slavery,* 106.

26. Pessen, "How Different," 1119; Nevins, *Ordeal of the Union,* 2:554; on the breakdown of the two-party system, see esp. Holt, *Political Crisis.* See also Ashworth, *Slavery, Capitalism, and Politics.*

27. On southern masters as economically modern but politically conserva-tive, see Mark M. Smith, *Mastered by the Clock;* Genovese, *Slaveholders' Dilemma.*

PART I

1. Huger quoted in Bowman, "Antebellum Planters and *Vormärz* Junkers," 783. Ulrich B. Phillips hinted at the association between authority and aurality when he referred to the "patrician's quietude" of white southerners. See Joyner, *Shared Traditions,* 193. On hearing by nineteenth-century British elites, see Bailey, *Popular Culture,* esp. 205–6.

2. Although the "South is a noisy and buggy place in the summer," the noise of cicadas is still considered a "soft summer song." See Brinson, "Cicadas' Soft Summer Song." Silence, even in an anechoic chamber, is impossible because one always hears the sounds of one's body. See Schafer, *Tuning,* 256–57; Truax, *Handbook,* 8–9. For slaveholders, silence was part of a continuum, and it was in their quieter moments that they felt most comfortable. For some interest-ing observations on southern oral loudness and the southern fear of silence, see Gretlund, "Southern Silence?"

3. See Schafer, *Tuning,* 48–52.

4. Ailenroc, *White Castle of Louisiana,* 18–19; Thornwell, *National Sins,* 33. My thinking here has been shaped in part by Fox-Genovese and Genovese, "Divine Sanction of Social Order," esp. 212–13. On the sound of time, see Mark M. Smith, *Mastered by the Clock.* On the slaveholders' alternate route to modernity, see Genovese, "Marxian Interpretations of the Slave South," esp. 119; Genovese, *Slaveholders' Dilemma,* esp. 5. On reform movements and hu-manitarianism in the South, see Quist, *Restless Visionaries.* For an interpretation stressing slaveholders' embrace of individualist and organic social values, see Young, *Domesticating Slavery.*

5. Anderson, *Imagined Communities;* Conner Diary, July 4, 1827, 13–14, SHC. For the firing of cannon in the antebellum South, see, for example, Valen-tine Diaries, vol. 3, July 4, 1838, SHC. For the evolution of July Fourth celebra-tions in North Carolina, see Fletcher Melvin Green, *Democracy in the Old South,* 111–56.

6. Schafer, *Tuning,* 52. See also Bertelson, *Lazy South,* for an implicit equation between tranquility and laziness. On pastoral tensions, see Grammer, *Pastoral and Politics,* and generally, Raymond Williams, *Country and the City,* esp. 16–

19. On the southern literary mind symbolizing "opposition to modernity in an image of pastoral permanence," see Simpson, *Dispossessed Garden*, 70.

CHAPTER I

1. Avirett, *Old Plantation*, 31, 45, 49, 58.

2. Eppes, "Negro of the Old South," 2–5, 19, Pine Hill Plantation Papers, FSU; [Charles Ball], *Fifty Years in Chains*, 160.

3. Valentine Diaries, Apr. 16, 1852, 12:18, SHC; Habersham Diary, Aug. 1, 1832, 28, SHC. For similar memories, see Chivers, "Memories of My Childhood," in his *Virginalia*, 68. For evidence that northerners found "the pattering of the falling rain as a night song," see Cathell Diary, Mar. 1, 1856, 195, SHC.

4. Clyde N. Wilson, *Carolina Cavalier*, 1, 53–55, 212; B. B. M. [John Reuben Thompson], "Impressions of Italy," *Southern Literary Messenger* 25 (Nov. 1857): 364, 366–67, 368–69. See also Burns, "1st Quarterly Oration," Burns Papers, SCL. For an argument that European travel consolidated U.S. national identity at the expense of sectional consciousness, see Kilbride, "Privileged Southerners." See also O'Brien, *Rethinking the South*, 84–111.

5. Valentine Diaries, Oct. 27, 1838, 3:29, SHC; Harriet Randolph, Leon Co., Fla., to Lucy Beverly, New London, Va., Sept. 8, 1829, in Whitfield and Chipman, "Florida Randolphs," FSU. For the English parallels, see Raymond Williams, *Country and the City*.

6. Stella Hull to Rev. Seth Hart, Aug. 8, 1806, 1–2; Stella Hull to Ruth Hart, May 25, 1808, 2; Stella Hall [Hull], New Smyrna, Fla., to Ruth Hart, June 6, 1804, 2–3, all in Hull Letters, MMC; Yarbrough, *Boyhood and Other Days*, 13. Those who lost "the equilibrium of a tranquil mind" experienced a dissonance, a disharmony that suggested insanity. See the remarks in the Ogden Journal, Jan. 30, 1834, MMC. Peace and quiet was commonly thought essential for the ill. Maria Baker Taylor of South Carolina feared for her ill son on January 27, 1864, "as he needs quiet." Three days later his condition had worsened. "He seemed very sensitive to noise," she wrote. A day later he was dead. See Taylor Diary, 6, MMC.

7. Richard J. Calhoun, *Witness to Sorrow*, 39, 45, 49, 99, 100–101.

8. Cecilia, "Sketches of Summer Visits," *Southern Literary Messenger* 23 (Nov. 1856): 366; Poe, "Man of the Crowd," in his *Tales*, 223.

9. Thomas R. R. Cobb to William R. Mitchell (1858), in Cobb Papers, SHC.

10. Cocke Diary, Aug. 1, 5, 4, 7, 1827, 14–15, 24–25, 20–21, 36–37, Cocke Papers, SHC. See, too, the similar sentiments expressed by Jane Carolina North on New York City, in O'Brien, *Evening When Alone*, 197.

11. Cocke Diary, Aug. 11, 17, 1827, 49, 82, Cocke Papers, SHC. Note, too, O'Brien, *Evening When Alone*, 7–14.

12. Courtesy Cards, FSU. These cards were used north and south.

13. Martineau, *Retrospect*, 36, 154, 173, 174; Pettigrew quoted in Clyde N. Wilson, *Carolina Cavalier*, 22; Bailey, *Popular Culture*, chap. 9. It would be a mis-

take to assume that members of the same class did not level the charge of noisy against one another, especially when behavior transgressed the bounds of gentility. See, for example, Reid Diary, June 3, 1833, SHC.

14. George Howe, *Endowments, Position and Education of Woman*, 5, 9, 10–11.

15. Cocke Diary, Sept. 27, 1827, 166–67, Cocke Papers, SHC. Harriet Randolph, Spring Garden, Va., to Mrs. T. E. Randolph, Lynchburg, Campbell, Va., May 29, 1829, in Whitfield and Chipman, "Florida Randolphs," FSU. For examples of southerners describing children in particular as noisy, see Reid Diary, June 12, 1833, SHC.

16. Conner Diary, June 24, 1827, 11, SHC. Because women were custodians of southern quietude, even their admonitions to silence bawdy and rude men were considered acceptable. Hearing "some gentlemen" slight the reputation of a young Tallahassee woman in 1845, for example, Mary Bates, a northern schoolteacher who had moved south, felt it proper to tell "these gentlemen to *hush*" (Bates to Dana, FSU). See also Fox-Genovese, *Within the Plantation Household*.

17. Conner Diary, n.d. and June 24, 1827, 42, 10, SHC. Recent work has shown that elite Virginia women did speak out on political matters but by wrapping their public voices in the feminine idioms of harmony, compromise, moral benevolence, and during the secession crisis, southern nationalism. See Varon, *We Mean to be Counted*.

18. Reid Diary, Feb. 22, 1833, SHC.

19. Lynch Journal, Feb. 8, 1837, FSU.

20. Avirett, *Old Plantation*, 173; Hughes, *Thirty Years a Slave*, 36, 40.

21. "Walterboro" [William Gilmore Simms], "Angling," and "E" [William Gilmore Simms], "The Cypress Swamp," *Southern Literary Gazette*, Oct. 15, Sept. 15, 1829, 257, 211; G. S. S., "Sketches of the South Santee," 316.

22. North, *Economic Growth*, 258; Wright, *Political Economy*, 110. See also Genovese, *Political Economy of Slavery*, esp. 49; Bateman and Weiss, *Deplorable Scarcity*, esp. 163–65; and Fox-Genovese, *Within the Plantation Household*, 70–88. On the acoustic effects of the industrial revolution generally, see Schafer, *Tuning*, 53.

23. Benjamin F. McPherson, "I Must Tell," 42, AU.

24. Martineau, *Retrospect*, 210–11. These were quintessential rural sounds. See Schafer, *Tuning*, 63.

25. Claybrooke to Claybrooke, Overton Family Papers, Swem; Hodgson, *Letters from North America*, 224–25.

26. Quoted in Mark M. Smith, *Mastered by the Clock*, 46.

27. "Harvest Labors in Germany, and Working Cows," *Farmers' Register*, Oct. 1834, 276. On the romantic construction of the sound of dinner bells on a Louisiana plantation, see Ailenroc, *White Castle of Louisiana*, 29.

28. Concerning a commitment to the historical past and nature's laws, see "Letter First," *Farmers' Register*, Jan. 1836, 536.

29. *Plantation and Farm Instruction*, 5, 13, 15, Cocke Papers, Virginia Historical Society, Richmond. The 1852 edition of the same title was identical except for the information on sound and time; see the copy held at FSU. See also "Man-

agement of Negroes," *Southern Planter*, Feb. 1851, 41; Stevens, *Sketch of Prof. John Le Conte*. On the science and variables governing the speed of sound, see Truax, *Handbook*, 134–35.

30. A Planter, "Notions on the Management of Negroes, &c.," *Farmers' Register*, Dec. 1836, 494–95; *Am. Sl.*, supp. ser. 1, vol. 3, Ga. Narrs., pt. 1, 87; William W. Brown, *Narrative*, 14; "Slaveholding Christians," *North Star*, Oct. 26, 1849. Hugh Davis was but one of many men who used the horn to regulate plantation bodies. See Weymouth T. Jordan, "Management Rules," 56, 58. Drivers on some West Indian plantations "blew their shells to summon the negroes" instead of using bells. See Commons et al., *Documentary History*, 2:138.

31. *Am. Sl.*, supp. ser. 2, vol. 10, Tx. Narrs., pt. 9, 4250–51; Gregg to Hammond, Hammond Papers, SCL.

32. Cruz, *Culture on the Margins*, 56 (quotation), and see also 45–66. Also on slave singing and plantation productivity, see Joyner, *Shared Traditions*, 39.

33. Ailenroc, *White Castle of Louisiana*, 12.

34. Dickens, *American Notes*, 80; McKay quoted in Takagi, *"Rearing Wolves,"* 91; Weymouth T. Jordan, "Management Rules," 59; "Management of Negroes," *Southern Planter*, Feb. 1851, 41.

35. Eppes, "Negro of the Old South," 5–6, 12–13, 17, Pine Hill Plantation Papers, FSU.

36. John Cumming, "Music in Its Relation to Religion," *Southern Literary Messenger* 23 (Dec. 1856): 437; Clinkscales, *On the Old Plantation*, 105.

37. Tyler to "My dear Brother" and "My dear Sister," Tyler Papers, Swem; Truax, *Handbook*, 33.

38. Viator, "The Night Funeral of a Slave," *De Bow's Review*, Feb. 1856, 220; Grayson quoted in McDonnell, "Money Knows No Master," 32. On the mundane and spiritual sounds of plantation life, see Abrahams, *Singing the Master*, esp. chaps. 3–4; Genovese, *Roll, Jordan, Roll*, 574–75.

39. Avirett, *Old Plantation*, 143–44; Hughes, *Thirty Years a Slave*, 52; A Mississippi Planter, "Management of Negroes upon Southern Estates," *De Bow's Review*, June 1851, 625.

40. Baker, *America Perceived*, 115; Tyler to "My dear Mama," Tyler Papers, Swem; Conner Diary, Aug. 10, 1827, 20, SHC; Valentine Diaries, July 4, 1851, 11:89, SHC.

41. Avirett, *Old Plantation*, 75, 82.

42. Stanford M. Lyman, *Selected Writings*, 195.

43. *Ray v. Lynes*. Laws against noisy blacksmithing in English towns were passed as early as the thirteenth century. See Schafer, *Tuning*, 190.

44. Valentine Diaries, Apr. 20, 1837, 1:17, SHC.

45. Commons et al. *Documentary History*, 2:182; Reid Diary, Sept. 18, 1833, SHC.

46. Colonel H. W. Walter, "Slave Laws of the Southern States: Mississippi," *De Bow's Review*, Dec. 1851, 618; "Art. XII—Manufactures and Mines," ibid., June 1853, 621.

47. William Gregg, "Domestic Industry—Manufactures at the South," ibid., Feb. 1850, 137, 140.

48. William Gregg, "Southern Patronage to Southern Imports and Domestic Industry," ibid., July 1860, 79.

49. J. D. B. De Bow, "The Cause of the South," ibid., July 1850, 120; Siegel, *Roots of Southern Distinctiveness,* 143.

50. Siegel, *Roots of Southern Distinctiveness,* 137–38. See also Hubbs, "Guarding Greensboro," esp. 56, 67. On Gregg, see Downey, "Planting a Capitalist South," esp. 159–60.

51. Commons et al., *Documentary History,* 2:331; Conner Diary, n.d., 1827, 40, SHC; Hubbs, "Guarding Greensboro," 20; Winthrop D. Jordan, *Tumult and Silence,* 22.

52. *Southern Literary Gazette* quoted in Rosengarten, *Tombee,* 193; "Country Life," *De Bow's Review,* Nov. 1860, 616. Northerners inadvertently testified to the growing decibel level of southern towns, for they sometimes found southern cities as noisy as any northern town. On the way to Louisville in April 1852 Henry L. Cathell lamented, "This is the first opportunity that I have had to pen anything in peace and quiet since leaving Nashville." But Louisville was scarcely a haven of quietude: "too much hustle, dirt and city like" (Cathell Diary, Apr. 23, 1852, 153–54, SHC).

53. Grayson's poem quoted in McKivigan, *Roving Editor,* 306–7, 308.

54. See Hahn, *Roots of Southern Populism,* 37. On the "sublime" nature of the sounds, see *Edgefield Advertiser,* Feb. 11, 1836, quoted in Downey, "Planting a Capitalist South," 111. See also McLean to McLean, Coit-McLean Papers, SCL.

55. See Mark M. Smith, *Mastered by the Clock,* esp. 92; *General Rules Applying to all Persons in the Employment of the Columbia and Greenville Railroad Company* (n.d., n.p., ca. 1840s).

56. See, for example, Diary of William C. Sommerville [?], Aug. 6, 1817, Swem.

57. Tyler, *Merchant of Former Times,* 3–5, Swem; "Charleston," in *Poems of Henry Timrod,* 97. On market bells, see Mark M. Smith, *Mastered by the Clock,* chap. 2.

58. "The Telegraph," *De Bow's Review,* Mar. 1854, 252–53. See, too, Schafer, *Tuning,* 165; Mark M. Smith, *Mastered by the Clock,* 76–78.

CHAPTER 2

1. Gorn, "'Gouge and Bite,'" 22, 27–29. On aspects of quietude in the ritual of the duel, see John Lyde Wilson, *Code of Honor,* 11.

2. Quoted in McCurry, "Two Faces of Republicanism," 1261–62. See also Ford, *Origins of Southern Radicalism;* Genovese, "Yeoman Farmers."

3. See Bolton, *Poor Whites of the Antebellum South,* esp. 12, 42–47, 90. We have known for some time, of course, that the extension of the franchise and the abolition of property qualifications for voting principally in the 1830s applied

as much to southern as to northern states. Both regions shared in a republican-democratic ethos for white men. Elites in both sections expressed reservations about unvirtuous, irresponsible, and overly passionate political participation by those who had recently been enfranchised, but this worry was more pronounced in the South because of the implications such democratic activity held for slavery. See the classic statement by Fletcher Melvin Green, *Democracy in the Old South*, 65–86. On the profound conservatism of antebellum South Carolina, see Sinha, *Counter-Revolution of Slavery*.

4. See, for example, McCurry, *Masters of Small Worlds*, 105.

5. According to Grimsted, "only about a third of Southern riots parallel the kinds that occurred in the North, and this percentage would shrink to about one-fourth if one excluded the cities of Louisville, St. Louis, and New Orleans." See Grimsted, *American Mobbing*, 101 (quotation), 86, 114–34.

6. Lynch Journal, Dec. 1836 [*sic*], FSU.

7. Hubbs, "Guarding Greensboro," 29; Reid Diary, Feb. 18, 1833, SHC. On the frontier, also see Flint, *Recollections of the Past Ten Years*, 97.

8. Lynch Journal, Feb. 8, 1837, FSU. The duality of Native Americans' sounds—alternate silence and hideous yelling—that haunted colonial Americans continued into the antebellum period. Witness the problems in early Virginia. Attacks by Native Americans necessitated colonists' use of public alarms, but this itself threatened public safety by offering stealthy Indians moments of opportunity. Because "the only means for the discovery of their plotts is by allarms, of which no certainty can be had in respect of the frequent shooting of gunns in drinking," read a mid-seventeenth-century Virginia statute, shooting "any gunns at drinkeing" was banned, although "marriages and ffuneralls onely excepted" (Hening, *Statutes at Large*, 401–2).

9. Capt. J. B. Mason, Jacksonville, Fla., to Mary S. Mason, Warren, R.I., Nov. 10, 1836, Mason Letters, and Reed to Reid, MMC.

10. "Diary from Jan. 1, 1805–Dec. 31, 1805," Aug. 13, 1805, 50, Carl Mavelshagen, trans., Cherokee Mission Records, FSU; Whitehead, "Childhood and Youth," 37, MMC; Benjamin F. McPherson, "I Must Tell," 9–10, 34, 39, AU. Ordinances against excessive drinking were, in fact, related directly to noise prevention, since it was widely believed drunks of any race were capable of making "a terrible noise howling and barking like a dog." Liquor and the lack of self-discipline of those who used it to excess were viewed as likely to lead to noise. In this instance, the quotation describes a white man. See "Diary of Moravian Mission, Springplace, Jan. 1–Dec. 31, 1822," Feb. 6, 1822, 298, Carl Mavelshagen, trans., Cherokee Mission Records, FSU. On drunk Indians as noise producers, see Ogden Journal, 1834, 6, MMC; Foreman, *Traveler in Indian Territory*, 215.

11. Conner Diary, Oct. 6, 1827, 32, SHC.

12. "Seminole War," MMC.

13. "Removal of the Seminoles," MMC.

14. Foreman, *Traveler in Indian Territory*, 33, 39; Stella Hall [Hull], New Smyrna, Fla., to Ruth Hart, June 6, 1804, 1, Hull Letters, MMC.

15. Conner Diary, n.d., 1827, 40, SHC.

16. Hamilton, *Cry of the Thunderbird,* 173. Guns were even more devastating. The report of a gunshot shortens dramatically what acousticians have termed the "attack" of a sound. Unlike natural sounds, the attack of which is relatively long, that of a gunshot is short and does not give the ear sufficient time to protect itself. Indeed, it is physically impossible for the bone structure in the ossicles or middle ear to react in time. Like Arctic Inuits, whose increasing reliance on the gun and the accompanying noise trauma seem to have contributed to their deafness at high frequencies and an unusually high degree of sensorineural hearing loss, antebellum Native Americans who came to rely increasingly on firearms as integral to their existence probably suffered a loss in hearing and, as a result, an impaired ability to hunt and to defend themselves. Ancient hunting techniques, for example, which relied on Native Americans' abilities either to mimic the sounds of animals or to remain silent, were changed with the introduction and use of the gun. When some southeastern Indians were "within a mile from home, each of the hunters loads his gun with one shot, and they are fired off, one by one, so that each family by counting the shots, can tell which group of hunters is returning." See Truax, *Handbook,* 9, 29–31, 61, 87; Counter, *Electromagnetic Stimulation of the Auditory System;* Counter, *North Pole Legacy,* esp. 9–10. Quotation in Hamilton, *Cry of the Thunderbird,* 46. More generally, the attack sounds of guns affected anyone who hunted regularly, not just Native Americans.

17. McKinstry, *Code of Ordinances,* 149; Dohrn-van Rossum, *History of the Hour,* 245–51; Mark M. Smith, *Mastered by the Clock,* chap. 2. See also Lewis, *Charter and Code of Ordinances,* 147–48.

18. "Ordinance Number One," June 13, 1853, St. Augustine Papers, F.3, MSS 74-1a, box 156, FSU.

19. Stuart Diary, 21, SHC.

20. Reid Diary, Feb. 27, 1833, SHC; Richard J. Calhoun, *Witness to Sorrow,* 61; Valentine Diaries, June 2, 1851, 11:81, SHC. See also Grimsted, *American Mobbing,* 101, 199, 229. For a different interpretation, see Luraghi, *Rise and Fall,* 69.

21. Thomas R. R. Cobb, New Orleans, to William R. Mitchell, Marion, May 9, 1858, 1–2, Cobb Papers, SHC; Daybook of Rev. William McCormick, Sept. 8, 1860, 9, MMC.

22. McKinstry, *Code of Ordinances,* 106, 135, 170.

23. Ibid., 154–55, 158; [Jonathan Walker], *Trial and Imprisonment,* 30, 68. See also Rousey, *Policing the Southern City;* Takagi, "Rearing Wolves," 64–65.

24. "Ordinance No. 8," Nov. 29, 1822, St. Augustine Papers, F.1, MSS 74-1a, FSU. See also Takagi, "Rearing Wolves," 57.

25. "An Ordinance Relating to the Police of the City of St. Augustine," Mar. 10, 1845, St. Augustine Papers, F.2, MSS 74-1a, FSU. See also Thomas Roper, "An Ordinance to Regulate the Wages of Licensed Porters, and Other Day Labourers; to fix their Stands, and for Other Purposes," Sept. 4, 1801, and Charles B. Cochran, "An Ordinance in Addition to an Ordinance entitled 'An Ordinance to Establish Certain Regulations for the Port of Charleston, and to

Define the Harbour-Master's Powers and Duties,'" July 1, 1806, in Edwards, *Ordinances of the City Council of Charleston*, 225, 361.

26. McKinstry, *Code of Ordinances*, 82, 110, 159.

27. Ibid., 161–62. Doubling the number for civil disorder probably confused more than it clarified. If sentinels tried to alert officials to a riot south of Dauphin and west of Franklin, this required four taps, which doubtless led some to run to the quadrant north of Dauphin and east of Franklin to put out a fire, not combat a riot.

28. Mary A. Stephenson to Perry, CWF. On the use of the town bell to signal curfew for slaves and the rather lax enforcement of it, see Taylor, *Reminiscences*, 31.

29. Georgia statute quoted in Lawrence M. Friedman, *Crime and Punishment*, 89. On drums, see Schafer, *Tuning*, 165; John F. Grimké, "An Ordinance to prohibit the frequent beating of Drums, in the City of Charleston," Nov. 12 1787, in Edwards, *Ordinances of the City Council of Charleston*, 57; Wood, *Black Majority*, esp. 271–81.

30. Quoted in Levine, *Black Culture*, 43.

31. Grier, *Culture and Comfort*, 99 (quotation), and see also 36–37. Of Charleston in 1849, James Johnston Pettigrew remarked, "The streets are sandy and unpaved, except a few" (quoted in Clyde N. Wilson, *Carolina Cavalier*, 32). On the changes in building materials in, for example, Greensboro, Alabama, see Hubbs, "Guarding Greensboro," 13. On the presence of carpets in elite colonial Williamsburg homes, see memoranda by Mr. Short, July 12, 1960, and by Mr. Graham, Nov. 7, 1966, Jan. 9, 1969, all in "Carpets," CWF. See also Glassie, *Patterns in the Material Folk Culture*; Lane, *Architecture of the Old South*; Emily Thompson, "Mysteries of the Acoustic"; Schwartz, "Noise and Silence," 1, 4–5.

32. Lamar to "My Dear Sister," University of Georgia, Athens; Bonsall to Hammond, Hammond Papers, SCL; Schwartz, "Noise and Silence," 5.

33. Schafer, *Tuning*, 54–55, 173–75; Mark M. Smith, *Mastered by the Clock*, chap. 2. On the colonial context, see Rath, "Sounding the Chesapeake," 5–8, 20.

34. See, for example, "Notes on St. Paul's Parish," 5, MMC; Claybrooke, Travel Account, Overton Family Papers, Swem; Cordley, *Pioneer*, 14, 20. On the bell denoting the preacher's mouth and the Christian symbolism, see Schafer, *Tuning*, 174.

35. Daybook of Rev. William McCormick, Oct. 17, 1860, 18, MMC.

36. Cocke Diary, Aug. 6, 17, 1827, 28–29, 90, Cocke Papers, SHC.

37. Mathews, *Religion in the Old South*, xvi; McCurry, *Masters of Small Worlds*, 112, 117, 127, 139, 170; Richard J. Calhoun, *Witness to Sorrow*, 69. See also Loveland, *Southern Evangelicals*, and by way of comparison, the excellent discussion in Schmidt, *Hearing Things*, esp. 65–77.

38. Schafer, *Tuning*, 51–52.

39. Ibid., 10.

40. "Travels of Henry Bryson," Dec. 8, 1826, Miscellaneous Collection, AU.

41. Holifield, *Gentlemen Theologians*, 16; Early Diary, bk. 1, May 27, Sept. 21, June 19, July 11, 1807; bk. 3, Aug. 3, 1808, May 12, 1811; bk. 7, Mar. 5, 1811, SHC.

42. Early Diary, bk. 1, Aug. 20, 1807, SHC.

43. Ibid., bk. 1, Sept. 20, 1807; bk. 3, July 23, 24, 25, Aug. 11, 1808; bk. 1, Sept. 22, Oct. 22, 1807.

44. Olive, *Wonders of the Age*, 23, 63, 67.

45. Valentine Diaries, Nov. 4, 1851, 11:138–39, and Sept. 13, 1853, 13:111, SHC. See, too, the description and characterization of the "tumult" and "howling" at an interracial evangelical service in Georgia by Frederick Olmsted in his *Journey in the Seaboard Slave States*, 459.

46. Jones quoted in Cruz, *Culture on the Margins*, 83.

47. Cruz, *Culture on the Margins*, 85–87.

48. Meade to [a Bishop], Swem.

49. John Cumming, "Music in Its Relation to Religion," *Southern Literary Messenger* 23 (Dec. 1856): 435, 439. Contemporary examples of such worries may be found in "Journal of my travels," May 28, 1833, Swem. The debate between inwardly quiet and outwardly loud spirituality was replayed at the end of the century. See Schwartz, "Noise and Silence," 8–10. In this sense the slaveholders embraced the rationalism of the Enlightenment's visualism and rejected what Schmidt has called the explosive aurality of popular Christianity, which was far less concerned with the interpretation of printed scripture than with establishing a personal, heard relationship with God. See Schmidt, "From Demon Possession to Magic Show."

50. Holifield, *Gentlemen Theologians*, 18, 39, 43, 45, 49.

51. Conner Diary, Aug. 30, 1827, 30–33, SHC.

52. "South Carolina Democracy," *National Era*, Oct. 26, 1848.

CHAPTER 3

1. Blight, "'Analyze the Sounds'"; Lovell, *Black Song;* Levine, *Black Culture,* 7; Genovese, *Roll, Jordan, Roll,* 324; Cruz, *Culture on the Margins,* esp. chap. 2.

2. As Cruz explains, to even vaguely southern abolitionist ears the power of slave singing, while generating emotional sympathy, was insufficient to make such whites anything more than incidental listeners or to combat the larger proslavery environment in which they lived. Such was the case with Mary Boykin Chestnut. See Cruz, *Culture on the Margins,* 29–30.

3. Josiah C. Nott, M.D., *Two Lectures on the Natural History of the Caucasian and Negro Races* (1844), in Faust, *Ideology,* 223; Steve Jones, *In the Blood,* esp. 187–90.

4. Samuel A. Cartwright, "Diseases and Peculiarities of the Negro Race," *De Bow's Review,* July, Sept. 1851, 66, 331–36.

5. Levine, *Black Culture,* 43; Habersham Diary, Apr. 1, 1832, 24, SHC. See also Ailenroc, *White Castle of Louisiana,* 28.

6. Pendleton/Anderson District, South Carolina Department of Archives and History, Columbia. See also McDonnell, "Money Knows No Master."

7. Eppes, "Negro of the Old South," 84, Pine Hill Plantation Papers, FSU; Willie Lee Rose, *Documentary History*, 515. See also Cruz, *Culture on the Margins*, 11, 27–31, 43–66; White and White, "'At intervals I was nearly stunned by the noise he made.'" The racialization of sound continues, principally through the critique of hip-hop and rap music. See Tricia Rose, *Black Noise*.

8. Cathell Diary, Dec. 25, 1851, 60, SHC.

9. Gross, "Pandora's Box," 307, 308, 315. Note also Baynton, *Forbidden Signs*.

10. *State v. Boyce;* Fisher, "Ideology and Imagery in the Law of Slavery," 51.

11. A Small Farmer, "Management of Negroes," *De Bow's Review*, Oct. 1851, 371.

12. Avirett, *Old Plantation*, 92, 93, 133; Clinkscales, *On the Old Plantation*, 10; Florida Territory Bill, MMC.

13. Hughes, *Thirty Years a Slave*, 13; Loguen, *As a Slave*, 115–16.

14. Pennington, *Fugitive Blacksmith*, 7; *Am. Sl.*, supp. ser. 1, vol. 11, N.C. and S.C. Narrs., 189; [Charles Ball], *Fifty Years in Chains*, 246.

15. Loguen, *As a Slave*, 124–27.

16. Brown and Taylor, *Gabr'l Blow Sof'*, 80–81.

17. Faust, "Slavery in the American Experience," 3; John Brown, *Slave Life in Georgia*, 72. See also Cruz, *Culture on the Margins*, 51–53.

18. [Charles Ball], *Fifty Years in Chains*, 248, 249, 252, 260. See also Faust, "Slavery in the American Experience," 3.

19. [Charles Ball], *Fifty Years in Chains*, 89, 189, 193. Some masters undoubtedly did not like the sound of their own whips. But some did. As one former slave recalled, his master's "whip had a cracker on the end of it, which Mr. Merritt liked to hear snap. If it broke or came off, he hitched up his horse and went to town to get another so that he might hear it crack when he whipped the negroes" (*Am. Sl.*, supp. ser. 1, vol. 4, Ga. Narrs., pt. 2, 623).

20. Kocher and Dearsyne, *Shadows in Silver*, 81; Gibbs to Brown, CWF; Fisher, "Ideology and Imagery in the Law of Slavery," 65; Brown and Taylor, *Gabr'l Blow Sof'*, 80. See also "Grand Pa's Ghost Slave Stories"; Pennington, *Fugitive Blacksmith*, 32–34; Stroyer, *My Life in the South*, 69.

21. *Am. Sl.*, supp. ser. 1, vol. 11, N.C. and S.C. Narrs., 132; Brown and Taylor, *Gabr'l Blow Sof'*, 96; [Charles Ball], *Fifty Years in Chains*, 256.

22. Williams quoted in Mark M. Smith, *Mastered by the Clock*, 130; *Am. Sl.*, supp. ser. 2, vol. 10, Tx. Narrs., pt. 9, 4098–99; Stroyer, *My Life in the South*, 98–99; [Charles Ball], *Fifty Years in Chains*, 187–88.

23. Hughes, *Thirty Years a Slave*, 23–24; William W. Brown, *Narrative*, 15–16; [Charles Ball], *Fifty Years in Chains*, 40–41. See also Darius Lyman, *Leaven for Doughfaces*, 146–50.

24. Loguen, *As a Slave*, 107, 109, 263.

25. Hughes, *Thirty Years a Slave*, 70; *Am. Sl.*, supp. ser. 1, vol. 4, Ga. Narrs., pt. 2, 458, also 507, 559.

26. Douglass quotation from Blight, "'Analyze the Sounds,'" 3–4; Randolph, *Sketches*, 54–55. See also Brown and Taylor, *Gabr'l Blow Sof'*, 33.

27. William H. Channing, "A Day in Kentucky," *Liberty Bell* (1843), 63; Brown and Taylor, *Gabr'l Blow Sof'*, 23. See also Loguen, *As a Slave*, 230–31; Cruz, *Culture on the Margins*, 52.

28. Eppes, "Negro of the Old South," 5–6, 12–13, 17, 114–15, Pine Hill Plantation Papers, FSU. For a comparative analysis of a variety of northern and southern harvest celebrations and the argument that slaves and free blacks injected a "greater spirit" into theirs, see Abrahams, *Singing the Master*, 77, 64–76.

29. Douglass quotations from Blight, "'Analyze the Sounds,'" 3, 4. On the construction of some slave songs as "wild" and difficult to fathom by Douglass and the use of song by slaves to mock other bondpeople not pulling their weight during work, see Cruz, *Culture on the Margins*, 25–26; Levine, *Black Culture*, 10–11. On the privileging of music as the art of sound in culture, see Kahn, "Introduction," 3. For a particularly good example of work stressing workers' use of music as resistance, see E. P. Thompson, "Rough Music." On singing and African American culture, see, for example, the emphasis in Levine, *Black Culture*, 54–55; Stuckey, *Slave Culture*; Rosenbaum, *Shout Because You're Free*; Joyner, *Shared Traditions*, esp. 38–39, 192–207; White and White, "'Us Likes a Mixtery'"; Abrahams, *Singing the Master*. More sensitive to varieties of noise and the power of silence as well as the rise of black acoustic public space is Cruz, *Culture on the Margins*, 20, 43–66, 69–98. Note also Wyatt-Brown, *Southern Honor*, esp. 440–42. On music as resistance generally, see Attali, *Noise*. For brief suggestive remarks on the hidden transcript of silence as resistance, see Scott, *Domination and the Arts of Resistance*, esp. 3.

30. *Am. Sl.*, supp. ser. 1, vol. 4, Ga. Narrs., pt. 2, 350; Joyner, *Shared Traditions*, 69. See also testimony of Joe McCormick, *Am. Sl.*, supp. ser. 1, vol. 4, Ga. Narrs., pt. 2, 391. Although by 1860 more than half of all slaves lived on plantations with twenty or more bondpeople, the situation on small farms with fewer slaves may have been different. Yet as Abrahams notes, "While those who lived on smaller farms would have less of a community at home, their isolation was diminished through participation in activities with slaves on other plantations" (Abrahams, *Singing the Master*, 45). For healthy reminders on the power of the slaveholders' authority, see Coclanis, "Slavery, African-American Agency, and the World We Have Lost"; Kolchin, "Reevaluating the Antebellum Slave Community."

31. *Am. Sl.*, supp. ser. 1, vol. 11, N.C. and S.C. Narrs., 10, and supp. ser. 2, vol. 10, Tx. Narrs., pt. 9, 4280.

32. Lydia Maria Child, "Charity Bowery," *Liberty Bell* (1839), 41–42.

33. Levine, *Black Culture*, 41–42. See also Genovese, *Roll, Jordan, Roll*, 236; Cruz, *Culture on the Margins*, 108–9. On the "riot of sound," see Raboteau, *Slave Religion*, 222. Other work on African American religious expression that emphasizes its audible components includes Frey and Wood, *Come Shouting to Zion*; Spencer, "Rhythms of Black Folks"; Stuckey, *Slave Culture*.

34. Loguen, *As a Slave*, 250, 277. African American religion inherited many of its aural dimensions from West Africa. There were carryovers, and evidence to that effect is abundant. See, for example, Stuckey, *Slave Culture*; Levine, *Black*

Culture. See also Stoller, "Sound in Songhay Cultural Experience," esp. 562–64; Carothers, "Culture, Psychiatry, and the Written Word"; Schafer, *Tuning,* 11.

35. *Am. Sl.,* vol. xvii, Va. Narrs., 11–12. The name is given as Minnie Folkes in Perdue, Barden, and Phillips, *Weevils in the Wheat,* 92–96.

36. *Am. Sl.,* vol. xvii, Va. Narrs., 29; Brown and Taylor, *Gabr'l Blow Sof',* 3; Perdue, Barden, and Phillips, *Weevils in the Wheat,* 217.

37. Gomez, *Exchanging Our Country Marks,* 260; Perdue, Barden and Phillips, *Weevils in the Wheat,* 196. Unlike historians who attribute, albeit conjecturally, the upturning of pots to some mystical and as yet unsubstantiated association "between pots and drums and river spirits in Africa," Gomez sensibly maintains that "there may be a purely functional as opposed to spiritual explanation"; see Gomez, *Exchanging Our Country Marks,* esp. 259–62, 260 (quotation). For other examples, see Mark M. Smith, "Time, Sound, and the Virginia Slave," 29–60.

38. I am indebted to Frank Fahy, professor emeritus of engineering acoustics of Southampton University's Institute of Sound and Vibration Research for suggesting this possibility; see personal communication to the author, Mar. 11, 1999. On "masking effects," see Flindell, "Fundamentals of Human Response to Sound," 151–53.

39. Brown and Taylor, *Gabr'l Blow Sof',* 34.

40. Washington, *Up from Slavery,* 3. See, too, the descriptions in Joyner, *Shared Traditions,* 75–76. Yeomen's houses were sometimes little better insulated. See McCurry, *Masters of Small Worlds,* 73.

41. Cordley, *Pioneer,* 52. Slave cabins could be made of hewn post oak and covered with cypress, with close plank floors and elevated a few feet above the ground. Cabins could be "about 200 feet apart." See A Mississippi Planter, "Management of Negroes upon Southern Estates," *De Bow's Review,* June 1851, 621–27. A self-styled Small Farmer made almost identical provisions with houses elevated two feet above the earth; see A Small Farmer, "Management of Negroes," *De Bow's Review,* Oct. 1851, 369–72.

42. Brown and Taylor, *Gabr'l Blow Sof',* 108; Hughes, *Thirty Years a Slave,* 25.

43. Brown and Taylor, *Gabr'l Blow Sof',* 64.

44. *Am. Sl.,* supp. ser. 2, vol. 10, Tx. Narrs., pt. 9, 4088, 4070, and supp. ser. 1, vol. 4, Ga. Narrs., pt. 2, 345, and vol. 11, N.C. and S.C. Narrs., 64; Washington, *Up from Slavery,* 11. On shoes and sound, see Schafer, *Tuning,* 164.

45. Raboteau, "Slave Autonomy and Religion." The best discussion of slave religion, which, incidentally, is also sensitive to its audible and inaudible dimensions, remains Genovese, *Roll, Jordan, Roll,* esp. 233–41, 236–37. For Tubman's ability to walk soundlessly, see Davis, *Women, Race, and Class,* 23. Runaways' ability to muffle sound and the importance of being quiet at the right time are revealed with wonderful precision in Pennington, *Fugitive Blacksmith,* 32–37. On hunting, see Marks, *Southern Hunting in Black and White,* esp. 26–28.

46. *Am. Sl.,* supp. ser. 1, vol. 11, N.C. and S.C. Narrs., 43, 46.

47. [Charles Ball], *Fifty Years in Chains,* 162–64, 169.

48. Loguen, *As a Slave,* 37, 44–45.

49. Ibid., 300, 45, 71, 72, 222. On whispers during slavery, see Washington, *Up from Slavery*, 2, 8.

50. [Charles Ball], *Fifty Years in Chains*, 376 (his moccasins), 309, 312–15; *Am. Sl.*, supp. ser. I, vol. II, N.C. and S.C. Narrs., 206. On planters' use of horns for hunting, see Clinkscales, *On the Old Plantation*, 110; Avirett, *Old Plantation*, 184.

51. [Charles Ball], *Fifty Years in Chains*, 316, 326, 336, 349.

52. Ibid., 385, 424. See also Bradford, *Harriet Tubman*, 35, 38, 54. Witness, too, Louis Hughes's attempt to escape. After Hughes managed to make it to the port of West Franklin, Indiana, "a gentleman passenger came to me hurriedly and whispered to me to go down stairs." And so "I went out as quietly as I could, and was not missed" (Hughes, *Thirty Years a Slave*, 82–83). Then there were the rogue slaves, who eluded capture and who loitered, existing as maroons, around plantations. "Essex" who "lived in the swamps" near Augusta, Georgia, merely "laughed at the ringing of the farm bells he heard." But his status was not without its price. Silence and craft stood between him and his capture. See Clinkscales, *On the Old Plantation*, 18, 19. Slaves also learned to talk fluently and persuasively when running away. Consequently, slaveholders described escapees as possessing a "bewitching and deceitful Tongue" (from the *Virginia Gazette*, Apr. 21, 1774, quoted in Commons et al., *Documentary History*, 2:38).

53. See Haney, "In Complete Order," 45–47. In 1822 Charlestonians, in the wake of the Vesey plot, scrutinized their methods for keeping urban bondpeople in check and found them wanting. In particular the absence of a good, full-time city guard was cause for concern, not least because members of the guard as it was constituted traded with slaves and allowed them "to pass unmolested through our streets after the bell has rung" (quoted in Commons et al., *Documentary History*, 2:113).

54. Dickens, *American Notes*, 68–69. See also the instances noted in Haney, "In Complete Order," 48–49.

55. "Ordinance Number One," June 7, 1853, St. Augustine Papers, F.3, MSS 74-1a, box 156, FSU.

56. Chesnut, *Diary from Dixie*, 147; Clinkscales, *On the Old Plantation*, 104.

57. Eppes, "Negro of the Old South," 51, Pine Hill Plantation Papers, FSU; Douglass cited in Cruz, *Culture on the Margins*, 43.

58. Martineau, *Retrospect*, 238.

59. McKivigan, *Roving Editor*, 300–301. On skittishness and self-imposed silence, see the thoughtful remarks in Winthrop D. Jordan, *Tumult and Silence*, esp. I, 26.

60. "Miscellaneous," *Provincial Freeman*, Feb. 17, 1855.

61. "Selections," reprinted in *North Star*, Feb. 11, 1848. Antislavery sympathizer James Redpath commented on the silent walking of slaves. He, too, found it ominous and disconcerting. See McKivigan, *Roving Editor*, 194.

62. McKivigan, *Roving Editor*, 231.

63. Martineau, *Retrospect*, 251.

64. Lydia Maria Child, "The Black Saxons," *Liberty Bell* (1841), 42; Hubbs, "Guarding Greensboro," 92, 95.

65. "Seminole War," MMC.

66. Commons et al., *Documentary History,* 2:44; [Charles Ball], *Fifty Years in Chains,* 230.

PART II

1. Karl Marx, *German Ideology,* 173.

2. See Kimberly K. Smith, *Dominion of Voice,* esp. 42–44, 56–61, 118–27.

CHAPTER 4

1. "Changes by Miss Phoebe Carey," *National Era,* Dec. 28, 1848; "When the Christmas Bells Are Ringing," *Harper's Weekly,* Dec. 26, 1857, 823; "Christmas—Gathering Evergreens," *Harper's Weekly,* Dec. 25, 1858, 820.

2. *Harper's Weekly,* Oct. 30, 1858, 691. See also Cheever, *Autobiography,* 27. Likewise with music, which sometimes collapsed space and time. Recalled British traveler Adam Hodgson while in Natchez, Mississippi, in May 1820, "In the society of some of these families I passed a few days very agreeably; and while listening to some of our own favourite melodies on the harp and piano-forte, I could have fancied myself on the banks of the Lune or the Mersey, rather than on those of the Mississippi" (Hodgson, *Letters from North America,* 185; on sounds familiar when traveling abroad as reminders of home, see 171).

3. Cornish Diary, July 4, 1837, 2:83, and July 4, 1838, 3:75, SHC. New Bedford, Massachusetts, lawyer Edwin L. Barney similarly bemoaned that his "city government even refused to appropriate a few hundred dollars to ring the bells" on July 4, 1858; see Barney Diary, July 5, 1858, MHS. On the contested nature of national celebrations, see Waldstreicher, *In The Midst of Perpetual Fetes,* 187, 314–15. For reasons that remain unclear, July Fourth celebrations in the North during the 1830s appear to have declined in frequency and intensity, although they became popular once again in the following two decades. See Fletcher Melvin Green, *Democracy in the Old South,* 134.

4. "Saturday Night," *Colored American,* Oct. 17, 1840; Henry Ingersoll Bowditch, "Slavery and the Church," *Liberty Bell* (1843), 5.

5. *Independent Mechanic,* Oct. 12, 1811, 2; *First Baptist Church in Schenectady v. Schenectady & S.R. Co.; First Baptist Church in Schenectady v. Utica & S.R. Co.* On property and nuisance laws, see Schwartz, "Hush." For a legal injunction against noise made in church, see *Owen v. Henman.*

6. "Censor, No. 1," *Independent Mechanic,* May 11, 1811, 2; "Censor, No. 2," ibid., May 18, 1811, 1, and Aug. 24, 1811, 2. See also Pope, *Tour through the Southern and Western Territories,* 102–3. Even sounds of industry could be deemed

noise at night. In fact many nocturnal sounds—from steam engines to dogs barking—were considered noise in a Connecticut court in 1862, not least because sleep was disrupted. See *Woolf v. Chalker.*

7. Chapin, *Sketches of the Old Inhabitants,* 214; The Lounger, "To My Neighbor across the Passage," *Harper's Weekly,* June 6, 1862, 770–71.

8. *Independent Mechanic,* May 18, 1811, 3; from the *Christian Advocate and Journal,* "The Gentleman at Church," *Colored American,* Oct. 13, 1838. See also Chapin, *Sketches of the Old Inhabitants,* 293.

9. *Harper's Weekly,* Feb. 26, 1859, 131.

10. Nathaniel Peabody Rogers, *National Era,* Oct. 21, 1847; Cathell Diary, Nov. 28, 1851, 14, SHC; Martha Russell, untitled, *National Era,* Sept. 19, 1849. On tanbark in streets, see Upton, "City as Material Culture," 60.

11. Chapin, *Sketches of the Old Inhabitants,* 32; Martineau, *Retrospect,* 49. In addition to churches, sanctuaries of silence included courts, because cases had to be heard accurately if justice was to be carried out. See *Independent Mechanic,* May 4, 1811, 2.

12. "On Gaming," *Independent Mechanic,* Nov. 30, 1811, 2. See also Richards Diary, Jan. 15, 1847, Walcott Family Papers, MHS; Habersham Diary, July 16, 1831, 7, SHC; Schafer, *Tuning,* 258; Schwartz, "Noise and Silence," 5.

13. "The Sword and the Press," *Frederick Douglass's Paper,* Nov. 6, 1851; "Dumbness in Libraries," *Harper's Weekly,* Oct. 31, 1863, 691.

14. *Problem of American Destiny,* 9–10.

15. Baker, *America Perceived,* 140. British elites sometimes found Americans noisy. Harriet Martineau said of a wedding in Boston in the 1830s, "The noise became such as to silence all who were not inured to the gabble of an American party, the noisiest kind of assemblage, I imagine (not excepting a Jew's synagogue), on the face of the globe. I doubt whether any pagans in their worship can raise any hubbub equal to it" (Martineau, *Retrospect,* 65).

16. Richard, "Notes and Observations," 25, MMC. See also Campbell to "Dear Brother and Friends," 1, MMC. On drunken soldiers, see Martineau, *Retrospect,* 87; Bancroft Diary and Scrapbook, Sept. 9, 1861, AU. Noisy foreigners got away with aural murder, according to northern travelers to the Far East. In Burma, "such a din as [Muslim prayers] make about our ears of an evening would get them a berth for the night in a Boston watch-house" ("An Oriental Glimpse," *National Era,* Aug. 5, 1847). See also "The Man About Town: My Hotel," *Harper's Weekly,* Mar. 2, 1857, 179.

17. Sellepa, "On Music," *Independent Mechanic,* Aug. 10, 1811, 3; Selah, "Of the Aeolean Harp," ibid., Dec. 14, 1811, 1. See also Meranze, *Laboratories of Virtue,* 160; Cmiel, *Democratic Eloquence.*

18. Urania Rus, "Casting Pearls Before ——," *Harper's Weekly,* May 8, 1858, 291. See also Kasson, *Rudeness and Civility;* Persons, *Decline of American Gentility,* esp. 29–50.

19. Joan Hoff Wilson, "Illusion of Change," 422. See also Lawrence M. Friedman, *Crime and Punishment,* 228, 233–34; Stansell, *City of Women,* esp. 171–216; Blackmar, *Manhattan for Rent,* esp. 170–72.

20. Young Ladies' Seminary, *Prospectus*, 20; Welter, "Cult of True Womanhood," esp. 160, 154, 158, 161, 162, 164–65, 172; "Woman," *Godey's Lady's Book*, Aug. 1831, 110; Gilman, *Recollections of a Southern Matron*, 256–57; Witheral, "How May an American Woman Best Show Her Patriotism?," 313. See also Sigourney, *Whisper to a Bride*, 44; "Good Wife," 229; Kahn, *Noise, Water, Meat*, 30–31.

21. M'Creery, "Time—Castle of Mayo," *Independent Mechanic*, July 13, 1811, 4; Hatheway Diary, Mar. 1, 1846, FSU; "Literary Notices," *Frederick Douglass's Paper*, May 20, 1852; "Home for Reformers," *North Star*, Aug. 21, 1848; "Poetry," *Colored American*, Dec. 22, 1838.

22. John Greenleaf Whittier, "Songs of Labor and Other Poems," *National Era*, Oct. 3, 1850; McMillen, "Mothers' Sacred Duty," 339; Martineau, *Retrospect*, 43. See also advertisement for Kathie Brande, *A Fireside History of a Quiet Life*, in *Harper's Weekly*, Jan. 17, 1857, 48; Blackmar, *Manhattan for Rent*, 131–33.

23. *Village Record*, May 28, 1823; "Let Us Call to Remembrance Former Days," *Independent Mechanic*, Apr. 6, 1811, 3.

24. "Epitaph on a Scolding Wife," *Independent Mechanic*, May 4, 1811, 4; "Censor, No. VI," ibid., July 6, 1811, 1. It should also be noted that while northern middle-class women were admonished to make homes quiet, they demanded nothing less from their male and female servants. In this way class complicated gender.

25. Eppes, "Through Some Eventful Years," 29, Pine Hill Plantation Papers, FSU; *Independent Mechanic*, Oct. 12, 1811, 2. See also [Jonathan Walker], *Trial and Imprisonment*, 19, and Kasson, *Rudeness and Civility*, esp. 147–81.

26. "A Friend to Females," *Independent Mechanic*, Jan. 4, 1812, 2; Friend to Females, "Taking Infants to Church," ibid., Dec. 21, 1811, 2. On men's construction of children as noisy and women's responsibility for quieting them, see L. Maria Child, "Home and Politics," *National Era*, Nov. 16, 1848.

27. Cornish Diary, Dec. 24, 1836, 2:67, SHC; "Shakerism Again," *Independent Mechanic*, May 25, 1811, 2.

28. Quotations in Sellers, *Market Revolution*, 211, and see also 202–36. See also Curtis D. Johnson, *Islands of Holiness*, esp. 113–33. Few northern popular preachers even conceived of the following sentence, which Moravian preacher George F. Bahnson penned following a service in Lancaster County, Pennsylvania, on February 2, 1845: "I spoke needlessly loud for which I was sorry—it does no good but may do harm." A diary entry on April 13 echoed this sentiment: "Spoke again much too loud—oh that the Lord wd enable me to break off this habit—useless for others, injurious to myself" (Bahnson Diary, Feb. 2, Apr. 13, 1845, SHC).

29. Watson quoted in Cruz, *Culture on the Margins*, 84. See also Schmidt, "From Demon Possession to Magic Show."

30. Georgina C. Munro, "The Life Ransom," *North Star*, Oct. 3, 1850; "The Trump of Jubilee," *Liberty Bell* (1841), 48. For virtually identical aural constructions of the Choctaws in 1820 through British ears, see Hodgson, *Letters from North America*, 237.

31. Hamilton, *Cry of the Thunderbird*, 47, 82.

32. Ibid., 100, 37, 107, 115. Native American scouts were the most acoustically sensitive of all, according to Chief Luther Standing Bear of the Sioux. "No movement," he maintained, "was too slight for the scout to ignore, and no sound too meaningless to go unheeded. . . . The calls of the sand or prairie cranes were announcers of the always important weather changes; a croaking frog proclaimed a tiny marsh or hidden spring." When human ears failed, scouts observed creatures more sensitive to molecular vibration. When looking for buffalo, scouts examined the "commonly called tumble-bug" because the "two horns on the top of the insect's head were movable in all directions, but were invariably pointed toward the buffalo herd, probably attracted in that direction by stamping hoofs too distant for even sensitive human ears to detect" (ibid., 104).

33. Martineau, *Retrospect*, 93; "Reports of the Secretary of War," *National Era*, Dec. 16, 1847.

34. "Up to the time of the Industrial Revolution," remarks Schafer, "the sound of the blacksmith's hammer was probably the loudest sound a solo human hand ever produced" (Schafer, *Tuning*, 56–58).

35. "From the Richmond *Enquirer*," *Independent Mechanic*, June 29, 1811, 1; ibid., Nov. 30, 1811, 2.

36. Quotation in Way, *Common Labor*, 18, 44; *Detroit Advertiser* quoted on 131. For the visual assessments of the canal as progress, see Sheriff, *Artificial River*, 9–51.

37. Gleaner, "That's Music," *Independent Mechanic*, Feb. 8, 1812, 1.

38. Mason A. Green, *Springfield*, 556, 559, 609.

39. "Stanzas by Miss Alice Carey," *National Era*, Dec. 16, 1847; John F. Watson, *Annals of Philadelphia*, 1:36–37.

40. Cornish Diary, Aug. 30, 1836, 2:39, SHC; "From Gleason's Pictorial," *Provincial Freeman*, June 3, 1854. Listen also to the sounds of New York City's commerce and industry coming alive of a morning in the 1820s as recounted in Milbert, *Picturesque Itinerary*, 20–21.

41. Sellers, *Market Revolution*, 71; John F. Watson, *Annals of Philadelphia*, 1:225–26. See also the similar sentiments expressed in 1802 quoted in McCoy, *Elusive Republic*, 198.

42. Cornish Diary, 1837–38, vol. 3, SHC. Older cities had established noise regulations much earlier. In 1664 Boston's general court passed an ordinance against singing and making a noise in "any place of public entertainment." In 1677 the "making [of] any noyse or otherwise, or during the daytime" was prohibited on the Sabbath. See Benton, *Story of the Old Boston Town House*, 156, 167.

43. John F. Watson, *Annals of Philadelphia*, 1:260. The incidence of false alarms was high. For references to them, see, for example, Bahnson Diary, Mar. 7, 1845, SHC. For examples of the prevention in Worcester, Massachusetts, of the firing of guns and the igniting of fireworks, which were both noisy and aurally misleading, see *Ordinances of the City of Worcester*, 55, 57.

44. Upton, "City as Material Culture," 60; Emily Thompson, "Mysteries of

the Acoustic." See also Schafer, *Tuning,* 164. Ohio's constitutional convention held in "the large room of the Cincinnati College building" in 1850 was infamous for its acoustic poverty, although steps had been taken to remedy the problem. "You know by experience," wrote one commentator in 1850, "that it is one of the very worst rooms to he[a]r or speak in, ever devised by the ingenuity of an architect; but the committee of the Convention have made some improvements, which make it more tolerable. A vestibule with inside cloth-covered door opening without noise, the entire floor carpeted" had helped a little. Some men plainly understood that sound could be absorbed. See P., "Letter from Cincinnati," *National Era,* Dec. 12, 1850. See also Beecher, *Treatise on Domestic Economy,* esp. 338–43; Webster and Parkes, *American Family Encyclopedia,* 256; and on the effects of wood paneling on the "acoustic properties" of a theater, see "Wallack New Theatre," *New York Herald,* Sept. 19, 1861.

45. Kulik, "Beginnings of the Industrial Revolution," 264, 271; quotation on Philadelphia in Meranze, *Laboratories of Virtue,* 229.

46. Cathell Diary, Nov. 25, 1851, 7, SHC.

47. *Harper's Weekly,* Jan. 28, 1860, 64, and Mar. 3, 1860, 142; see also Feb. 8, 1862, 96. Colonial antecedents abound. Boston's officials in the mid-seventeenth century paid someone four pounds annually to ring the town bell "at nine o'clock at night and half-past four in the morning." In 1650 a clock was established. In 1645 officials decided that the "said town of Salem shall provide two good drums" to be used to alert inhabitants of an enemy's approach. In the 1650s the town watch or patrol was charged "that they Silentlie but vigilantlie walke theire seuerall turnes in the seuerall quarters and partes of the Towne." By the 1660s a town bell was used to regulate the times of town markets, and the ringer of the market bell was paid "12d p. yeare." See Benton, *Story of the Old Boston Town House,* 8, 41, 70, 74. On rural schools that had bells with which to call students to study, see L. Maria Child, "Home and Politics," *National Era,* Nov. 16, 1848.

48. "Notes on the Arts and Sciences," *Harper's Weekly,* Feb. 14, 1857, 111; "Chime of Church-Bells," ibid., May 26, 1860, 324; "Amalgam Bells," ibid., Apr. 25, 1863, 272.

49. Hatheway Diary, Nov. 30, 1845, FSU.

50. David N. Johnson, *Sketches of Lynn,* 286–87; Bruegel, "'Time That Can Be Relied Upon,'" 553. See also John F. Watson, *Annals of Philadelphia,* 1:41–42. On the alarm system used in Philadelphia in the 1820s, see Milbert, *Picturesque Itinerary,* 276.

51. Habersham Diary, Aug. 4, 1831, 11–12, and Cornish Diary, Sept. 22, 1834, 1:25, and Feb. 18, 1837, 2:72, SHC.

52. Poore, *Perley's Reminiscences,* 44.

53. N. B. Gordon, *Diary* (1829–30), in *NEMV,* 290; McLeod, "Philadelphia Artisan," 25; *Independent Mechanic,* Nov. 30, 1811, 2.

54. *Records from the Life of S. V. S. Wilder* (New York, 1865), in *NEMV,* 246. On Gregg, see Downey, "Planting a Capitalist South," 166–68.

55. George S. White, *The Moral Influence of Manufacturing Establishments*

(1836), in *NEMV*, 347–51, 355–60, 366; Zachariah Allen, *The Practical Tourist* (1832), in *NEMV*, 7.

56. Thomas, "Romantic Reform in America," 672; Dix, *Remarks on Prisons*, 21. See, too, the arrests made for disturbing the peace, in Washington, D.C., Police Department Ledger and Daybook, 1862–65, LC; Haskell, "Capitalism and the Origins of the Humanitarian Sensibility," esp. 108–11.

57. Dickens, *American Notes*, 30–31, 63. On manufacturing work in antebellum prisons, see McLeod, "Philadelphia Artisan," esp. 8–12. Generally, see Foucault, *Discipline and Punish*.

58. Martineau, *Retrospect*, 123–25, 137. On the use of the system in a New York prison in the 1820s, see Milbert, *Picturesque Itinerary*, 19.

59. Lawrence M. Friedman, *Crime and Punishment*, 79–82, 156–59.

60. Meranze, *Laboratories of Virtue*, 188–89, 194–95, 169, 173–74, 110, 239. See also Upton, "City as Material Culture," 62–63.

61. Meranze, *Laboratories of Virtue*, 218–19.

62. Ibid., 141, 133. See also Bender, *Imagining the Penitentiary;* Upton, "City as Material Culture," 63–68.

63. Meranze, *Laboratories of Virtue*, 253, 262. Like slaves, northern inmates used the institutionalized sounds of prison life to escape. In 1789 twenty-two inmates escaped from a Philadelphia prison and managed to get out of the main door thanks to one prisoner, who having donned a turnkey's hat and coat, ventured to the prison's outer doors, where "as usual with the turnkey, [he] rattled the keys as a signal for the keeper of the outer door to come and let them out" (ibid., 92). See also Rothman, *Discovery of the Asylum;* Ignatieff, *Just Measure of Pain*.

64. Lawrence M. Friedman, *Crime and Punishment*, 156. See also Teeters and Shearer, *Prison at Philadelphia;* de Beaumont and de Tocqueville, *On the Penitentiary System;* Gray, *Prison Discipline in America;* Ayers, *Vengeance and Justice;* Upton, "City as Material Culture," 66; and esp. Hindus, *Prison and Plantation*, 163–69, 171–72, 210–17.

65. Ralph Waldo Emerson, "Nature," *North Star*, Feb. 25, 1848; John Smith the Younger, "Bon Sejour," *National Era*, Aug. 19, 1847; Cruz, *Culture on the Margins*, 115–17. See also Nathaniel Hawthorne, "The Great Stone Face," *National Era*, Jan. 24, 1850.

66. G. B., "Letter from the Editor," *National Era*, Sept. 14, 1848.

67. All elites found solace in nature's rhythm, and in this respect all were caught in the romanticization of the pastoral. On consuming the quietude of northern springs and spas—particularly for the sick—see Milbert, *Picturesque Itinerary*, 173–74. One northerner in 1822 noted how, during a trip to New York, his "first time of hearing the Whipoorwill; eighty one miles from Boston" (West Travel Journal, May 22, 1822, Walcott Family Papers, MHS).

CHAPTER 5

1. I can think of no better work for giving a feel of the political and (albeit incidental) acoustic texture of the antebellum North than the fine study by Wilentz, *Chants Democratic*.

2. See Fletcher Melvin Green, *Democracy in the Old South*, esp. 66–81. On southern democratic participation among nonslaveholding southern whites, see particularly Thornton, *Politics and Power*; Ford, *Origins of Southern Radicalism*.

3. North, *Economic Growth*, 110.

4. Figures are from James M. McPherson, "Antebellum Southern Exceptionalism," 236–38.

5. According to Schafer, a steam engine produces a level of 85 dBA; printing works, 87 dBA; a boiler works and hammering, 118 dBA. The level of a modern jet taking off, by way of comparison, is 120 dBA; a rocket launching, 160 dBA. I offer this information simply to convey some idea of the loudness of industrialization. The cognitive interpretation of the timbre of each activity is another matter. Note that the A after dB indicates that the lower frequencies of the sound recorded were discriminated against by a weighting in the measuring device in a manner roughly the same as the human ear's discrimination against low-frequency sounds. The notation "dBB" indicates less discrimination, while "dBC" represents virtually flat response to the sound that is being measured. See Schafer, *Tuning*, 39.

6. Ibid., 43, 78–79.

7. Carolyn F. Ware, *Launching the Industry* (1931), in *NEMV*, 34–35; David Wilkinson, *Reminiscences* (1846), in *NEMV*, 85; Truax, *Handbook*, 4–5.

8. Data from Inter-University Consortium for Political and Social Research, *Study 0003*.

9. Davies, *American Scenes*, 177; "Art. III—The Southern States," *De Bow's Review*, Jan. 1850, 46.

10. See, for example, Thoreau, *Week on the Concord and Merrimack*, esp. 50, 65, 112, 211, 256.

11. The process of what Schama has called "making metaphors" and symbols "more real than their referents" is explained in "Sound Matters: An Essay on Method" at the end of this study. See, though, Schama, *Landscape and Memory*, 61; Hibbitts, "Making Sense of Metaphors"; Mark M. Smith, "Listening," 88–90. On structural, objective similarities between North and South, see Pessen, "Egalitarian Myth"; Pessen, "How Different."

12. James M. McPherson, "Antebellum Southern Exceptionalism," 233, 234.

13. Baker, *America Perceived*, 142. Indeed, individuals who braved the transatlantic passage were sensitive to northern sounds upon arrival. Even hard-of-hearing Harriet Martineau commented on this in the 1830s after she arrived in New York: "All the sounds, except human voices, were quite unlike all we had heard for six weeks. One of my companions took the sound of the catydid for a noise in her head for many hours after" arrival (Martineau, *Retrospect*, 35–36).

14. Philipson, *Reminiscences by Isaac M. Wise*, 18. In *The Jungle* Upton Sin-

clair made a similar point by highlighting how Eastern European immigrants heard the soundscape of industrializing Chicago as modern in the late nineteenth century. The advanced technological base of Chicago generated an aural world that was qualitatively and quantitatively different from that experienced by the largely rural immigrants. Unsure of how to understand the sounds, they fell back on rural comparisons, but the sheer scale of the noise was unmistakably that of a busy, modern, industrial world. As Sinclair put it, "Then the party became aware of another strange thing. This, too, like the color, was a thing elemental; it was a sound, a sound made up of ten thousand little sounds. You scarcely noticed it at first—it sunk into your consciousness, a vague disturbance, a trouble. It was like the murmuring of the bees in the spring, the whisperings of the forest; it suggested endless activity, the rumblings of a world in motion. It was only by an effort that one could realize that it was made by animals, that it was the distant lowing of ten thousand cattle, the distant grunting of ten thousand swine" (Sinclair, *The Jungle,* 25). See, too, the 1877 testimony concerning the "roar of the train" to Norwegian immigrant Andreas A. Hjerpland, in Zempel, *In Their Own Words,* 13. Note as well the suggestive remarks in Cashman, "Sight and Sound of Industrial America."

15. Cornish Diary, Sept. 23, 1835, 1:84, SHC; Thoreau, *Week on the Concord and Merrimack,* 50, 211, 65, 256, 112; Hankins and Silverman, *Instruments and the Imagination,* 109. See also Schafer, *Tuning,* 34–35, 36, 78–79; Truax, *Handbook,* 136; "Agriculture," *Provincial Freeman,* Sept. 2, 1854; Nash, *Wilderness and the American Mind,* 90–92; McIntosh, *Thoreau as Romantic Naturalist,* 283–86. For numerous examples of the northern incorporation of the sounds of modernity, see Leo Marx, *Machine in the Garden,* esp. 17–18, 211–12.

16. *Delaware County Republican,* Feb. 23, 1849.

17. "Letter from London," *National Era,* Dec. 12, 1850. Northern industrialists saw their industrialization as qualitatively different from and superior to the English models, which produced noise and blighted slums and an impoverished, wretched workforce. For a brief but useful discussion of these themes, see *NEMV,* xxiii–xxxiii.

18. Cornish Diary, Aug. 25, 1837, 3:12, SHC; diary entry, July 3, 1849, Lancaster County Historical Society, Lancaster, Pa.; David Wilkinson, *Reminiscences* (1846), in *NEMV,* 87; "Rambling Epistles from New York, by John Smith the Younger, No. 1, the City of New York," *National Era,* Oct. 12, 1848; Sprague Manufacturing Company, "Bankruptcy Inventory and Appraisal" (1830), in *NEMV,* 218; Zachariah Allen, *History of Cotton Manufacture in America* (1861), in *NEMV,* 145.

19. *Delaware County Republican,* Apr. 1, 1857, and Nov. 21, 1851.

20. Ibid., Mar. 19, 1858.

21. Ibid., Mar. 21, 1856, July 29, 1859, and Sept. 13, 1861; David N. Johnson, *Sketches of Lynn,* 158, 160, 162.

22. "Frederick Douglass' Address," *North Star,* Aug. 4, 1848.

23. *Delaware County Republican,* Feb. 8, 1856. See also Sears, *Sacred Places,* 202–3.

24. "Notes by a Resident," *North Star,* June 16, 1848; "Engine Building in Galt," *Provincial Freeman,* May 13, 1854.

25. "The Passion for City Life," *Frederick Douglass's Paper,* Nov. 20, 1851; Cornish Diary, Sept. 9, 1833, 1:2, SHC.

26. On mobocracy and democracy, see Kimberly K. Smith, *Dominion of Voice,* esp. 38–45, 56–63.

27. Kulik, "Beginnings of the Industrial Revolution," 3.

28. McLeod, "Philadelphia Artisan," 35, 37.

29. Testimony of Carl Conrad, in Garraty, *Labor and Capital in the Gilded Age,* 20.

30. McLeod, "Philadelphia Artisan," 56.

31. Kulik, "Beginnings of the Industrial Revolution," 6–7, 24.

32. Ibid., 35, 43.

33. Ibid., 45–46. It should also be remembered that even small changes in technology altered the soundscape. David Johnson remembered "the heavy tread of fishing boots" in antebellum Lynn: "There were no rubber boots then." The introduction of "friction matches" in Lynn in 1833 added a new sound, but one that was quieter than striking flint. See David N. Johnson, *Sketches of Lynn,* 130, [479].

34. Kulik, "Beginnings of the Industrial Revolution," 105, 111.

35. Ibid., 147.

36. See *NEMV,* [202], 240–41; Kulik, "Beginnings of the Industrial Revolution," 370–71. Because factory owners and the middle classes often lived at some distance from industrial centers and because "a sound level drops 6 dB for every doubling of the distance from the source," the middle class had quieter lives even though they were often within earshot of the throb of progress. See Truax, *Handbook,* 65. In some instances it is inappropriate to talk simply of the "factory soundscape," for factories, made up of different rooms with different operations, probably had a variety of acoustic environments. The main mill in Pawtucket at Belper, for example, was made of stone, and each "of the floors was divided into specialized rooms, each used for particular processes in the production of cotton tarn." Until about 1810 when iron became more common, "wood predominated in the textile power-generation system" (Kulik, "Beginnings of the Industrial Revolution," 136).

37. "Linwood factory," *Delaware County Republican,* Dec. 30, 1859; Kulik, "Beginnings of the Industrial Revolution," 138. On the factory bell as regulator of time and work, see Commons et al., *Documentary History,* 5:196. On the running of machinery within factories—gears running "silently and smoothly as belts"—see Montgomery, *Practical Detail of the Cotton Manufacture,* 62.

38. Samuel Ogden, *Thoughts, What Probable Effect the Peace with Great Britain Will Have on the Cotton Manufactures of This Country* (1815), in *NEMV,* 317.

39. Kulik, "Beginnings of the Industrial Revolution," 128.

40. Milbert, *Picturesque Itinerary,* 215.

41. Kulik, "Beginnings of the Industrial Revolution," 134, 283, 286, 355, 357, 363–64. For ways in which disciplined evangelicalism promoted social order

among workers, see Paul E. Johnson, *Shopkeeper's Millennium;* Sellers, *Market Revolution,* 202–36.

42. David N. Johnson, *Sketches of Lynn,* 92, 100–101; McLeod, "Philadelphia Artisan," 165, 167.

43. Kulik, "Beginnings of the Industrial Revolution," 342, 314.

44. Philip S. Foner, *Factory Girls,* 5–6, 7.

45. Ibid., 77.

46. Ibid., 114, 132.

47. Thomas Mann, *Picture of a Factory Village* (1833), in *NEMV,* 339–41.

48. Philip S. Foner, *Factory Girls,* 146.

49. Ibid., 179–80.

50. N. B. Gordon, *Diary* (1829–30), in *NEMV,* 303–5; *Pawtucket Chronicle,* Oct. 18, 1828, in *NEMV,* 266, 265. See also E. P. Thompson, "Time, Work-Discipline, and Industrial Capitalism."

51. David N. Johnson, *Sketches of Lynn,* 3, 5, 16, 340.

52. Woloch, *Early American Women,* 246–47; Cott, *Root of Bitterness,* 129. This and similar examples are used by Herbert Gutman, although to rather different effect. See Gutman, "Work, Culture, and Society," 28–29, but see, too, McDonnell, "'You Are Too Sentimental.'"

53. Schafer, *Tuning,* 74–76. On the threshold levels of acoustic trauma and the causes of permanent hearing loss, see Truax, *Handbook,* 5–6, 16, 86, 146.

54. Philip S. Foner, *Factory Girls,* 88–89.

55. Commons et al., *Documentary History,* 5:189, 318, 117.

56. Mason A. Green, *Springfield,* 486–87.

57. "The Attempted Rescue of Burns," *Provincial Freeman,* June 3, 1854; Kimberly K. Smith, *Dominion of Voice,* esp. 56, 61–63.

58. "Debate in the Senate," *National Era,* Apr. 27, 1848.

59. David N. Johnson, *Sketches of Lynn,* 127; Walton Felch, *The Manufacturer's Pocket-Piece* (1816), in *NEMV,* 326, 330.

60. *Manufacturers' and Farmers' Journal,* May 31, 1824, in *NEMV,* 485–86. On workers' use of noise to protest their exploitation, see also John Adams testimony, *Thompsonville Manufacturing Company v. William Taylor et al.* (1836) in *NEMV,* 508.

61. McLeod, "Philadelphia Artisan," 93.

62. Walton Felch, *The Manufacturer's Pocket-Piece* (1816), in *NEMV,* 326, 330.

63. "Domestic Correspondence," *National Era,* Apr. 8, 1847; "Suffrage: From the Philadelphia *Republic,*" *North Star,* Dec. 8, 1848; Whitman, "Starting from Paumanok," in his *Leaves of Grass,* 17.

64. Cathell Diary, Dec. 1, 1851, 18, SHC.

65. N. Sizer, "Biographical Sketch of Rev. Henry Ward Beecher," *Frederick Douglass's Paper,* Apr. 8, 1852.

66. "Free Soil Convention," *National Era,* Aug. 17, 1848. See also Sheidley, *Sectional Nationalism.* The reference to Cass describes his speech in Cleveland in the summer of 1848 where local Democrats criticized him on his sympathy for extending slavery into the territories. Apparently the haranguing was so loud

that Cass said simply, "Sir, the noise and confusion which pervades this assembly will prevent my being heard on the important topics to which you have called my attention." Following a few more remarks, Cass left. See "General Cass Bothered," *National Era*, June 29, 1848. Democrats apparently resented such silence.

CHAPTER 6

1. Donald L. Robinson, *Slavery in the Age of Revolution*, 80. This was particularly the case during the Revolution. John Hancock's 1774 oration on the Boston Massacre portrayed the British as noisemakers and the colonists as quiet and harmonious. "Our streets nightly resounded with the noise of riot and debauchery," explained Hancock. But worse was to come: "Hence, the rude din of arms which broke in upon your solemn devotions in your temples. . . . Hence, impious oaths and blasphemies so often tortured your unaccustomed ear." Americans, suggested Hancock, were different because they did not share the noise of the Old World, where the din of war and the cacophony of disorder were commonplace. The British were constructed in aural terms during the Revolution in an effort to tighten the definition of national selves and noisy others. See Hancock, "Boston Massacre Oration," 100–108. Jürgen Habermas's depiction of the creation of a public sphere in eighteenth-century Europe—a place where privileged private citizens came together to form a public in an effort to debate in a rational but sometimes passionate manner governmental policy—applied equally to the United States in the eighteenth and nineteenth centuries. The nature of such open conversations and attendant social spaces necessarily gave vent to voice. As the public expanded at the behest of liberal dictates, it grew louder still, literally and metaphorically. It was this development that made the public political and the political aural. See Habermas, *Structural Transformation of the Public Sphere*. See also, Kimberly K. Smith, *Dominion of Voice*.

2. Joseph Harmer, "To the Public" and "Mr. Editor," *Independent Mechanic*, Apr. 6, 1811, 1; "The Passing Times," ibid., Apr. 20, 1811, 2; John Smith the Younger, "No. 5, The Raw Western Member," *National Era*, Feb. 18, 1847.

3. Reid Diary, Feb. 1, 1833, SHC; Hubbs, "Guarding Greensboro," 136. On literal and figurative boisterous antebellum political noise north and south, see Grimsted, *American Mobbing*, 181–217. Unionists in other southern states portrayed secessionists as political noisemakers. "Georgia," for example, in a letter to the editor of Savannah's *Morning News*, castigated secessionists' "thunder tones" for creating unnecessary "alarm" among the people; see *Savannah Morning News*, 1–2, 4, MMC.

4. "Movements in the Southern States," *National Era*, Apr. 5, 1849.

5. Eppes, "Through Some Eventful Years," 116, Pine Hill Plantation Papers, FSU.

6. Valentine Diaries, May 19, Oct. 29, 1852, 12:36, 134, SHC.

7. Ibid., Apr. 6, 1840, 4:71.

8. Ibid., Nov. 22, 1844, 7:71, and Jan. 4, 1851, 11:4. On the didactic, less popu-

list nature of Whig oratory, see Daniel Walker Howe, *Political Culture of the American Whigs,* esp. 27–30.

9. Richard J. Calhoun, *Witness to Sorrow,* 184, 186; Dorman, *Party Politics,* 144.

10. Mason A. Green, *Springfield,* 437; [Joel W. Jones,] "Brief Narrative," 92–93, MMC; Charles Francis Adams, *What Makes Slavery a Question of National Concern . . .* (1855), in Pease and Pease, *Antislavery,* 450; Holt, *Political Crisis,* 121; "The Election in Indiana," *National Era,* July 12, 1849. On the emotionalism of Know-Nothing rhetoric, see Grimsted, *American Mobbing,* esp. 219. On constructions of immigrants as corrupting of democracy, see Kimberly K. Smith, *Dominion of Voice,* esp. 61–66.

11. Cornish, Nov. 22, 1837, 3:23, SHC; *Cincinnati Daily Enquirer,* May 24, 1854.

12. "A Voice from the South," *National Era,* June 15, 1848; John Perkins, "Art. IV—The Public Lands and Land System of the United States," *De Bow's Review,* Aug. 1854, 161.

13. "Vermont Resolutions and the Senate," *National Era,* Jan. 17, 1850; Eppes, "Through Some Eventful Years," 21–26, Pine Hill Plantation Papers, FSU.

14. Clyde N. Wilson, "Economic Platform of John C. Calhoun," 85; Trask, "Constitutional Republicans of Philadelphia," 299, 403, 474.

15. Generally, see Holt, *Political Crisis.*

16. Haskell, "Capitalism and the Origins of the Humanitarian Sensibility"; Cruz, *Culture on the Margins,* 3, 36, 111. Consult also Kimberly K. Smith, *Dominion of Voice,* 184–235; Young, *Domesticating Slavery.*

17. Cathell Diary, Jan. 2, 1852, SHC. On positive and negative visual evaluations of the South by northerners, see Grant, *North over South.*

18. Edward Beecher, *Narrative of Riots at Alton . . .* (1838), in Pease and Pease, *Antislavery,* 271; Loguen, *As a Slave,* 359; Nathaniel Peabody Rogers, "Color-Phobia" (1838), in Pease and Pease, *Antislavery,* 317–19.

19. Webb, *Slavery and Its Tendencies,* 2–3; Gorn, "'Gouge and Bite,'" esp. 24–40. See also Commons et al., *Documentary History,* 2:166; Moore, "Jared Sparks," 425–31.

20. Pease and Pease, *Antislavery,* xliv; "Letters From Scotland, Letter IV," *Independent Mechanic,* Sept. 7, 1811, 1.

21. "From the Northern Budget, Negro Slavery, No. VIII," *Independent Mechanic,* June 29, 1811, 1; "From the Northern Budget, Negro Slavery, No. IX," ibid., July 6, 1811, 1; *Colored American,* Feb. 20, 1841.

22. Baker, *America Perceived,* 109. The South and its masters, the very word "aristocracy," said the *New York Tribune,* had "rather a bad sound" to northerners; see *New York Tribune,* Apr. 29, 1856, quoted in Eric Foner, *Free Soil, Free Labor, Free Men,* 68.

23. "Suffering in the States," *Provincial Freeman,* Jan. 20, 1854; "The Death of Thomas Clarkson," *National Era,* Jan. 28, 1847; John Pierpont, "The Chase," *Liberty Bell* (1843), 112. See also George W. Putnam, "New England," *National Era,* Nov. 8, 1849.

24. Arnold Buffum, *Lecture* . . . (1844) in Pease and Pease, *Antislavery,* 426; Caroline Weston, "The Church and the World," *Liberty Bell* (1839), 49. On the visual dimensions of abolitionist efforts to inspire support for their cause, see Clark, "'Sacred Rights of the Weak'"; Lapansky, "Graphic Discord."

25. David N. Johnson, *Sketches of Lynn,* 175, 178, 180. See also Hoffert, *When Hens Crow,* 60.

26. "Frederick Douglass' Address," *North Star,* Aug. 4, 1848. Note as well the brief but perceptive remarks in Blight, "'Analyze the Sounds,'" 4; Levine, *Black Culture,* 15.

27. [Theodore Dwight Weld], *American Slavery As It Is* (1839), in Pease and Pease, *Antislavery,* 99–100. On southern whipping and punishments, see Fogel and Engerman, *Time on the Cross,* esp. 145. Note also Gutman and Sutch, "Sambo Makes Good," which suggests that corporal punishment on southern plantations was more common but not to the extent that abolitionists suggested.

28. Frank P. Blair Jr., *The Destiny of the Races* . . . (1859), in Pease and Pease, *Antislavery,* 50, 52.

29. Darius Lyman, *Leaven for Doughfaces,* 45, 109. See also Bernard, *Retrospections of America,* 152.

30. "The Convention," *Colored American,* June 27, 1840; "Letter from Fernandina," MMC; "Reply to Mr. A. C. C. Thompson," *North Star,* Oct. 13, 1848. See also Cruz, *Culture on the Margins.*

31. Curtis, "The Crisis Over *The Impending Crisis,*" 166, 170, 176, 178–79; Leonard Bacon, "Slavery" (1846), in Pease and Pease, *Antislavery,* 113–14.

32. Curtis, "The Crisis Over *The Impending Crisis,*" 180–84, 191, 195. See also Grimsted, *American Mobbing,* 114.

33. "Extract," *North Star,* May 26, 1848; Pease and Pease, *Antislavery,* xlv; Grimsted, *American Mobbing,* 99; Webb, *Slavery and Its Tendencies,* 6. See also "Freeman's National Convention," *Colored American,* May 1, 1841; Ashworth, *Slavery, Capitalism, and Politics.*

34. Calvin Robinson, "Manuscript," MMC.

35. William H. Channing, "A Day in Kentucky," *Liberty Bell* (1843), 58–59.

36. Richard, "Notes and Observations," 3–6, 24, MMC. See also Baker, *America Perceived,* 71–72. There were northern visitors to the South whose ears were quite southern. David Ogden of New York, for example, appears to have heard the city of Mobile, Alabama, in 1834 as a native southerner did. Progress and the sounds of slavery joined in curious harmony: "The activity and businesslike appearance of Mobile makes quite a favorable impression on the mind of a stranger. . . . The wharves are lined with shipping and steamboats. . . . The Negroes with their merry ways singing . . . joyful in their bondage and happy in subjection." Such instances were exceptional. See Ogden Journal, Feb. 5, 1834, MMC.

37. See, for example, Warren, *Culture of Eloquence,* 18–19; Lehuu, *Carnival on the Page,* esp. 23. On literacy levels and the dissemination of print culture, see Sellers, *Market Revolution,* 366–71; John, *Spreading the News.*

38. *New York Plaindealer,* "Blessings of Slavery." Contrast the abolitionists' caricature of Kentucky with Boles, *Religion in Antebellum Kentucky;* Freehling, *Road to Disunion,* esp. 209, 465–68.

39. G. B. S., "The Nation's Destiny" and "Statesmen and Slavery," *North Star,* Mar. 17, Jan. 28, 1848; *Boston Daily Evening Transcript,* May 23, 1856; *Hartford Daily Courant,* May 24, 1854; Hale quoted in Eric Foner, *Free Soil, Free Labor, Free Men,* 56; Cordley, *Pioneer,* 66; American and Foreign Anti-Slavery Society, *Address to the Nonslaveholders . . .* (1843), in Pease and Pease, *Antislavery,* 156; *Albany Evening Journal,* May 23, 1854. On early tensions between the self-styled "calm" voices of northern conservative patricians and the excited emotionalism of abolitionists, see Bruce Mills, *Cultural Reformations,* esp. 30–54.

40. Quoted in "Contrasts of Slavery and Freedom," *National Era,* Apr. 19, 1849.

41. De Tocqueville, *Democracy in America,* bk. 1, chap. 14.

42. "S. M. Janney's Review of Rev. William A. Smith's Address on Slavery," *National Era,* Oct. 18, 1849.

43. "Editorial Correspondence," *North Star,* Apr. 14, 1848.

44. "The Sound of Industry," *Provincial Freeman,* Jan. 20, 1854.

45. Meranze, *Laboratories of Virtue,* 299, 311.

46. J. E. Snodgrass, "The Childless Mother," *National Era,* May 6, 1847. Anti-slaveryites who lived in or visited the South reported both the region's acoustic otherness and its enormity to those who were too far away to hear literally. James Redpath heard the South through accents. While in Virginia he felt distanced from the "hum and mellifluous nasal melody of New England pronunciation." But for most antislaveryites, the sound of the South resided less in regional accents and more in the aurality of the South's specific configuration of social and economic relations. See McKivigan, *Roving Editor,* 210.

47. Elias Hicks, *Observations . . .* (1861), in Pease and Pease, *Antislavery,* 147. See also Lydia Maria Child, *An Appeal . . .* (1836), in Pease and Pease, *Antislavery,* 88, 90.

48. Moore, "Jared Sparks," 425, 426–27, 431. Sectional religious schisms were similarly aural. See "Journal of my travels," May 28, 1833, Swem. The critique and simultaneous construction of the noisy South by abolitionists extended to the southern elite. Critiques of proslavery clergy were couched in aural terms: "And then how beautifully pious the priesthood chime in with the patriotic dema-gogue, to denounce all political agitation of slavery, and instantly a loud baying rises to the ear in one incessant din" ("Letter to a Friend in the South, No. II," *Frederick Douglass's Paper,* Aug. 21, 1851).

49. Anthony Benezet, *Observations . . .* (1760), in Pease and Pease, *Antislavery,* 3; Davies, *American Scenes,* 59. See also S. G. Howe, "Scene in a Slave Prison," *Liberty Bell* (1843), 175; Martineau, *Retrospect,* 259; McKivigan, *Roving Editor,* 246; Walter Livezey Johnson, "Masters and Slaves in the Market of Slavery," esp. 256–57.

50. Reprinted in "Slavery and the Slave Trade in the District of Columbia,"

North Star, Sept. 28, 1849; "Progress," *National Era,* Apr. 19, 1849; Webb, *Slavery and Its Tendencies,* 4.

51. Adams, *South-Side View of Slavery,* 75; Loguen, *As a Slave,* 67, 70. On sale days, see Russell, "Slave Auctions on the Courthouse Steps," esp. 333–35, 346. See also Cathell Diary, Jan. 1, 1852, 67, SHC. Slave prisons, such as the one in New Orleans, were aurally little different from slavery per se. The prison was full of "the sound of hellish orgies, intermingled with the clanking of the chains of the more furious demons." This visitor, or rather, auditor "heard the snap of a whip, every stroke of which sounded like the sharp crack of a pistol"; poor creatures "writhed and shrieked" (S. G. Howe, "Scene in a Slave Prison," *Liberty Bell* [1843], 176–77).

52. Frederick Douglass, *The Constitution of the United States . . .* (1860), in Pease and Pease, *Antislavery,* 356; George S. Burleigh, "World-Harmonies," *Liberty Bell* (1843), 25–26; "Address," *Colored American,* Oct. 3, 1840; Greeley quotation in Joyner, *Shared Traditions,* 112.

CHAPTER 7

1. Recent germane literature on the link between capitalism and abolitionism includes Stachiw, "'For the Sake of Commerce,'" and O'Connor, "Slavery in the North." On the importance of the Slave Power to Republican ideology and a belief in wage labor uniting Democrats and Republicans, see Gienapp, "Republican Party and the Slave Power." On republicanism, consult Holt, *Political Crisis;* on wage labor as the cornerstone of Republican ideology, see Eric Foner, *Free Soil, Free Labor, Free Men.* On the common loathing of the Slave Power by Republicans and abolitionists, see Ashworth, *Slavery, Capitalism, and Politics.* For alternative emphases that focus less on slavery and more on ethnocultural factors but nonetheless reveal the Republican preoccupation with establishing cultural hegemony, see, for example, Silbey, *Partisan Imperative,* esp. 166–89. On sound imperialism, see Schafer, *Tuning,* 77. On the shift from northern regionalism to nationalism, see Grant, *North over South.*

2. "The Nation's Appeal to Britannia," *North Star,* Mar. 24, 1848; E. C. Ellis, "A Poem Written for the Rochester Anti-Slavery Bazaar," ibid., Jan. 12, 1849.

3. "Address," *Colored American,* Aug. 25, 1838; Mark M. Smith, *Mastered by the Clock,* chap. 6.

4. E. Godwin, "England and America," *Liberty Bell* (1841), 5, 10; Amos Augustus Phelps, *Lectures on Slavery . . .* (1834), in Pease and Pease, *Antislavery,* 79–80.

5. Beriah Green, *Things for Northern Men to Do . . .* (1836), in Pease and Pease, *Antislavery,* 184, 188–90; *Albany Evening Journal,* May 23, 1854.

6. Mason A. Green, *Springfield,* 509.

7. "Frederick Douglass' Address," *North Star,* Aug. 4, 1848; Douglass, "What to the Slave."

8. Douglass, "What to the Slave."

9. John Greenleaf Whittier, "The Hunters of Men" (1837), in Pease and Pease, *Antislavery*, 103.

10. John Greenleaf Whittier, "A Sabbath Scene" (1855), in Pease and Pease, *Antislavery*, 124–28.

11. "Emancipation in Kentucky," *National Era*, Mar. 15, 1849; Stanford M. Lyman, *Selected Writings*, 92.

12. "An Appeal," *National Era*, Aug. 24, 1848.

13. Hoffert, *When Hens Crow*, 1, 2–3. See also Julie Roy Jeffrey, *Great Silent Army;* "Censor, No. VI," *Independent Mechanic*, July 6, 1811, 1. For the congruence between this aurality and what Isenberg has styled "visual politics," see her thoughtful *Sex and Citizenship*, esp. 41–74.

14. Hoffert, *When Hens Crow*, 62, 97, 101, and see also 71, 93.

15. "Woman and Her Pastor," *Liberty Bell* (1842), 161–62.

16. See, for example, American Anti-Slavery Society, "Declaration . . . ," *The Abolitionist* 1 (Dec. 1833), in Pease and Pease, *Antislavery*, 66.

17. Schafer, *Tuning*, 175–76. On the historical melting of bells for war, see "Rushton Memorial Carillon," Miscellaneous Collection, AU.

18. "The Ohio Black Laws—No. 2," *North Star*, June 2, 1848; "Tremendous Excitement in New Bedford," ibid., Mar. 20, 1851. See also the report concerning the founding of a "go-ahead" abolitionist newspaper, *The Tocsin of Liberty*, in 1841 in *Colored American*, Oct. 30, 1841. On the characterization of northern free blacks as quiet, even during a celebration, see Barney Diary, Aug. 1, 1858, MHS.

19. Douglass, "What to the Slave."

20. Frontispiece of *Liberty Bell*, 1839; Martin Robinson Delany, *The Condition,* . . . (1852), in Pease and Pease, *Antislavery*, 324.

21. See Hughes, *Thirty Years a Slave*, 7.

22. Darius Lyman, *Leaven for Doughfaces*, 116. On the *Liberty Bell* ringing out "the clearest notes of personal, civil, and spiritual liberty," see May, *Some Recollections*, 231.

23. Lydia Maria Child, "The Black Saxons," *Liberty Bell* (1841), 24; "The Trump of Jubilee," ibid., 45, 46–47, 52–53.

24. James Houghton, "A Voice from Erin," ibid. (1842), 59–60.

25. Pease and Pease, *Antislavery*, xxxviii; M. W. C., "Sonnet," *Liberty Bell* (1839), v–vi.

26. John Pierpont, "The Liberty Bell," *Liberty Bell* (1842), 1–3.

27. Pease and Pease, *Antislavery*, xxxi–xxxii, xl–xli; Richards Diary, Dec. 13, 1846, Walcott Family Papers, MHS. See also Cruz, *Culture on the Margins*, 6–7; Curtis D. Johnson, *Islands of Holiness*, esp. 130–32; William Lee Miller, *Arguing about Slavery*, esp. 83–84; Sellers, *Market Revolution*, 202–36; Hammond Commonplace Book, "Does religion prompt to a Monastic life?," Palfrey Papers, MHS; Lawrence J. Friedman, *Gregarious Saints;* Goodman, *Of One Blood*, esp. 67.

28. James Forten Jr., *An Address . . .* (1836), in Pease and Pease, *Antislavery,* 239; "Selections from the Liberator," *North Star,* Sept. 21, 1849.

29. Cruz, *Culture on the Margins,* 26; [Theodore Dwight Weld], *American Slavery As It Is* (1839), in Pease and Pease, *Antislavery,* 95; Beriah Green, *Things for Northern Men to Do . . .* (1836), in Pease and Pease, *Antislavery,* 184; *Ithaca Chronicle,* Mar. 1, 1854.

30. C. L. R., "The Spirit Voice, Or Liberty Call to the Disfranchised," *Colored American,* Aug. 7, 1841.

31. Richard C. Webb, "A Word from Ireland," *Liberty Bell* (1843), 19–20.

32. "Fifteenth Annual Meeting of the American Anti-Slavery Society," *North Star,* June 1, 1849; *Colored American,* May 18, 1839.

33. "A Good Example," *Freedom's Journal,* July 11, 1828; James Russell Lowell, "The Prejudice of Color" (1845), in Pease and Pease, *Antislavery,* 314; William H. Channing, "A Day in Kentucky," *Liberty Bell* (1843), 60; Stephen Symonds Foster, *The Brotherhood of Thieves . . .* (1843), in Pease and Pease, *Antislavery,* 140; New England Anti-Slavery Society, "Annual Report, Extracts," *The Abolitionist* 1 (Jan. 1830), in Pease and Pease, *Antislavery,* 62. See also Cruz, *Culture on the Margins,* 6–7.

34. James A. Thome, "Speech . . ." (1834), in Pease and Pease, *Antislavery,* 92–93.

35. For the broader argument, see Fox-Genovese and Genovese, "Divine Sanction of Social Order," esp. 213. On aural habituation, see Truax, *Handbook,* 117, 119. On the broad social basis of abolitionism and the variety of people who by 1838 made up the 140,000 members of the more than 1,300 societies, see Goodman, *Of One Blood,* esp. 66–67.

36. Pease and Pease, *Antislavery,* xxvii; "Africa in America," *Southern Literary Messenger* 22 (Jan. 1856): 10–11, 1–2; Dewees, *Great Future of America and Africa,* 102–3.

37. Richard J. Calhoun, *Witness to Sorrow,* 123; Hubbs, "Guarding Greensboro," 106. The October 1834 election riots in Philadelphia no doubt reaffirmed slaveholders' worry about the excess of democratic passion being introduced in northern society. See Grimsted, *American Mobbing,* esp. 199. Although riots of this sort were not unknown in the early republic, the 1834 disturbances have been seen as inaugurating a class-conscious and intensified form of rioting. See Gilje, *Road to Mobocracy;* Weinbaum, *Mobs and Demagogues.*

38. See Schafer, *Tuning,* 77. Note also Silbey, *Partisan Imperative,* 166–89; Simpson, *Mind and the American Civil War,* 33–69.

39. Fisher, "Ideology and Imagery in the Law of Slavery," 64; American and Foreign Anti-Slavery Society, *Address to the Non-Slaveholders of the South . . .* (1843), in Pease and Pease, *Antislavery,* 158–59. On the effect of the transport and communication revolutions in heightening sectional consciousness, see Huston, "Property Rights in Slavery."

40. "Speech of Mr. Clingman, of North Carolina, on the Question of Slavery," *National Era,* Jan. 20, 1848. See also Thomas E. Jeffrey, *Thomas Lanier*

Clingman. On the importance of oratory to a region where literacy rates were quite low, see Zboray, *Fictive People,* 196–98 (Helper quotation on 198); Gaines, *Southern Oratory.*

41. *Mississippian,* Mar. 31, 1854.

42. Thornton, *Politics and Power,* 220, 224.

43. A Citizen of Alabama, "James Russell Lowell and His Writings," *De Bow's Review,* Mar. 1860, 272–78.

44. L. C. B. [of Westmoreland, Va.], "The Country in 1950 [*sic*], or the Conservatism of Slavery," *Southern Literary Messenger* 22 (June 1856): 426–27, 428, 430, 432, 433–44. See also Huston, "Property Rights in Slavery."

45. Henry Field James, *Abolitionism Unveiled,* 80.

46. *Abbeville Banner,* July 24, 1856; Joseph A. Turner, "What Are We To Do?," *De Bow's Review,* July 1860, 70, 71; A Mississippian, "Our Country—Its Hopes and Fears," *De Bow's Review,* July 1860, 83, 84, 85. William Gilmore Simms chastised Sumner not just for what he said but for his retreat from the "language of wisdom" into popular, demagogic oratory that "disturb[s] harmony." Southern oratory, by contrast, was supposed to appeal not to emotions but to the rational mind. See Warren, *Culture of Eloquence,* 155–60 (quotation on 161). On the changes in northern oratorical practices, see Clark and Halloran, *Oratorical Culture in Nineteenth-Century America.*

47. "The Hullabaloo about the Forts," *Mercury,* Dec. 19, 1860; "From the Tallahassee *Patriot,*" *National Era,* May 27, 1847; "Vermont Resolutions and the Senate," *National Era,* Jan. 17, 1850.

48. Thomas R. Dew, "Abolition of Negro Slavery," *American Quarterly Review* 12 (1832), in Faust, *Ideology,* 27, 30, 58.

49. *Greenville Southern Enterprise,* Nov. 22, 1860. Virtually identical arguments were made in *Journal of the Convention of the People of South Carolina,* 461–66, justifying the state's secession.

50. Henry Hughes, *Treatise on Sociology: Theoretical and Practical* (1854), in Faust, *Ideology,* 246–47, 248. Others agreed. "The joyous sound of the Drivers Bugle," wrote James Burchell Richardson of South Carolina in 1824, "is rejoicing to summon the workmen to rise to labor; to rouse the drowsy, & stimulate the industrious" (James B. Richardson, Jamesville, to William Henry Burchell Richardson, Nov. 25, 1824, reel 14, frame 47, p. 2, Richardson Papers, DU). Scholars, sometimes unwittingly, agree with slaveholders' assessments of free wage labor. Fisher, for example, recently referred to "the brutal and tumultuous wage labor system used in the industrializing North" (Fisher, "Ideology and Imagery in the Law of Slavery," 52).

51. John C. Calhoun, "Slavery as a Positive Good," 211.

52. *Richmond Enquirer,* Mar. 13, 1857. See also *Charleston Mercury,* May 30, 1856; Stanford M. Lyman, *Selected Writings,* 87.

53. Eppes, "Through Some Eventful Years," 55, Pine Hill Plantation Papers, FSU; William Harper, *Memoir on Slavery, Read Before the Society for the Advancement of Learning of South Carolina at Its Annual Meeting at Columbia, 1837* (1838), in Faust, *Ideology,* 80, 122–23, 133. See also James Henry Hammond, *Two Let-*

ters on Slavery in the United States, Addressed to Thomas Clarkson, Esq. (1845), in Faust, *Ideology*, 180, 191.

54. *Charleston Mercury,* May 30, 1856.

55. "From Columbia," *Yorkville Enquirer,* Jan. 17, 1861.

PART IV

1. See the very helpful essays by Roark, "Behind the Lines"; Faust, "'Ours as well as that of the Men'"; Kolchin, "Slavery and Freedom in the Civil War South."

2. Faust, *Creation of Confederate Nationalism.* We know from the work of psychoacousticians that prolonged exposure to noises that threaten the customary soundmarks of a particular locale can inspire opposition to those noises (witness modern antinoise legislation), sometimes encourage accommodation to alien sounds that then become normalized, or often fatigue individuals and communities to the point of grudging acceptance and capitulation to the noises. See Handel, *Listening,* esp. chap. 6; J. D. Miller, "Effects of Noise on People"; J. D. Miller, "General Psychological and Sociological Effects of Noise"; J. D. Miller, "General Physiological Effects of Noise"; Kryter, *Effects of Noise on Man,* esp. 270–77.

3. On the scale and modernity of the war, see Hagerman, *American Civil War,* esp. xi–xvii.

CHAPTER 8

1. Ross, "Ssh!" See also James M. McPherson, *Battle Cry,* 520. For what we do know about the perception of battle sounds and noises by Union soldiers, we are indebted to the few but thoughtful observations in Hess, *Union Soldier in Battle,* esp. 15–18, 28, 46, 112–13. The sounds of the Civil War are, on the whole, better reflected in book titles than they are in the analysis presented in texts. On book titles, see, most obviously, Cozzens, *This Terrible Sound.*

2. Schafer, *Tuning,* 50; *Freedom's Journal,* Sept. 7, 1827; "Extract of a Letter to the Editor of the Enquirer," *Independent Mechanic,* Dec. 7, 1811, 1. See also Woods, *History of Tactical Communication Techniques.*

3. "The War Begun," *New York Herald,* Apr. 14, 1861.

4. H. W. Longfellow, "The Arsenal at Springfield," *North Star,* Oct. 5, 1840.

5. "Highly Important Events," *New York Herald,* July 1, 1862; Hess, *Union Soldier in Battle,* 15–16.

6. "The Evacuation of Corinth," *New York Herald,* June 5, 1862; Linsley Diary, May 16, 1864, MMC. Also on the sounds of human pain during the war, see Hess, *Union Soldier in Battle,* 28; W. C. King and W. P. Derby, eds., *Camp-Fire Sketches and Battle-field Echoes* (W. C. King, 1887) in Meltzer, *Voices from the Civil War,* 70.

7. Bancroft Diary and Scrapbook, Sept. 29, 1861, AU. Note also Hess, *Union Soldier in Battle,* 17.

8. Bancroft Diary and Scrapbook, Sept. 30, 1861, AU; "Interesting from Gen. Burnside," *New York Herald,* Jan. 11, 1863. Camp Victory, near Richmond, sounded the same. At ten o'clock at night "all was now comparatively quiet. . . . The click of the axes against the stately sons of the forest, or the crash of falling timber, mingled now with the crack of a distant rifle in the advanced pickets through the stillness of a lonely night" ("Highly Important Events," *New York Herald,* July 1, 1862).

9. "The Evacuation of Corinth," *New York Herald,* June 5, 1862; Linsley Diary, Feb. 23, 1862, Oct. 17, 1863, MMC; "The Disaster," *New York Herald,* Dec. 17, 1862; Bancroft Diary and Scrapbook, June 4, 1862, AU; Fellman, *Citizen Sherman,* 179. See also Longstreet Diary, Jan. 15, 1864, MMC.

10. Barrett, *Reminiscences,* 23; Bancroft Diary and Scrapbook, Dec. 13, 1862, Sept. 29, 1861, AU. Everett, "Gettysburg Oration." For examples of aural sensitivity in other battles, some from the ancient world, see Schafer, *Tuning,* 50.

11. "The End," *New York Herald,* Apr. 5, 1865; "Sherman," ibid., Aug. 6, 1864. See also Hess, *Union Soldier in Battle,* 16.

12. "Important from Virginia," *New York Herald,* July 30, 1862; Bancroft Diary and Scrapbook, July 21, 1861, AU; Barrett, *Reminiscences,* 17; "The Disaster," *New York Herald,* Dec. 17, 1862.

13. "Sherman: Operations to July 3," *New York Herald,* July 8, 1864; "The Evacuation of Corinth," ibid., June 5, 1862.

14. W. W. Hall, *Take Care of Your Health,* MMC; Bancroft Diary and Scrapbook, Aug. 5, 6, 7, 1861, AU; Barrett, *Reminiscences,* 8, 15–16.

15. "The Evacuation of Corinth," *New York Herald,* June 5, 1862.

16. Orville Repton [?] to Mary Repton [?], 3, 4, SCL.

17. "Highly Important Events," *New York Herald,* July 1, 1862; "News From Gen. Pope's Army," ibid., Aug. 10, 1862; Robert Dickinson to "Amanda," 2, SCL; Hubbs, "Guarding Greensboro," 277. Generally on visual and aural dimensions to warfare, see Woods, *History of Tactical Communication Techniques.*

18. "Additional Details of the Battle of Pittsburgh," *New York Herald,* Apr. 9, 1862. For a different interpretation stressing the lower register of Union shouts compared to Confederate, see Hess, *Union Soldier in Battle,* 18.

19. "Highly Important Events," *New York Herald,* July 1, 1862; "The Army before Yorktown," ibid., Apr. 26, 1862. On the morale-boosting effects of songs and music used by Union soldiers during the war, see Hess, *Union Soldier in Battle,* 112–13. For a sampling of Civil War songs, see Meltzer, *Voices from the Civil War,* 58–61. On shouting across lines to aggravate the enemy, see Moffatt to "Dear Sister," 3, SCL.

20. "Sheridan!" *New York Herald,* May 17, 1864; "Sherman: Operations to July 3," ibid., July 8, 1864; "The Evacuation of Corinth," ibid., June 5, 1862; "The Battle of Fair Oakes," ibid., June 6, 1862.

21. Truax, *Handbook,* 76, 103; A. William Mills, "Auditory Localization," 310; Ross, "Ssh!"; Chamberlain to "My dear friends," 4, MMC. Aural miscues were

not peculiar to the Civil War. False alarms plagued Americans at all times, and earlier wars created similar problems. See, for example, Whitehead, "Childhood and Youth," 37–38, MMC. On echolocation during First Bull Run, see Burns to "Dear Tim," 2, Burns Papers, SCL.

22. J. A. Campbell, Smyrna Camp-Ground, Ga., July 7, 1864, Special Field Orders No. 46, Union correspondence, orders, and returns relating to operations in the Atlanta Campaign, from July 1, 1864, to Sept. 8, 1864—#4, *Official Records of the Union and Confederate Armies*, ser. 1, vol. 38/5 [S#76]; H. B. Scott, Folly Island, S.C., Feb. 7, 1864, Union correspondence, orders, and returns relating to operations in South Carolina, and the Georgia Coast, from Jan. 1 to Feb. 29, 1864—#1, ibid., vol. 35/1 [S#65]; "Hooker's Army," *New York Herald*, May 9, 1863; "Correspondence of Mr. S. M. Carpenter," *New York Herald*, May 8, 1863; "Important from Charleston," *New York Herald*, Aug. 15, 1863.

23. "The Siege of Yorktown," *New York Herald*, Apr. 22, 1862; Benjamin F. McPherson, "I Must Tell," 99, AU; "Sherman," *New York Herald*, Aug. 6, 1864. See also Bancroft Diary and Scrapbook, Aug. 16, 1862, AU.

24. "Captivity in Rebeldom," *New York Herald*, Feb. 10, 1864; John Walker, *Cahaba Prison*.

25. "The Battle of Perryville," *New York Herald*, Oct. 13, 1862; "The Disaster," ibid., Dec. 17, 1862; Snedeker Diary, May 29, 31, June 2, 1863, Civil War Collection, AU.

26. "The Campaign in the Southwest—Its Tremendous Importance," *New York Herald*, Jan. 5, 1863; "A Great Victory," ibid., July 6, 1863. See also "Sherman," ibid., June 14, 1864; L. Q. W., "Letter from the South, No. 5," MMC; Linsley Diary (Feb. 1862?), MMC.

27. "Charleston," *New York Herald*, Aug. 28, 1863; "The Rebellion," ibid., July 18, 1861; "The Situation," ibid., July 31, 1864.

28. *Charleston Mercury*, July 25, 1861; William L. Jones (48th Regt., N.C. Troops), Petersburg, Va., to "Dear Brother," June 19, 1862, 2, Civil War Papers, FSU. Southern soldiers compared the tremendous noise of battle at Gettysburg to natural phenomena. One man mentioned an earthquake. See Clyde N. Wilson, *Carolina Cavalier*, 196. Mechanical metaphors were used more sparingly. "The truth is," remarked a sergeant in an Alabama company in 1861, "a good soldier is a machine, worked at the option of the superior officers, and not always allowed to exercise the brain, like a clock when it has run down rests perfectly quiet" (Hubbs, "Guarding Greensboro," 158). See also Keith Civil War Diary, inside cover and 18–19, SCL.

29. "Another Account of the Two Days Fighting on the Chickamauga," *Charleston Mercury*, Oct. 1, 1863.

30. Dill Diary, May 12, Nov. 23, 1861, Apr. 5, Dec. 26, 1862, Jan. 4, Feb. 24, May 22, June 10, 23, July 4, 1863, MMC.

31. Memoranda of E. J. Vann, July 1864, 1, Civil War Papers, FSU.

32. "Notes of the War," *Charleston Mercury*, Oct. 23, 1861; "Important News," *New York Herald*, Apr. 7, 1863.

33. John Wesley James, "Incidents in the History of the Civil War," 7, Miscel-

laneous Collection, AU. See also "Your friend Alexander [Beck]," Camp, 20th Geo. Regt., to Miss E. J. Smith, June 8, 1862, 1, Letters of Confederate Soldiers, FSU. Hunting skills probably played a role in these and many other maneuvers. See Marks, *Southern Hunting in Black and White,* esp. 18–28. Civilians likewise listened. One white woman reportedly colored herself black in an effort to penetrate Sherman's Atlanta and "capture the details of his purposes against her people." But first she had to find him. She did so by "following the sound of an occasional cannon shot" (Yarbrough, *Boyhood and Other Days,* 43). See also Benjamin F. McPherson, "I Must Tell," 78, AU.

34. "The Great Battle of Chickamauga—A Connected Account," *Charleston Mercury,* Sept. 30, 1863; *Charleston Mercury,* July 25, 1861; William L. Jones (48th Regt., N.C. Troops), Petersburg, Va., to "Dear Brother," June 19, 1862, 2, Civil War Papers, FSU.

35. Memoranda of E. J. Vann, July 1864, 1, Civil War Papers, FSU.

36. "Charleston," *New York Herald,* Aug. 28, 1863.

37. "Your friend Alexander [Beck]," Camp, 20th Geo. Regt., to Miss E. J. Smith, June 8, 1862, 1, Letters of Confederate Soldiers, FSU; Benjamin F. McPherson, "I Must Tell," 79, AU; Hubbs, "Guarding Greensboro," 226; "From the Army of Northern Virginia," *Charleston Mercury,* Apr. 26, 1864; "Camp Pickens, Manassas Station," *Charleston Mercury,* June 17, 1861.

38. Civil War Poem, MMC. See also Jackson, "Memoir of the Battle," 2, MMC; "Correspondence of Mr. S. M. Carpenter," *New York Herald,* May 8, 1863. Such jitters were nothing new. For years soldiers had listened for bullets. Capt. J. B. Mason, for example, reported how "the Indians balls have whistled by our heads" during the Seminole Wars in Florida; see Capt. J. B. Mason, Tallahassee, to Sarah Ann Mason, Warren, R.I., Feb. 24, 1840, Mason Letters, MMC. On battle noise deafening a Union soldier for two days, see Hess, *Union Soldier in Battle,* 15.

39. "The Ohio Penitentiary," *Charleston Mercury,* Sept. 8, 1863. See also the insistence on the "universal reign of quiet and decorum" experienced by another Confederate prisoner, in Barbiere, *Scraps from the Prison Table,* 117, 106, 131–32, and, conversely, the hideous noise of the Union prison in Maryland recounted in Diary of J. L. McCrorey, June 6, 1864, 2, SCL. On the cacophony experienced by Griffin Frost in a variety of Union prisons, see Frost, *Camp and Prison Journal,* 24, 30, 268.

40. "The United States Army," *New York Herald,* Oct. 30, 1863; "The War," ibid., Apr. 15, 1861; "The News," ibid., Apr. 13, 1861; "Anticipated Attack on the Brooklyn Navy Yard," ibid., Jan. 22, 1861.

41. "For Our Country and For Glory," ibid., Apr. 22, 1861; Thompson Diary, Feb. 10, Mar. 18, 1864, Civil War Collection, AU.

42. "From the West," *Charleston Mercury,* Sept. 22, 1862.

43. "The Invasion," *New York Herald,* June 30, 1863.

44. "Our Baltimore Correspondence," ibid., July 1, 1863. See also "From the West," *Charleston Mercury,* Sept. 22, 1862; "The Rebel Raid," *New York Herald,* July 8, 1864.

45. S. M. Carpenter, "The Rebel Invasion," *New York Herald,* June 25, 1863.

46. Orville Repton [?] to Mary Repton [?], 1, SCL.

47. Drake, *Union and Anti-Slavery Speeches,* 81–82, 93, 123; J. N. McElroy, Elk River, July 25, 1861, Union correspondence, orders, and returns relating to operations in Maryland, Eastern North Carolina, Pennsylvania, Virginia (except southwestern), and West Virginia, from Jan. 1, 1861, to June 30, 1865—#6, *Official Records of the Union and Confederate Armies,* ser. 1, vol. 51/1 [S#107].

48. Mary A. Livermore, *My Story of the War* (Worthington, 1888), in Meltzer, *Voices of the Civil War,* 107; "The Celebration of the Fourth of July," *New York Herald,* July 6, 1863. On the ringing of bells to commemorate Union victories and the winding down of the war, see "The Situation," *New York Herald,* Apr. 12, 1865.

49. *New York Herald,* July 8, 1863; John A. Andrew, Boston, Mass., to "His Excellency A. Lincoln," Feb. 1, 1865, Union correspondence, orders, and returns relating to operations in northern and southeastern Virginia, North Carolina (Jan. 1–31), West Virginia, Maryland, and Pennsylvania, from Jan. 1, 1865, to Mar. 15, 1865—#15, *Official Records of the Union and Confederate Armies,* ser. 1, vol. 46/2 [S#96]; "The Situation," *New York Herald,* Feb. 3, 1865.

50. "The Christmas Festival," *New York Herald,* Dec. 25, 1864; "The Sleighing Carnival," ibid., Feb. 25, 1863. See also ibid., Jan. 13, 14, 28, 1861, Jan. 5, 1865. Only rarely during the war were the same sounds heard in the Confederacy. Snow fell in "the vicinity of Richmond and Petersburg" on March 24, 1864, and "the jingle of sleigh bells was heard." See "The Situation," ibid., Mar. 31, 1864.

51. "Death of the President!!" ibid., Apr. 15, 1865; "The Situation," ibid., Apr. 20, 1865; "The Funeral Train," ibid., Apr. 22, 1865; "The Situation," ibid., Apr. 23, 25, 1865; "Our Grief," ibid., Apr. 17, 1865.

52. Meltzer, *Voices of the Civil War,* 83, 87; *New York Herald,* July 16, 20, 1863. Note, too, the voice of the "mob" at the beginning of the war as depicted in Ivey to "My Dear Sister," 3, SCL. See also Gallman, *North Fights the Civil War,* 147–50.

53. George Fitzhugh, "Conduct of the War," *De Bow's Review,* Jan./Feb. 1862, 139, 140, 142.

CHAPTER 9

1. J. D. B., "Notes of the War," *Charleston Mercury,* Aug. 29, 1861.

2. Manigault to "Mon Cher Pere," Mar. 22, 1861, 1, 2; Nov. 24, 1861, 1, 4, Manigault Papers, SCL. See also Morton to Bonham, SCL.

3. Manigault to "Mon Cher Pere," Dec. 5, 1861, 1–3, Manigault Papers, SCL.

4. Ransom, *John Ransom's Andersonville Diary,* 151.

5. Mundy to "My dear Sister," 2–3, SCL; "From Virginia," *Charleston Mercury,* Jan. 1, 1862. See also Puckett, "'. . . The chains which had bound us so long were well nigh broken,'" 9, 25. Some southern sympathizers in the North simi-

larly kept silent for fear of ostracism or attack. On this and antiwar Democrats, see Gallman, *North Fights the Civil War,* 141, 144–46.

6. On the Confederacy's rapid entry into industrial modernity, see, for example, Luraghi, "Civil War and the Modernization of American Society," esp. 244; DeCredico, *Patriotism for Profit;* Dew, *Ironmaker to the Confederacy;* and the literature cited in Roark, "Behind the Lines," esp. 210–11.

7. Eppes, "Negro of the Old South," 76–77, and Eppes, "Through Some Eventful Years," 156, Pine Hill Plantation Papers, FSU.

8. "Siege Matters—Three Hundred and Eighty Eighth Day," *Charleston Mercury,* Aug. 1, 1864.

9. "Our Richmond Correspondence," ibid., June 24, 1861; "The News," *Richmond Enquirer,* May 13, 1864.

10. George Fitzhugh, "Conduct of the War," *De Bow's Review,* Jan./Feb. 1862, 139; "H.," *Yorkville Enquirer,* Aug. 12, 1863; "Notes of a Recent Tour in the South," *New York Herald,* Nov. 12, 1861; Faust, *Creation of Confederate Nationalism,* 52–54. See also Grimsted, *American Mobbing,* esp. 101.

11. "The Firing Yesterday Morning," *Charleston Mercury,* June 19, 1861.

12. Dawson, *Confederate Girl's Diary,* 43, 79, 116; "Notes of the War," *Charleston Mercury,* Aug. 28, 1861. See also the attempts to maintain domestic quietude in the absence of a husband's authority and the increasing noisiness of children as recounted in [Amanda Dickinson] to "My Dear Husband," SCL.

13. "Army Correspondence," *Charleston Mercury,* June 28, 1864.

14. Black Diary, Oct. 4, 1862, FSU.

15. Dawson, *Confederate Girl's Diary,* 136, 146.

16. Anonymous to "Dearest Mattie," SCL; "Happy New Year!," *Charleston Mercury,* Jan. 1, 1863; W. T. J. [Company E, 20th Regt., Georgia Volunteer Infantry], to "Miss Jennie Smith," Feb. 21, 1862, 1, Letters of Confederate Soldiers, FSU.

17. "Siege Matters," *Charleston Mercury,* Mar. 10, 1864; "The Lesson taught by the Fall of Vicksburg and Port Hudson," ibid., July 24, 1863.

18. Bryce, *Personal Experience,* 9; *Daily Southern Guardian,* Sept. 17, 1863; Capt. Benjamin Wesley Justice to his wife, May 20, 1864, quoted in Power, *Lee's Miserables,* 14.

19. "From the Army at Atlanta," *Charleston Mercury,* July 20, 1864; "Merry Christmas," ibid., Dec. 25, 1862.

20. "Meade's Army," *New York Herald,* Oct. 9, 1863; "Savannah," ibid., Feb. 16, 1865.

21. "Charleston," ibid., Feb. 28, 1865; "News from Port Royal," ibid., Dec. 20, 1861; "The Coming of Autumn and Winter," *Charleston Mercury,* Aug. 29, 1862; Anonymous to "Dear Parents & Sister," 1, SCL.

22. *Am. Sl.,* supp. ser. 1, vol. 4, Ga. Narrs., pt. 2, 508; Taylor, *Reminiscences,* 33; *Am. Sl.,* supp. ser. 1, vol. 3, Ga. Narrs., pt. 1, 116, and supp. ser. 2, vol. 10, Tx. Narrs., pt. 9, 3996; Hurmence, *Before Freedom,* 48.

23. *Am. Sl.,* vol. 13, pts. 3 and 4, Ga. Narrs., 75; supp. ser. 1, vol. 4, Ga. Narrs., pt. 2, 346–47; supp. ser. 2, vol. 10, Tx. Narrs., pt. 9, 4061.

24. Washington, *Up from Slavery*, 21.

25. See the examples given in Levine, *Black Culture*, 51–52.

26. *Am. Sl.*, supp. ser. 1, vol. 3, Ga. Narrs., pt. 1, 261; Hurmence, *Before Freedom*, 12–13.

27. Hughes, *Thirty Years a Slave*, 142–44. See also the similar account in "Story of a Black Refugee," 7, SCL.

28. Hughes, *Thirty Years a Slave*, 146, 150–51.

29. Eppes, "Negro of the Old South," 86, 87, Pine Hill Plantation Papers, FSU; *Am. Sl.*, supp. ser. 1, vol. 5, Ga. Narrs., pt. 2, 568. On the "general chaos and confusion produced by the war" and the possibility of "a servile insurrection" in the South, see "Southern Slavery—The Black Race—the Dangers of a Protracted War," *New York Herald*, Nov. 22, 1863.

30. *Am. Sl.*, vol. 12, Ga. Narrs., pts. 1 and 2, 274–75.

31. "The Value of Church Bells," *Charleston Mercury*, Apr. 3, 1862; "Letter from Richmond," ibid., Nov. 28, 1862. By way of comparison, see Corbin, *Village Bells*, 13–14. Bell metal (an alloy of 77 percent copper to 23 percent tin) is, by design, quite brittle—"it can be broken on the palm of the hand up to 1″ in thickness by a sharp tap with a 2 lb. Hammer," notes the Whitechapel Foundry. Indeed, its brittleness gives the bell its tone, in part. Brittle bell metal could translate into brittle Confederate cannon. See *Whitechapel*, 4, and "Liberty Bell," 2, "Bell Founders," CWF; Boland, *Ring in the Jubilee*, 40–41.

32. (Elder), William Henry to Archbishop John Mary Odin, New Orleans, Louisiana, Mar. 16, 1862, University of Notre Dame Archives, South Bend, Ind.; G. T. Beauregard, Jackson, Tenn., to Father Mullon, Saint Patrick's Church, New Orleans, La., Mar. 20, 1862, Confederate correspondence, orders, and returns relating to operations in Kentucky, Tennessee, North Mississippi, North Alabama, and Southwest Virginia from Mar. 4 to June 10, 1863—#3, *Official Records of the Union and Confederate Armies*, ser. 1, vol. 10/2 [S#11]. On bell donation, see "The Mississippi," *New York Herald*, Sept. 17, 1863.

33. "The Situation," *New York Herald*, Jan. 22, 1865, Sept. 29, 1864; "The Porter-Butler Expedition," *Charleston Mercury*, Jan. 3, 1865. Small bells were not worth the effort to melt. See Malet, *Errand to the South*, 51.

34. "Important from Richmond," *New York Herald*, Feb. 10, 1864; Snedeker Diary, June 2, 1863, Civil War Collection, AU.

35. Yarbrough, *Boyhood and Other Days*, 44.

36. Richard J. Calhoun, *Witness to Sorrow*, 226. St. Michael's bells, housed in the church since 1764, had been removed from the city before. They were seized at the capture of Charleston during the Revolutionary War, sent back to England, and later returned. During the Civil War the bells were removed to Columbia but perished during the burning of the city. See Hughes to Tucker, "Bell Founders," CWF. See also the mistaken belief that the bells were sent to Columbia "to be melted into cannon," in Malet, *Errand to the South*, 33.

37. Drayton to "My dear Commodore," 1–2, SCL. Confederates did not want to hear Union bells of victory, especially when they recalled that "300 bells tolled for the fate of [John] Brown." See Barbiere, *Scraps from the Prison Table*, 41.

38. "Desperation of the Rebels—Their Deficiency in Arms," *New York Herald,* Apr. 20, 1862; "The New Defensive System of the Rebellion," ibid., Apr. 7, 1862. See as well the report in ibid., Apr. 8, 1862.

39. "Sale of Rebel Bells in Boston," *Harper's Weekly,* Aug. 16, 1862, 515.

40. "The News," *Richmond Enquirer,* May 13, 1864; "The Enemy's Dash on Petersburg," *Charleston Mercury,* June 13, 1864. When bells were scarce, then "the screeching and screaming of locomotive whistles" was also used to rouse citizens, as in Charleston in 1863. See "Charleston," *New York Herald,* Aug. 28, 1863. Early in the war the *Charleston Mercury* explained how a funeral procession would be coordinated: "A telegraph operator will be stationed at the Northeastern Railroad Depot to notify the different quarters of the city of the arrival of the remains of the honored dead. . . . From the moment of the arrival the Fire Alarm Telegraph Bells will toll minute bells. If a fire occurs during the time, the usual signals or alarms will be given" ("The Funeral Procession," *Charleston Mercury,* July 25, 1861).

41. "Important News," *New York Herald,* Apr. 7, 1863; Julia and Edward K. Pritchard, to Van Bergen, Miscellaneous Collection, AU; Schafer, *Tuning,* 60; John Cumming, "Music in Its Relation to Religion," *Southern Literary Messenger* 23 (Dec. 1856): 436; Le Conte quoted in Mark M. Smith, *Mastered by the Clock,* 45; [Kate Cumming], *Journal,* 247; "Last Honors to the Heroes of Manassas," *Charleston Mercury,* July 27, 1861.

CHAPTER 10

1. *Am. Sl.,* supp. ser. 1, vol. 11, N.C. and S.C. Narrs., 45; Hurmence, *Before Freedom,* 154; Brown and Taylor, *Gabr'l Blow Sof',* 58.

2. Brown and Taylor, *Gabr'l Blow Sof',* 57; Hurmence, *Before Freedom,* 24; *Am. Sl.,* supp. ser. 2, vol. 10, Tx. Narrs., pt. 9, 4068, and supp. ser. 1, vol. 3, Ga. Narrs., pt. 1, 51; William W. Brown, *Negro in the American Rebellion,* 114. See also *Am. Sl.,* supp. ser. 2, vol. 10, Tx. Narrs., pt. 9, 4046–47.

3. Dancy, Memoirs, 12, MMC.

4. Eppes, "Negro of the Old South," 107, Pine Hill Plantation Papers, FSU.

5. Edward Ball, *Slaves in the Family,* 345; *Am. Sl.,* supp. ser. 2, vol. 10, Tx. Narrs., pt. 9, 4105, 4095, and vol. 16, Va. Narrs., 4. On the malleability of the bell as symbol and slaves' use thereof at the end of the war, see also the brief but perceptive remarks in McDonnell, "Work, Culture, and Society," 146. As another ex-slave remarked, "Answering bells is played out" (quoted in Hunter, *To 'Joy My Freedom,* 4).

6. *Am. Sl.,* supp. ser. 2, vol. 10, Tx. Narrs., pt. 9, 4095, and supp. ser. 1, vol. 3, Ga. Narrs., pt. 1, 87.

7. Hurmence, *Before Freedom,* 40–41.

8. *Am. Sl.,* supp. ser. 2, vol. 10, Tx. Narrs., pt. 9, 4166; supp. ser. 2, vol. 10, Tx. Narrs., pt. 9, 4219; supp. ser. 1, vol. 11, N.C. and S.C. Narrs., 189; Washington, *Up from Slavery,* 22, 42; Hurmence, *Before Freedom,* 127.

9. Eppes, "Negro of the Old South," 84, Pine Hill Plantation Papers, FSU; Hudgins to "Uncle Frank," Swem; George Fitzhugh, "Cui Bono?—The Negro Vote," *De Bow's Review*, Oct. 1867, 292.

10. Eppes, "Negro of the Old South," 91, Pine Hill Plantation Papers, FSU.

11. Ibid., 115, 117; Hart to "Dear Father," June 15, 1869, 1, Hart Letters, MMC.

12. Eppes, "Negro of the Old South," 115, 117, 132, Pine Hill Plantation Records, FSU; *Daily South Carolinian*, Nov. 14, 1864.

13. LeConte, *When the World Ended*, 113–14.

14. Clinton and Silber, *Divided Houses*, 311; "Revised and Amended Prescript," 20–21, 24, MMC; Zuczek, *State of Rebellion*, 91. See also Trelease, *White Terror*, 5. On the continued use of political noise to drown out opponents during debate, see the instances of Democratic silencing of Republican rivals during the 1876 campaign in South Carolina offered in Zuczek, *State of Rebellion*, esp. 172.

15. Page, Articles of agreement, and Wise, Articles of agreement, SCL; Commons et al., *Documentary History*, 2:182; Leigh quoted in Juncker, "'Over the Water,'" 8.

16. Newman, *Code of the City of Atlanta*, 108–9. See also Lewis, *Charter and Code of Ordinances*, 95, 103. The New South's children were admonished to be quiet, particularly in school lest noise disrupt their learning. In 1884, for example, the Board of Public Instruction in Escambia County, Florida, endorsed a series of "Rules and Regulations" governing students' behavior in class. Prominent among the rules was the following admonition: "During the regular exercises of the School, whether of study or resitation, [*sic*] pupils are required to abstain entirely from communication with one another by speaking, writing or signs, without special permission of the teacher." Antebellum factory workers were familiar with such prohibitions. See "Rules and Regulations!," MMC.

17. Lewis, *Charter and Code of Ordinances*, 153–58, 129, 146, 147. Other continuities remained. Atlanta's policemen were entrusted by an 1873 statute to "guard the city from fire . . . by ringing the fire bell and crying 'fire'" (Newman, *Code of the City of Atlanta*, 121). Paupers, drunks, and the irreligious were all deemed likely to disturb the peace, and those who would "disturb or disquiet" both "any congregation met for religious worship" and "the neighborhood" were arrested. See Lewis, *Charter and Code of Ordinances*, 172–73.

18. Reprinted in "Examination of the Colored Schools," *De Bow's Review*, May 1866, 560; *Rouse & Smith v. Martin & Flowers*.

19. [George Fitzhugh], "Art. IV—Commerce, War and Civilization," *De Bow's Review*, Sept. 1866, 256; "The Truth Forcibly Uttered," *Yorkville Enquirer*, July 31, 1862.

20. On the bourgeois and probusiness nature of conservative northern Republicans during Reconstruction, see, for example, Bensel, *Yankee Leviathan*; Simpson, *Mind and the American Civil War*, esp. 70–95.

21. "Under the Shadow of the Dragon," *Overland Monthly*, Nov. 1883, 452; Turner, *Gazetteer of the St. Joseph Valley*, 61; Arnett, *Centennial Thanksgiving Ser-*

mon; Meader, *Merrimack River,* 186; Debar, *West Virginia Hand-Book,* 11; Tuttle, *Illustrated History of the State of Wisconsin,* 657.

22. "Editor's Table," *Appleton's Journal,* Sept. 26, 1874, 410; *Andreas' History of the State of Nebraska.*

23. Eric Foner, *Reconstruction,* 460; Edwin Q. Bell, "In Lieu of Labor," *De Bow's Review,* July/Aug. 1867, 73.

24. Eric Foner, *Reconstruction,* 477, 478, 490. See also Stanley, *From Bondage to Contract,* 85.

25. See Hobsbawm, *Age of Capital;* Eric Foner, *Reconstruction,* esp. 514–15; Eric Foner, *Politics and Ideology,* esp. 170–75; Stanley, *From Bondage to Contract,* 70.

26. Stanley, *From Bondage to Contract,* 103, 178. See also Eric Foner, *Reconstruction,* 520–21. On antebellum precedents, see Isenberg, *Sex and Citizenship.*

27. Smilor, "Toward an Environmental Perspective." Consult also Schwartz, "Noise and Silence," 2–6; the injunction issued against a roller coaster in 1885 in *Schlueter v. Billingheimer;* Schwartz, "Hush"; Tuley, "Ordinance for Revising and Consolidating," 76, 77, 89, on the prohibitions against disorderly gaming and activities pursued by Chicago's working classes.

28. *Sawyer v. Davis.* This overturned the earlier case of *Davis v. Sawyer.* For a summary of the cases, see John C. Williams, "Annotation," 1271. A few cases ruled in favor of communities and against factories. See esp. *Colgate v. New York C. & H.R.R. Co.*

29. *Sparhawk v. Union Pass. Ry. Co.* For the earlier case, see *First Baptist Church in Schenectady v. Schenectady & S.R. Co.* There were exceptions to this. If heavy machinery physically damaged an adjoining business through vibration, it was deemed noise (*Demarest v. Hardham; Goodall v. Crofton*), and the "bleating of calves kept overnight at a slaughter house, to be slaughtered in the morning, to the serious annoyance of a family dwelling" was enjoined, presumably because the noise was deemed threatening to familial integrity; see *Bishop v. Banks,* quotation in *American Digest* 37:1554. For a case where the tapping associated with silversmithing was considered noise because it interfered with what appears to have been the quietude of a middle-class Philadelphia neighborhood, *see Wallace v. Auer.* For an early-twentieth-century legal case in which the sound generated by locomotive shunting was deemed not to be a nuisance even when churchgoers complained that it interfered with their religious worship, see *Taylor et al. v. Seaboard Air Line Ry.* Legal efforts ensued to silence the voice of the laboring class. In 1876 the Massachusetts Supreme Court interpreted workers' silence about conditions of work and wages as legal assent to the legitimacy of the contract. Since this was the case, elites could hardly blame workers for not assenting through silence. Indeed, the legislation encouraged workers to speak out against exploitation. See Stanley, *From Bondage to Contract,* 66.

30. Tuley, "Ordinance for Revising and Consolidating," 90, 93, 4, 5, 75, 105, 143. In the 1870s Chicago prohibited street peddlers. Exceptions were made, mostly to accommodate farmers who came to the city to sell "produce for mar-

ket." Clearly, though, the statute was designed to free the streets from congestion and keep the noise level down. See ibid., 89, 93, and Stanley, *From Bondage to Contract*, 108.

31. Tuley, "Ordinance for Revising and Consolidating," 80, 100.

32. Ibid., 25.

33. Ibid., 80, 78, 101. Technological advances inevitably forced cities to redefine false signals. Electronically communicated alarms now had to be protected, and so Chicago passed an ordinance to that effect in 1865 and 1872. See ibid., 102.

34. Ibid., 111, 112, 143. For examples of deaths and accidents caused by trains passing through towns and villages and the apparent deafness of the victims to the sound of the engine's whistles and bells, see *Village Record*, Aug. 7, 1866, Apr. 19, 1870; *Delaware County America*, Apr. 27, Dec. 21, 1864, Sept. 8, Oct. 20, 1865.

35. Silber, *Romance of Reunion*, 81–82. On health and the feminine South, see ibid., chaps. 1–3. See also Rodgers, *Work Ethic*, 94–124. European soundscapes seem to have lured the more affluent. John Carver Palfrey, for example, was charmed by England's "quiet & orderly streets" and being "wakened by bells" in Zurich during his trip in 1871. He also commented on a Venetian parade in which "the orderly, quiet multitude walk up & down" (Palfrey Journal, June 4 and undated entries, Palfrey Papers, MHS).

36. Stanley, *From Bondage to Contract*, 142; "Tropical Florida," "Key West," and "About Florida," all in MMC. On antebellum precedents on the healthy aspects of living in Florida, see Sewall, *Sketches of St. Augustine*, esp. 50–53.

37. See, "About Florida," MMC.

38. Campbell, *Winter Cities in Summer Lands*, 80, and [Bloomfield], *Bloomfield's Illustrated Historical Guide*, 30, FEC.

39. Campbell, *Winter Cities in Summer Lands*, 14, FEC.

40. *Florida! Its Climate, Productions, and Characteristics*, 34, FEC; Silber, *Romance of Reunion*, 85.

41. Campbell, *Winter Cities in Summer Lands*, 3, 8, FEC.

42. *Florida! Its Climate, Productions, and Characteristics*, 34, FEC.

43. White [?], *Jacksonville, Florida, and Surrounding Towns*, 71, and *Florida! Its Climate, Productions, and Characteristics*, 49, 58, 64, 84, 87, FEC. Progress, especially of the industrial kind, required better railroads, which in turn demanded "blasting . . . rock and dirt," just as it had years earlier. See L. Q. W., "Letter from the South. No. 1," MMC.

44. *Florida: Beauties of the East Coast*, 1–4, 6–8; West Florida Land Co., *Sub-Tropical Exposition Leaflet*; Campbell, *Winter Cities in Summer Lands*, 13 ("carriage rolls"), all in FEC. As a "quaint and ancient town," St. Augustine in 1871 was necessarily home to waters "gently murmuring." See E. B. K., "Glance at St. Augustine," MMC.

45. E. B. K., "Glance at St. Augustine," MMC; Campbell, *Winter Cities in Summer Lands*, 16, FEC; L. Q. W., "Letter from the South. No. 4," MMC.

46. Campbell, *Winter Cities in Summer Lands,* 47, FEC.

47. Ibid., 80; L. Q. W., "Letter from the South. No. 3," MMC.

48. Ceryx, "Pen Sketches," 1, MMC. See also Ambrose B. Hart, Lake City, Fla., to "Dear Father," Dec. 18, 1866, 2, Hart Letters, MMC.

49. Hill Diary, Feb. 8, 1877, 8, 11, and Huston to Davidson, 3–4, MMC.

50. George Franklin Thompson, Journal, 2, MMC. That Thompson, not the landlord, got the time wrong is clear from p. 5 of the journal. See also *Atlantic Monthly,* Feb. 1866, 237.

51. Thorpe, "Life in Virginia," 7, 17, 19, 31, 39–41, Swem.

52. L. Q. W., "Letter from the South. No. 2," MMC; Hart to "My dear brother Wille," Apr. 19, 1868; to "My dear Mother," May 11, 1868, 1; to "Dear Father," June 15, 1869; to "Dear *little* Emily," Feb. 6, 1867, 3; "Negro Character (Special)," July 31, 1870, 2; and Hart to "My very dear Mother," Oct. 11, 1872, 1, all in Hart Papers, MMC.

53. Hart to "My dear Sister Polly Carp," May 1868, 1, Hart Papers, MMC.

SOUND MATTERS

1. On the "paralinguistic" desirability of historical soundscape studies, see Rath, "Sounding the Chesapeake," 3. On the excessive focus on music as the preeminent form of sound, see Kahn, "Introduction," 3. On the necessity of treating seen and heard worlds (and, in fact, all senses) as non-oppositional and thereby avoiding a retreat to classical and medieval tendencies that considered the senses hierarchically, see Kahn, *Noise, Water, Meat,* 3–4; Lowe, *History of Bourgeois Perception,* esp. 6–8; Classen, *Worlds of Sense,* esp. 2–4; Feld, "Waterfalls of Song," 94–96; Idhe, *Listening and Voice,* esp. 20–22; Howes, *Varieties of Sensory Experience.* The literature on music is voluminous, but on American music and its social context, see Crawford, *American Musical Landscape;* Heintze, *American Musical Life;* Malone, *Southern Music, American Music.* Also useful for its theoretical insights is Attali, *Noise.* On the vocal world in colonial America, see Kamensky, *Governing the Tongue.* Neither is this study preoccupied with nineteenth-century auditory communications technologies. On this matter, consult Sterne, "Audible Past."

2. On the ascendancy of the visual in Western thought and culture, the primacy of hearing in non-Western societies, and our fetish with the ocular, see Stoller, "Sound in Songhay Cultural Experience," 559–70; Foucault, *Birth of the Clinic,* esp. 3; Schafer, *Tuning.* Although generalizations about cultures are notoriously dangerous, some observers have nevertheless attempted to delineate precisely between the cultural geography of Western and non-Western meanings of sound. Much of this work has focused on West African and Western European cultural constructions of the heard world. See, for example, Stoller, "Sound in Songhay Cultural Experience," esp. 560; Ridington, "In Doig People's Ears." But note also Schmidt, *Hearing Things,* 20–37. On calls by linguists and theo-

rists, consult Ong, *Presence of the Word;* Zuckerkandl, *Sound and Symbol.* On recent and forthcoming work by historians, see Bruce R. Smith, *Acoustic World;* Kahn, *Noise, Water, Meat;* Bailey, "Breaking the Sound Barrier," in his *Popular Culture,* chap. 9; Corbin, *Village Bells;* Hibbitts, "Making Sense of Metaphors"; Schwartz, "Beyond Tone and Decibel"; Kamensky, *Governing the Tongue;* Schwartz, "Hearing Aids"; Smilor, "Toward an Environmental Perspective"; Edward T. Hall, *Hidden Dimension,* esp. 42–45. For pioneering work on acoustemology in anthropology, see Feld, *Sound and Sentiment* and his excellent essay "Waterfalls of Song," 91–97. Richard Rath is engaged in a study titled "North American Soundways, 1600–1800." For other work and references to the matter in related disciplines, see the very helpful overview in Kahn, "Big Bang of Auditory Culture." For a recent and particularly good example of the history of deafness in the United States, see Baynton, *Forbidden Signs.* Why this relatively sudden flurry of listening to the past is happening now is not entirely clear. As Kahn has suggested, "These questions can be approached only after the sonic dust settles" (Kahn, "Big Bang of Auditory Culture"). See also the suggestive remarks offered in Upton, "City as Material Culture," esp. 56–59, 64–68. For my part, I came to the study of heard pasts less out of a fascination with new linguistic or theoretical analyses and more through my earlier study of time consciousness in the nineteenth century and, indirectly, via the much earlier and influential work of E. P. Thompson. While I hesitate to speculate on why others seem to have made the shift from an implicit recognition of past sounds (embedded deeply in historical narratives) to the explicit treatment of sound (as in Bruce Smith's work, for example), I, like Thompson and several historians of time consciousness, came to listen more closely to the past simply because it is impossible to understand how time was communicated in the eighteenth and nineteenth centuries without acknowledging that time was as much heard as it was seen. Indeed, in the antebellum South, time had to be heard by the enslaved not least because they were often denied visual access to the watch. See Mark M. Smith, *Mastered by the Clock;* E. P. Thompson, "Time, Work-Discipline, and Industrial Capitalism"; Mark M. Smith, "Old South Time."

3. Rath, "Sounding the Chesapeake," 3. See also Kahn, *Noise, Water, Meat,* esp. 2–19.

4. Kahn, *Noise, Water, Meat,* 4, 5; Bruce R. Smith, *Acoustic World,* 26–29. On the Enlightenment's impact on the sense of hearing and a supposed growing distrust of aurality, see Schmidt, "From Demon Possession to Magic Show" and *Hearing Things.* On hearing as conveying authenticity, see Ong, *Presence of the Word,* esp. 111, 124, 309; Lowe, *History of Bourgeois Perception,* 8. On antebellum orators, see Warren, *Culture of Eloquence.* Note also Corbin, *Village Bells,* xviii–xx; Goody and Watt, "Consequences of Literacy." Soundscape theorist R. Murray Schafer claims one should "formulate an exact impression of a soundscape [rather] than of a landscape," an argument that applies to the historical reclamation of the sounds of the past particularly. "To give a totally convincing image of a soundscape," he maintains, "requires the capturing of thousands of

sounds. Indeed, because the historical work in detecting these sounds recorded in writing and images is so laborious, we can only be relieved that the subject is beginning to attract the attention of historians." Future work, it hardly seems necessary to add, is not merely desirable; it is essential if historians of the American experience are even to begin to complement visual history with the aural. See Schafer, *Tuning*, 7–10; Truax, *Handbook*, 6. See also Schwartz, "Noise and Silence," 2. On metaphor and metonymy, see Tilley, *Metaphor and Material Culture*, esp. 4–6. On the sometimes murky relationship between metaphor and metonymy and for an overview of the argument that metonymy is a type of metaphor, see Fass, *Processing Metonymy and Metaphor*, esp. 46–47.

5. Schafer, *Tuning*, 3–4. The working definition of soundscape here is "An environment of sound (sonic environment) with emphasis on the way sound is perceived and understood by the individual, or by a society. . . . A soundscape is shaped by both the conscious and subliminal perceptions of the listener" (Truax, *Handbook*, 126).

6. Schwartz, "Noise and Silence," 1, 11. The word "noise" is probably derived from *noxa*, a dispute or wrangle, or *nausea*, thus suggesting its pejorative origins. In technical terms a noise is opposed to a tone and is a sound produced by irregular, discordant vibrations. See "Noise," *Century Dictionary and Cyclopedia*.

7. Truax, *Handbook*, 7, 32–33, 39, 143.

8. See Schafer, *Tuning*, 191–94; Truax, *Handbook*, 21–24; Barron, *Tyranny of Noise*, 50.

9. On the problems with quantitative attempts to measure noise, see Truax, *Handbook*, vii–viii.

10. Ibid., 79. Loudness is the subjective impression of the intensity of a sound. On this and the difficulty with measuring loudness, see ibid., 71–72, 93.

11. Gleaner, "That's Music," *Independent Mechanic*, Feb. 8, 1812.

12. J. D. Miller has summarized the problem of relating objective sound levels to subjective perceptions and evaluations of those levels in this way: "Whether a sound is classed as noise depends in part on the quality of auditory experience it produces. If there were a comprehensive system to describe the quality of auditory experience produced by complex sounds, one could imagine a model that would relate auditory experience to physical measurements of the sounds and another model which would in turn relate the quality of the auditory experience to the annoyance produced by the sound." "Unfortunately," laments Miller, "our knowledge is not so extensive as to allow a model of this kind." The absence of such models is a reflection of the utter complexity and contingency of social values and not just objective decibel levels in determining the aesthetics of a particular sound or complex of sounds. Certainly, decibel levels have a role to play here (especially in the aggregate), but it is the skein of social values (which are inevitably historical) that gives particular sounds their social meaning. See J. D. Miller, "General Psychological and Sociological Effects of Noise," 652. On the interrelatedness of sound as abstract and actual, see Kahn, *Noise, Water, Meat*, 3–4.

13. Truax, *Handbook,* 126. See also Truax, "Soundscape Studies." For wise words on context and sound, see Winthrop D. Jordan, *Tumult and Silence,* esp. 2–3, 20–23.

14. Note particularly Kahn, *Noise, Water, Meat;* Bruce R. Smith, *Acoustic World,* 8, 222, 267. Both historians, though, are clearly influenced by a wide range of other ways of understanding the acoustic. On psychoacoustics, see, for example, Handel, *Listening;* Kryter, *Effects of Noise on Man;* J. D. Miller, "General Psychological and Sociological Effects of Noise"; Solomon, "Search for Physical Correlates."

15. Handel, *Listening,* 163, 174; Feld, "Waterfalls of Song," 91. See also Austin, *Sense and Sensibilia.* People tend to hear objective sound levels in consistent ways. Take, for example, loudness. Experiments in psychoacoustics show that respondents generally agree that when a sound "such as a tone or a band of noise is raised in intensity by about 10 dB, it sounds twice as loud." To this must be added the matter of frequency—low-frequency sounds (below 900 Hz) are perceived as less loud than high-frequency sounds (900 to about 5,000 Hz) when both sounds are of equal physical intensity. Industrialization in all likelihood increased the number of high-frequency sounds, and residents in such areas, because of their day-to-day familiarity with such sounds, became attuned to them. Not so with visitors to sites of high-frequency sound. If they come from relatively quieter areas or, in fact, areas where decibels levels are roughly the same but of lower frequency, such visitors hear the high-frequency and high-decibel sounds as noise. See J. D. Miller, "General Psychological and Sociological Effects of Noise," 654.

16. There is no direct link between the physical characteristics of the sound—decibel levels, frequencies, and the like—and the way the sound is perceived. Consequently, acoustic engineers do not figure this definition of noise into their designs because, simply put, they cannot anticipate every perception or estimate the cognitive meaning of each sound. See J. D. Miller, "General Psychological and Sociological Effects of Noise," 655–57; Kryter, *Effects of Noise on Man,* 270–77.

17. Solomon, "Search for Physical Correlates," 492–97; J. D. Miller, "General Psychological and Sociological Effects of Noise," 658.

18. J. D. Miller, "General Psychological and Sociological Effects of Noise," 663–66.

19. Research into the psychology of acoustics, particularly the effects of sound on sleep patterns, is suggestive and revealing. Certain sounds, for example, become familiar, and it is likely that "environmental sounds only disturb sleep when they are unfamiliar." Hence there is an acculturation by individuals to their soundscape. But where are the lines drawn? One axis is rural-urban. As Miller remarks, "A rural person may have difficult[y] sleeping in a noisy [acoustically unfamiliar] urban area." Conversely, "An urban person may be disturbed by the quiet, the sounds of animals, and so on when sleeping in a rural area" (J. D. Miller, "General Psychological and Sociological Effects of Noise," 642).

20. Schafer, *Tuning,* 7–10, 152, 162; Truax, *Handbook,* 68. Schafer tends to perpetuate the visual/aural binary by not giving much explicit recognition to how hearing often complements seeing. See Feld, "Waterfalls of Song."

21. Tilley, *Metaphor and Material Culture,* 16, 8, and see also 33–34. See, too, Lakoff and Johnson, *Metaphors We Live By;* Beck, "Metaphor as Mediator"; Gadamer, *Truth and Method.*

22. Hibbitts, "Making Sense of Metaphors," pt. 1, 1; Schama, *Landscape and Memory,* 61.

BIBLIOGRAPHY

PRIMARY SOURCES

Manuscript Collections

Athens, Georgia
University of Georgia, Hargrett Rare Book and Manuscript Library
    John Basil Lamar, Swift Creek, to "My Dear Sister," Jan. 27, 1835,
        Howell Cobb Papers

Auburn, Alabama
Auburn University, Special Collections and Archives
    Civil War Collection
        John M. Bancroft Civil War Diary and Scrapbook, 1861–64, typescript
        Charles Henry Snedeker Diary, 1863–65, typescript by Mildred Britton,
            1966
        Andrew Thompson Diary, 1864–65, typescript
        Benjamin F. McPherson, "I Must Tell: An Autobiography," unedited
            manuscript
    Miscellaneous Collection
        Dwayne Cox, "The Alabama Supreme Court on Slaves," ser. 3, folder
            "Slavery"
        John Wesley James, "Incidents in the History of the Civil War," n.d.,
            typescript, ser. 3
        Julia and Edward K. Pritchard, Charleston, S.C., to Harmannus H.
            Van Bergen, Atlanta, Ga., July 27, 1976, ser. 1, Samford Carillon
            Dedication folder
        "Rushton Memorial Carillon, Samford University," Birmingham, Ala.,
            May 5, 1968, n.p., ser. 1, Samford Carillon Dedication folder
        "Travels of Henry Bryson, 1826–1827," typescript, ser. 3

Boston, Massachusetts
Massachusetts Historical Society
    Edwin L. Barney Diary, vol. 1, Jan. 1, 1858–June 30, 1871
    John Carver Palfrey Papers
        Mary Anne Hammond Commonplace Book, 1816
        John Carver Palfrey Journal, 1871
    Henry Pickering Walcott Family Papers

William B. Richards Diary, 1846–47
Samuel West Travel Journal, 1821–24

Chapel Hill, North Carolina
University of North Carolina, Wilson Library, Southern Historical Collection
Penelope Eliza Howard Alderman Diary, 1851–56
George F. Bahnson Diary, 1845–46
Henry L. Cathell Diary, 1851–52, 1856
Thomas R. R. Cobb Papers, 1855–58, typescript
Elizabeth Ruffin Cocke Diary, 1827, Harrison Henry Cocke Papers,
   1762–1876, folder 13
Juliana Margaret Courtney Conner Diary, 1827, typescript
John Hamilton Cornish Diary, vols. 1–3, 1833–38, bound typescript
John Early Diary, bks. 1–3, 7–9, 1807–14
Robert Habersham Diary, 1831–32, typescript
Robert Raymond Reid Diary, 1833–35, microfilm
James Stuart Diary, 1814, typescript by William Percy Gibbes
William D. Valentine Diaries, vols. 1–15, 1837–55

Columbia, South Carolina
South Carolina Department of Archives and History
Pendleton/Anderson District, Court of Magistrates and Freeholders, Trial
   Papers, Case 185, *State v. Lossan, Bas, Joe, Andrew, Amos, Aaron, Lewis,*
   Jesse, July 31, 1846
University of South Carolina, South Caroliniana Library
Anonymous, [Charleston, S.C.?], to "Dearest Mattie," [8? Nov. 1861]
Anonymous, Charleston Harbour, to "Dear Parents & Sister," June 10, 1865
J. H. Burns Papers
   J. H. Burns, "1st Quarterly Oration before the Calliopean and
      Polytechnic Society delivered by J. H. Burns, May 11th 1860.
      Subject[:] Italy"
   J. H. Burns, Bull Run, Va., to "Dear Tim" July 27, 1861
[Amanda Dickinson], Darlington, S.C., to "My Dear Husband"
   [Robert Dickinson], July 29, Aug. 31, 1862
Robert Dickinson [CSA, Captain, 21st S.C.], Camp Ford, Charleston, S.C.,
   to "Amanda," June 23, 1862
P[ercival] Drayton, Flag Officer of S[amuel] F[rancis] DuPont, to "My dear
   Commodore" [Gideon Welles], USS *Pawnee,* Stono, June 19, 1862
James Henry Hammond Papers
   S. Bonsall to James Henry Hammond, Barnwell Court House, Apr. 11,
      1833
   William Gregg to James Henry Hammond, Kalmia, Feb. 16, 1857
Edward E. Ivey, Philadelphia, to "My Dear Sister," [S.C.?], Apr. 15, 1861
Elliott Keith Civil War Diary, 1862
Diary of J. L. McCrorey, 2nd Lieut. Co. B. 4th S.C. Cavalry, 1864, typed
   transcription

Murdoch McLean to Sarah McLean, Mar. 19, 1855, Coit-McLean Papers
Louis Manigault Papers
  Louis Manigault, Gowrie (Savannah River) to "Mon Cher Pere"
    [Charles Manigault], Mar. 22, 1861; Nov. 24, 1861
  Louis Manigault, Gowrie (Argyle Island) to "Mon Cher Pere"
    [Charles Manigault], Dec. 5, 1861
Daniel S. Moffatt, Camp Flint Hill, [Va.], to "Dear Sister," Sept. 20, 1861
A. H. Morton, Abbeville District Court House to Milledge Luke Bonham,
    [Brigadier General, CSA], June 6, 1861, box 3, folder 71, Milledge Luke
    Bonham Papers
Johnson M. Mundy, Columbia, [S.C.], to "My dear Sister"
    [Mary E. Mundy], Feb. 20, 1861
Samuel G. Page, Articles of agreement between Samuel G. Page and
    freedpeople, Jan. 31, 1866, Marion District
Orville Repton [?], Co. D, 7th Regt., New Hampshire Volunteers, Morris
    Island, S.C., to Mary Repton [?], Contoocookville, N.H., Aug. 15, 1863
"Story of a Black Refugee," *New-York Semi-Weekly Tribune,* Friday, June 24,
    1864, Henry Fuller Papers
George Wise, Articles of agreement between George Wise and freedpeople,
    Newberry District, 1866; contract until "January 1867"

Durham, North Carolina
Duke University, William R. Perkins Library, Manuscript Department
  "The Church Bell," undated and unidentified newspaper clipping,
    M. J. Solomon Scrapbook, 1861–63, Savannah, Ga.
  Sarah E. J. Boyles Dilworth, "My Reminiscences of the Sixties (1861–65),"
    n.p., William Johnston Cocke Papers
  James Burchell Richardson Papers, 1803–65, in *Records of Ante-Bellum
    Southern Plantations from the Revolution through the Civil War,* edited by
    Kenneth M. Stampp (Frederick, Md.: University Publications of
    America, 1985), ser. F, pt. 2, microfilm

Gainesville, Florida
University of Florida, Smathers Library, Department of Special Collections
  Miscellaneous Manuscripts Collection
    "About Florida," *Watchman & Reflector,* Feb. 27, 1868, box 33
    R. S. C. Campbell, Port Fernandina, Amelia Island, to "Dear Brother
      and Friends," Jan. 31, 1812, box 27
    Ceryx, "Pen Sketches, No. 2. Jacksonville and the St. John's River,"
      Trinity Record (Washington, Pa.), box 33
    Valentine Chamberlain, Hilton Head, S.C., [7th Connecticut
      Volunteers], to "My dear friends," Oct. 10, 1862, box 7
    Civil War Poem to "Mrs. Page," Apr. 24, 1860 [*sic*], typescript, box 67
    Joseph B. Cottrell Diary, 1888[?], box 24
    James M. Dancy, Memoirs of the Civil War and Reconstruction, 1933,
      typescript, box 27

Joseph Dill Diary, May 1861–July 1863, typescript, box 26

E. B. K., "A Glance at St. Augustine," *Christian Register,* July 29, 1871, clipping, box 33

Margaret Seton Fleming Documents, 1876, box 74

Florida Territory Bill, 1845, "A Bill Entitled an Act Concerning Patrols," typescript, box 67

W. W. Hall, M.D., *Take Care of Your Health: Advice to Soldiers* (Boston: American Tract Society, n.d.), box 20

Ambrose B. Hart Letters, 1866–76, box 13

Erastus C. Hill Diary, 1877, typescript, box 26

F. W. Hoskins, "The Story of the St. Joseph Convention: The First Constitutional Convention Held at St. Joseph, Florida, December 3rd 1838," typed manuscript, box 78

Rev. Nathan Hoyt, "A Religious Revival in Tallahassee in 1843," box 19

L. D. Huston, Palatka, Fla., to J. D. Davidson, Mar. 17, 1874, box 7

Ambrose Hull Letters, 1804–28, box 3

Lawrence Jackson, "Memoir of the Battle of Olustee [Fla.], Feb. 20, 1864," 1929, box 27

[Joel W. Jones,] "A Brief Narrative of Some of the Principle Events in the Life of Joel W. Jones with a Few Observations, [1849]," vol. 2, box 78

"Key West," *New-York Tribune,* Mar. 18, 1853, clipping, box 33

"Letter from Fernandina, Fla.," *Boston Semi-Weekly Advertiser,* Mar. 1, 1865, clipping, box 33

Lt. James H. Linsley Diary, 1862–64, box 24

Cornelius H. Longstreet Diary, 1864–65, box 26

L. Q. W., "Letter from the South. No. 1," *Daily Patriot* (Washington, D.C.), Feb. 13, 1872, box 33

———, "Letter from the South. No. 2," *Daily Patriot* (Washington, D.C.), Feb. 14, 1872, box 33

———, "Letter from the South. No. 3," *Daily Patriot* (Washington, D.C.), Feb. 15, 1872, box 33

———, "Letter from the South. No. 4," *Daily Patriot* (Washington, D.C.), Feb. 19, 1872, box 33

———, "Letter from the South. No. 5," *Daily Patriot* (Washington, D.C.), n.d. [1872], box 33

"Manuscript Diary of a Boston Artist: Excursions, Fishing and Bird Hunting on a trip from New York to Key West and Cuba, Feb. 10 to May 31, 1851," n.p., box 74

Daybook of Rev. William McCormick, Aug. 21–Dec. 31, 1860, box 26

Capt. J. B. Mason Letters, 1836–40, typescript, folder 2, box 78

"Notes on St. Paul's Parish, Key West. From old Journals of the Diocese of Fla.," in "Notes on Fla. Episcopal Churches," by Corrine Robinson, n.d., box 44

David Ogden Journal, 1834–35, typescript, box 47

Rosalie Reed, St. Augustine, to Robert R. Reid, May 27, 1840, box 7

"Removal of the Seminoles from Florida," *National Intelligencer,* Nov. 7, 1853, clipping, box 33

"Revised and Amended Prescript of the Order of the ***," [1870], General James Patton Anderson Material, 1822–72, box 64

John C. Richard, "Notes and Observations upon the present condition of Florida," 1843, typescript, box 67

Calvin Robinson, "Manuscript of his residence in Jacksonville during the Civil War," n.d., typescript, box 51

"Rules and Regulations! For the Guidance of the Public Schools of Escambia County, Adopted by the Board of Public Instruction," Pensacola, Apr. 8, 1884, box 67

Savannah Morning News, n.d., box 3

"The Seminole War," *Family Magazine,* 1835, clipping, box 33

Maria Baker Taylor Diary, 1864–68, typescript, folder 1, box 26

George Franklin Thompson, Journal as Inspector with the Bureau of Freedmen, Refugees and Abandoned Lands, Dec. 1865–Jan. 1866, box 24

"Tropical Florida," St. Augustine, Aug. 1848, unidentified newspaper clipping, box 33

W. A. Whitehead, "Childhood and Youth of W. A. Whitehead 1810–1830," 1874, typescript, box 7

Nineteenth Century Promotional Literature, Florida Ephemera Collection, P. K. Yonge Library of Florida History

[Max Bloomfield], *Bloomfield's Illustrated Historical Guide. Embracing an Account of the Antiquities of St. Augustine, Florida, with Map* (St. Augustine, Fla., 1883), #1883

Reau Campbell, *Winter Cities in Summer Lands. A Tour Through Florida and the Winter Resorts of the South* (Cincinnati, 1885), #2585

Florida: Beauties of the East Coast. A Collection of Photographs, with Text by Mrs. H. K. Ingram (St. Augustine, 1893), #2504

Florida! Its Climate, Productions, and Characteristics. A Handbook of Important and Reliable Information for the Use of the Tourist, Settler and Investor. Prepared by John P. Varnum. Issued by the Passenger Department of the Jacksonville, Tampa and Key West Railway, the Trunk Line of the Florida Peninsula (1885), #2584

J. B. White [?], *Jacksonville, Florida, and Surrounding Towns, Containing a brief description of Florida; Her Climate, Health, Soil, Agricultural Products, Fruits, List of Post-Offices, Transportation Lines, Counties, and Other Useful Information* (Jacksonville, Fla.: J. B. White, 1889–90), #737

West Florida Land Co., *Sub-Tropical Exposition Leaflet of De Funiak Springs, Florida* (1888), #266

Lancaster, Pennsylvania
Lancaster County Historical Society
　Diary, 1849, box 2, folder 10, Eleanor Fulton/Presbyterian Collection,
　　MG-50, Manuscript Collection

Richmond, Virginia
Virginia Historical Society
　*Plantation and Farm Instruction, Regulation, Record, Inventory and Account
　　Book. For the Use of the Manager on the Estate of Philip St. George Cocke,
　　And for the better Ordering and Management of Plantation and Farm
　　Business in many Particulars. Second Edition, with additions* (Richmond:
　　J. W. Randolph, 1861), Philip St. George Cocke Papers, 1854–71

South Bend, Indiana
University of Notre Dame Archives
　(Elder), William Henry, Bishop of Natchez, Miss., to Archbishop John
　　Mary Odin, C.M.
　New Orleans, Louisiana, Mar. 16, 1862, calendar: 1862, accessed at
　　http://cawley.archives.nd.edu/calendar/cal1862c.html (6/1/00)

Tallahassee, Florida
Florida State University, Robert Manning Strozier Library, Special
　　Collections Department
　Miss Mary Bates, Tallahassee, Fla., to Charles Dana Jr., Woodstock, Vt.,
　　Apr. 11, 1845, typescript, MSS 0-16, box 137
　Capt. Hugh Black Diary, 6th Florida Infantry, [1862–63], MSS 0-22,
　　box 138, folder 2
　Laura Margaret Smith Brumby Diary, 1884–85, Brumby Family Papers,
　　typescript
　Cherokee Mission Records, 1800–1804, typescript, box 152
　Civil War Papers, 1862–64, MSS 0-61, box 140
　Courtesy Cards, Tallahassee, Fla., Papers, 1853, 1854, 1855, 1868, MSS 0-226,
　　box 157
　Frank Hatheway Diary, 1845–46, MSS 0-137, box 149
　Letters of Confederate Soldiers, 1862–64, typescript, MSS 0-70, box 141
　Batholomew Lynch Journal, Florida War, 1837–39, MSS 0-180, box 152
　Pine Hill Plantation Papers
　　Susan Bradford Eppes, "Negro of the Old South," typed manuscript,
　　　MSS 0-204, folder 4, box 368
　　———, "Through Some Eventful Years," typed manuscript, box 368
　*Plantation and Farm Instruction, Regulation, Record, Inventory, and Account
　　Book* . . . . (Richmond: J. W. Randolph, 1852), MSS 0-66, box 141
　St. Augustine Papers, 1822, 1845, 1853, typescript
　Randolph Whitfield and John Chipman, comps., "The Florida Randolphs
　　and Some Related Families," 1978, typescript, MSS 80-5, folder 1,
　　box 156

Washington, D.C.
Library of Congress, Manuscripts Division
  Washington, D.C., Police Department Ledger and Daybook, 1862–65,
    Daybook, 8th Precinct, microfilm

Williamsburg, Virginia
Colonial Williamsburg Foundation, John D. Rockefeller Jr. Library, Research
  Queries Files
  "Bell Founders"
    "Bruton Church Bells Found A-Pealing"
    Bill Hughes (Whitechapel Foundry) to George Holbert Tucker, July 10,
      1987
    "The Liberty Bell"
    Whitechapel: An Account of the Whitechapel Foundry
    "Whitechapel Bell Foundry Data"
  "Carpets"
  Patricia A. Gibbs to Nancy N. Brown, Sept. 2, 1986
  Patricia A. Gibbs to Thomas W. Dwyer, Feb. 10, 1976
  Mary R. M. Goodwin to Mr. Graham, "Posthorns," Aug. 21, 1958
  Mary R. M. Goodwin to Mr. John Graham, "Horse, Cart, Harness, or
    Wagon Bells," n.d.
  Mrs. Rutherfoord Goodwin to Miss Vivien Kellems, Dec. 15, 1972
  Mrs. Rutherfoord Goodwin to Mrs. E. F. Lay, July 29, 1964
  E. K. Meade Jr. to Mr. Riley, June 3, 1955
  A. P. Middleton to Mr. Alexander, May 21, 1954
  A. P. Middleton to Mr. Campioli, "The Use of Bells at the Palace," Dec. 27,
    1949
  S. P. Moorehead to A. E. Kendrew, "Enlivening the Streets," Oct. 27, 1955
  Lou Powers to Nicholas A. Pappas, "Cannon fire in eighteenth-century
    Williamsburg," Mar. 22, 1985
  Edward M. Riley, "Excerpts from the Virginia Gazette on Snow," Feb. 2,
    1966
  Edward M. Riley to Ms. Agnes P. Miller, Feb. 10, 1976
  E. M. Riley to Mr. Meade, July 7, 1955
  Mary A. Stephenson to Mrs. Carroll Perry, Mar. 9, 1953
  M. Stephenson to E. M. Frank, "Birds in the Landscape," n.d.
  "Streets"
    "Authenticity of Street Conditions"
    Mary Goodwin, "The Streets of Colonial Williamsburg, 1699–1800"
College of William and Mary, Earl Gregg Swem Library, Manuscripts and
  Rare Books Department
  "Journal of my travels as agent of the Presbyterian Education Society in the
    states of Ohio, Kentucky, and Indiana, 1833" (author unknown, possibly
    Rev. Franklin Y. Vail or Rev. John Spaulding), Travel Accounts
  William P. Hudgins, Texas, to "Uncle Frank," Benjamin Franklin Dew,

Newtown P.O., King and Queen Co., Va., Nov. 8, 1865, folder 3, Dew
Family Papers

Bishop William Meade, Millwood Fred., to [a Bishop], May 30, 1832, typed
transcript, box 1, folder 3, William Meade Papers

Overton Family Papers

John S. Claybrooke, Trenton, [N.J.], to Thomas W. Claybrooke,
Cherryville, Haywood Co., Tenn., Jan. 12, 1833, box 9, folder 52

Thomas W. Claybrooke, Travel Account, 1832–33, box 16, folder 15

Margaret Newbold Thorpe, "Life in Virginia by a Yankee Teacher,"
1866–67, positive photostat

John Tyler, *The Merchant of Former Times and the Merchant of Today*, lecture
given by John Tyler to the Petersburg Library Association, May 4, 1854,
pamphlet, box 3, folder 1, group A, Tyler Family Papers

Julia Gardiner Tyler Papers, group A, typescripts

Julia Gardiner Tyler, Sherwood Forest Plantation, Va., to "My dear
Brother," June 17, 1845

———, to "My dear Sister," June 19, 1845

———, White Sulpher Springs, Greenbriar Co., Va., to "My dear
Mama," Aug. 10, 1845

Public Documents

City of Boston. *Report on the Proposed Marginal Street, From Broad Street, at
Rowe's Wharf, to Commercial Street, at Eastern Avenue, with Remarks of
Members of the Board of Aldermen in Favor of Said Measure*. Boston: n.p.,
1868.

Edwards, Alexander. *Ordinances of the City Council of Charleston . . . (1804–1807)*.
Charleston, S.C.: W. P. Young, 1802 [*sic*].

Hening, William Waller, ed. *The Statutes at Large; Being a Collection of all the
Laws of Virginia, from the First Sessions of the Legislature, in the Year 1619*.
Vol. 1. New York: R. & W. & G. Bartow, 1823.

Inter-University Consortium for Political and Social Research. *Study 0003:
Historical Demographic, Economic, and Social Data: U.S., 1790–1970*. Ann
Arbor: ICPSR. http:fisher.lib.Virginia.EDU/cgi-local/censusbin/census/
cen.pl.

*Letters to His Excellency, George S. Boutwell, on His Veto of the Bill to Create The
Eastern Avenue Corporation, in the Session of 1851*. Boston: Office of the Daily
Courier, 1851.

Lewis, Reuben A., comp. *The Charter and Code of Ordinances of the City of
Mobile*. Mobile, Ala.: Advertiser & Register Office, 1866.

McKinstry, Alexander. *The Code of Ordinances of the City of Mobile, With the
Charter, and an Appendix*. Mobile, Ala.: S. H. Goetzel, 1859.

Newman, William T. (revisor). *The Code of the City of Atlanta, Containing the
Acts Incorporating the City, the Ordinances Adopted by the Mayor and Council,
the Acts Organizing the City Courts of Atlanta, and an Appendix Containing*

the *Rules for the Government of Council, and the Members of the city Council from 1848 to 1873, Inclusive.* Atlanta: W. A. Hemphill, 1873.

Official Records of the Union and Confederate Armies. Army Official Record Computer File, CD-ROM. Wilmington, N.C.: Broadfoot, 1995.

Ordinances of the City of Worcester; with the City Charter, and Other City Laws. Worcester, Mass.: Chas. Hamilton, 1854.

State of South Carolina. *Journal of the Convention of the People of South Carolina, Held in 1860, 1861, and 1862. Together with the Ordinances, Reports, Resolutions, etc.* Columbia, S.C.: R. W. Gibbes, 1862.

Tuley, Murray F., comp. "An Ordinance for Revising and Consolidating the General Ordinances of the City of Chicago," in *Laws and Ordinances Governing the City of Chicago.* Chicago: Bulletin Printing Co., 1873.

Legal Cases and Documents

Bishop v. Banks, 33 Conn. 118, 87 Am. Dec. 197 (Conn. 1865).
Colgate v. New York C. & H.R.R. Co. (1906), 51 Misc. 503, 100 NYS 650.
Davis v. Sawyer, 133 Mass. 289, 43 Am. Rep. 519 (Mass. 1882).
Demarest v. Hardham, 34 N.J. Eq. (7 Stew.) 469 (N.J. 1881).
First Baptist Church in Schenectady v. Schenectady & S.R. Co., 5 Barb. 79 (N.Y. 1848).
First Baptist Church in Schenectady v. Utica & S.R. Co., 6 Barb. 313 (N.Y. 1848).
Goodall v. Crofton, 33 Ohio St. 271, 31 Am. Rep. 535 (Ohio 1877).
Owen v. Henman, 1 Watts & S. 548, 37 Am. Dec. 481 (Pa. 1841).
Ray v. Lynes, 10 Ala. 63 (1846), 65–66.
Rouse & Smith v. Martin & Flowers, 75 Ala. 510 (1883), 510–16.
Sawyer v. Davis, 136 Mass. 239 (Mass. 1884).
Schlueter v. Billingheimer (Com. Pl.), 14 Wkly. Law Bul. 224 (Ohio, 1885).
Sparhawk v. Union Pass. Ry. Co., 54 Pa. (4 P. F. Smith) 401 (Pa. 1867).
State v. Boyce, 32 N.C. (10 Ired.) 541 (1849).
Taylor et al. v. Seaboard Air Line Ry., Supreme Court of N.C., Nov. 6, 1907.
Wallace v. Auer, 10 Phila. 356 (Pa. 1875).
Williams, John C. "Annotation: Bells, Carillons, and the Like as Nuisance." *American Law Reports 3d, Cases and Annotations,* 95:1268–72. Rochester, N.Y.: Lawyers Cooperative Publishing, 1979.
Woolf v. Chalker, 31 Conn. 121, 81 Am. Dec. 175 (Conn. 1862).

Contemporary Newspapers, Journals, and Periodicals

Abbeville (S.C.) Banner, 1856
Albany Evening Journal, 1854
Appleton's Journal of Literature, Science, and Art (New York, N.Y.), 1874
Atlantic Monthly (Boston, Mass.) 1866
Boston Daily Evening Transcript, 1856
Charleston Mercury, 1856–64
Cincinnati Daily Enquirer, 1854

Colored American (New York, N.Y.), 1837–41
Daily Georgian (Savannah, Ga.), 1828
Daily South Carolinian (Columbia, S.C.), 1864
Daily Southern Guardian (Columbia, S.C.), 1863
De Bow's Review (New Orleans, La.) 1850–66
Delaware County American (Media, Pa.), 1864–65
Delaware County Republican (Darby and Chester, Pa.), 1849–61
Farmers' Register (Richmond, Va.) 1834–36
Frederick Douglass's Paper (Rochester, N.Y.), 1851–52
Freedom's Journal (New York, N.Y.), 1827–28
Godey's Lady's Book (Philadelphia, Pa.), 1853
Harper's Weekly (New York, N.Y.), 1857–63
Hartford Daily Courant, 1854
Independent Mechanic (New York, N.Y.), 1811–12
Ithaca (N.Y.) Chronicle, 1854
Leslie's Illustrated Newspaper (New York, N.Y.), 1863
Liberty Bell (Boston, Mass.), 1839–58
Mississippian (Jackson, Miss.), 1854
National Era (Washington, D.C.), 1847–50
New York Herald, 1861–65
New York Tribune 1854
North Star (Rochester, N.Y.), 1848–51
Overland Monthly: Devoted to the Development of the Country (San Francisco,
    Calif.), 1883
Pennsylvania Gazette, 1728–65. Accessible Archives *Pennsylvania Gazette*
    Computer File on CD-ROM. Provo, Utah: Folio Corp., 1991– .
Provincial Freeman (Chatham, Canada West), 1854–55
Richmond Enquirer, 1857–65
Southern Enterprise (Greenville, S.C.), 1860
Southern Literary Gazette (Charleston, S.C.), 1829
Southern Literary Messenger (Richmond, Va.), 1856–57
Southern Planter (Richmond, Va.), 1851
Upland Union (Chester, Pa.), 1834
Village Record (West Chester, Pa.), 1823, 1866, 1870
Yorkville (S.C.) Enquirer, 1861–63

Contemporary Accounts

Adams, Nehemiah. *A South-Side View of Slavery; or, Three Months at the South,
    in 1854.* Boston: T. R. Martin, 1855.
Ailenroc, M. R. *The White Castle of Louisiana.* Louisville, Ky.: John P. Morton,
    1903.
Andreas' History of the State of Nebraska: Otoe County. Pt. 8. http://www.ukans.
    edu/carrie/kancoll/andreas_ne/otoe/otoe-p8.html (12/18/99).
Arnett, Benjamin. *Centennial Thanksgiving Sermon.* Online at "Thanksgiving

in American Memory," http://lcweb2.loc.gov/ammen/ndlpedu/features/
thanks/thanks.html (12/18/98).

Avirett, James Battle. *The Old Plantation: How We Lived in Great House and
Cabin Before the War.* New York: F. Tennyson Neely, 1901. Electronic ed.,
University of North Carolina, Chapel Hill, digitization project, 1998.

Bagnall, William R. *Samuel Slater and the Early Development of the Cotton
Manufacture in the United States.* Middletown, Conn.: J. S. Stewart, 1890.

Baker, William J., ed. *America Perceived: A View from Abroad in the Nineteenth
Century.* West Haven, Conn.: Pendulum Press, 1974.

[Ball, Charles]. *Fifty Years in Chains; or, The Life of an American Slave.* New
York: H. Dayton, 1859.

Barbiere, Joe. *Scraps from the Prison Table at Camp Chase and Johnson's Island.*
Doylestown, Pa.: W. W. H. Davis, 1868.

Barrett, O. S. *Reminiscences, Incidents, Battles, Marches, and Camp Life of the Old
4th Michigan Infantry in War of Rebellion, 1861 to 1864.* Detroit: W. S. Ostler,
1888.

Beecher, Catherine. *A Treatise on Domestic Economy.* New York: Schocken
Books, 1977.

Benton, Josiah Henry. *The Story of the Old Boston Town House, 1658–1711.*
Boston: privately printed, 1908.

Bernard, John. *Retrospections of America, 1797–1811.* New York: Harper and
Brothers, 1887.

Bobo, William M. *Glimpses of New-York City, by a South Carolinian (Who Had
Nothing Else to Do).* Charleston, S.C.: J. J. McCarter, 1852.

Bradford, Sarah. *Harriet Tubman: The Moses of Her People.* Gloucester, Mass.:
Peter Smith, 1981.

Brown, Alan, and David Taylor, eds. *Gabr'l Blow Sof': Sumter County,
Alabama, Slave Narratives.* Livingston, Ala.: Livingston Press, 1997.

Brown, John. *Slave Life in Georgia: A Narrative of the Life, Sufferings, and Escape
of John Brown, a Fugitive Slave, Now in England.* Edited by F. N. Boney.
Savannah, Ga.: Beehive Press, 1972.

Brown, William W. *Narrative of William W. Brown, a Fugitive Slave. Written
by Himself.* Boston: Anti-Slavery Office, 1847. In *Five Slave Narratives:
A Compendium,* edited by William Loren Katz. New York: Arno Press,
1968.

———. *The Negro in the American Rebellion, His Heroism and His Fidelity.*
Boston: Lee & Shepard, 1867.

Bryce, Mrs. Sarah Campbell. *The Personal Experience of Mrs. Campbell Bryce
During the Burning of Columbia, South Carolina, by General William T.
Sherman's Army, February 17, 1865.* Philadelphia, 1899.

Calhoun, John C. "Slavery as a Positive Good, Feb. 6, 1837." In *Readings in the
History of the American Nation,* edited by Andrew C. McLaughlin, 206–12.
New York: D. Appleton, 1914.

Calhoun, Richard J., ed. *Witness to Sorrow: The Antebellum Autobiography of
William J. Grayson.* Columbia: University of South Carolina Press, 1990.

Chapin, Charles Wells. *Sketches of the Old Inhabitants and Other Citizens of Old Springfield of the Present Century, and Its Historic Mansions of "Ye Olden Tyme."* Springfield, Mass.: Press of Springfield and Binding Co., 1893.

Cheever, Rev. Henry T., ed. *Autobiography and Memorials of Ichabod Washburn.* Boston: D. Lothrop, 1878.

Chesnut, Mary Boykin. *A Diary from Dixie.* Edited by Ben Ames Williams. Boston: Houghton Mifflin, 1949.

Chivers, Thomas Holley. *Virginalia: Songs of My Summer Nights.* New York: Eugene L. Schwaab, 1942.

Chopin, Kate. *Bayou Folk.* Boston: Houghton Mifflin, 1894. Electronic ed., University of North Carolina, Chapel Hill, digitization project, 1998.

Clinkscales, John George. *On the Old Plantation: Reminiscences of His Childhood.* Spartanburg, S.C.: Band & White, 1916. Electronic ed., University of North Carolina, Chapel Hill, digitization project, 1997.

Commons, John R., Ulrich B. Phillips, Eugene A. Gilmore, Helen L. Sumner, and John B. Andrews, eds. *A Documentary History of American Industrial Society.* Vol. 2, *Plantation and Frontier.* New York: Russell & Russell, 1958.

———. *A Documentary History of American Industrial Society.* Vol. 5, *Labor Movement.* New York: Russell & Russell, 1958.

Cordley, Richard. *Pioneer Days in Kansas.* Boston: Pilgrim Press, 1903.

Cott, Nancy F., ed. *Root of Bitterness: Documents of the Social History of American Women.* New York: Dutton, 1972.

[Cumming, Kate]. *The Journal of Kate Cumming, a Confederate Nurse, 1862–1865.* Edited by Richard Harwell. Savannah, Ga.: Beehive Press, 1975.

Davies, Ebenezer. *American Scenes, and Christian Slavery: A Recent Tour of Four Thousand Miles in the United States.* London: John Snow, 1849.

Dawson, Sarah Morgan. *A Confederate Girl's Diary.* Boston: Houghton Mifflin, 1913. Electronic ed., University of North Carolina, Chapel Hill, digitization project, 1997.

Debar, J. H. Diss. *The West Virginia Hand-Book and Immigrant's Guide.* Parkersburg, W.Va.: Gibbens Bros., 1870.

De Beaumont, Gustave, and Alexis de Tocqueville. *On the Penitentiary System in the United States and Its Application in France.* Philadelphia: Carey, Lea & Blanchford, 1833. Reprint, Carbondale: Southern Illinois University Press, 1964.

De Tocqueville, Alexis. *Democracy in America.* http://xroads.virginia.edu/~HYPER/DETOC/1_ch14.html (4/2/01).

Dewees, Jacob. *The Great Future of America and Africa; An Essay Showing Our Whole Duty to the Black Man, Consistent with our Own Safety and Glory.* Philadelphia: H. Orr, 1854.

Dickens, Charles. *American Notes, and Reprinted Pieces.* London: Chapman and Hall, n.d.

Dix, Dorothea Lynde. *Remarks on Prisons and Prison Discipline in the United States.* Montclair, N.J.: Patterson Smith, 1984.

Dorman, Lewy. *Party Politics in Alabama from 1850 through 1860.* Publication of the Alabama State Department of Archives and History. Historical and Patriotic Series No. 13. Wetumpka, Ala.: Wetumpkin Printing Co., 1935.

Douglass, Frederick. "What to the Slave is the Fourth of July?" July 5, 1852. http://douglass.speech.nwu.edu/doug_a10.htm (2/18/00).

Drake, Charles D. *Union and Anti-Slavery Speeches, Delivered during the Rebellion.* 1864. Reprint, New York: Negro Universities Press, 1969.

Everett, Edward. "Gettysburg Oration, Nov. 19, 1863." In his *Orations and Speeches on Various Occasions,* 4:622–59. Boston: Little, Brown, 1885.

Faust, Drew Gilpin, ed. *The Ideology of Slavery: Proslavery Thought in the Antebellum South, 1830–1860.* Baton Rouge: Louisiana State University Press, 1981.

Fitzhugh, George. *Cannibals All! Or Slaves without Masters.* Edited by C. Vann Woodward. Cambridge, Mass.: Belknap Press of Harvard University Press, 1988.

Flint, Timothy. *Recollections of the Past Ten Years, Passed in Occasional Residences and Journeyings in the Valley of the Mississippi . . . .* Boston: Cummings, Hilliard, 1826.

Foner, Philip S., ed. *The Factory Girls: A Collection of Writings on Life and Struggles in the New England Factories of the 1840s by the Factory Girls Themselves, and the Story, in Their Own Words, of the First Trade Unions of Women Workers in the United States.* Urbana: University of Illinois Press, 1977.

Foreman, Grant, ed. *A Traveler in Indian Territory: The Journal of Ethan Allen Hitchcock, Late Major-General in the United States Army.* Norman: University of Oklahoma Press, 1996.

Frost, Griffin. *Camp and Prison Journal, Embracing Scenes in Camp, on the March, and in Prisons.* Quincy, Ill.: N.p., 1867.

Garraty, John A., ed. *Labor and Capital in the Gilded Age: Testimony Taken by the Senate Committee upon Relations between Labor and Capital—1883.* Boston, Mass.: Little, Brown, 1968.

Garstin, Crosbie, ed. *Samuel Kelly: An English Eighteenth Century Seaman Whose Days Have Been Few and Evil, to Which Is Added Remarks etc. on Places He Visited During His Pilgrimage in This Wilderness.* London: Jonathan Cape, 1925.

Gilman, Caroline. *Recollections of a Southern Matron.* New York: Harper, 1838.

"A Good Wife. From a Sermon of Rev. Dr. Bishop, in the *National Preacher.*" *American Ladies' Magazine,* Apr. 1835, 228–30.

"Grand Pa's Ghost Slave Stories." http://scriptorium.lib.duke.edu/harris/harris-appendix/harris-ap04.html (6/6/99).

Gray, Francis C. *Prison Discipline in America.* London: J. Murray, 1848. Reprint, Montclair, N.J.: Patterson Smith, 1973.

Green, Mason A. *Springfield, 1636–1886: History of Town and City.* Springfield, Mass.: C. A. Nichols, 1888.

G. S. S. "Sketches of the South Santee." *American Monthly Magazine,* Oct.,
    Nov. 1836, 313–19, 431–42.

Hamilton, George, ed. *Cry of the Thunderbird: The American Indian's Own
    Story.* Norman: University of Oklahoma Press, 1972.

Hancock, John. "Boston Massacre Oration." Mar. 5, 1774. In *Three Centuries of
    American Rhetorical Discourse,* edited by Ronald F. Reid, 100–108. Prospect
    Heights, Ill.: Waveland Press, 1988.

Hodgson, Adam. *Letters from North America, Written During a Tour in the
    United States and Canada.* Vol. 1. London: Hurst, Robinson, 1824.

Holsey, L. H. *Autobiography, Sermons, Addresses, and Essays.* Atlanta: Franklin
    Printing and Publishing, 1898. Electronic ed., University of North Carolina,
    Chapel Hill, digitization project, 1999.

Howe, George. *The Endowments, Position, and Education of Woman. An Address
    Delivered Before the Hemans and Sigourney Societies of the Female High School
    at Limestone Springs, July 23, 1850.* Columbia, S.C.: I. C. Morgan, 1850.

Hughes, Louis. *Thirty Years a Slave. From Bondage to Freedom. The Institution of
    Slavery as Seen on the Plantation and in the Home of the Planter.* Milwaukee:
    South Side Printing Co., 1897. Electronic ed., University of North
    Carolina, Chapel Hill, digitization project, 1998.

Hull, John. "Some Passages of God's Providence About Myself and In
    Relation to Myself." *Transactions and Collections of the American Antiquarian
    Society* 3 (1857): 141–64.

Hundley, D. R. *Social Relations in Our Southern States.* New York: Henry B.
    Price, 1860. Electronic ed., University of North Carolina, Chapel Hill,
    digitization project, 1999.

Hurmence, Belinda, ed. *Before Freedom: Forty-Eight Oral Histories of Former
    North and South Carolina Slaves.* New York: Mentor, 1990.

James, Henry Field. *Abolitionism Unveiled; or, Its Origin, Progress, and
    Pernicious Tendency Fully Developed.* Cincinnati: E. Morgan and Sons, 1856.

Johnson, David N. *Sketches of Lynn; or, The Changes of Fifty Years.* Lynn, Mass.:
    Thos. P. Nichols, 1880.

Jordan, Weymouth T., ed. "The Management Rules of an Alabama Black Belt
    Plantation, 1848–1862." *Agricultural History* 18 (Jan. 1944): 53–64.

Kulik, Gary, Roger Parks, and Theodore Z. Penn, eds. *The New England Mill
    Village, 1790–1860.* Cambridge, Mass.: MIT Press, 1982.

LeConte, Emma. *When the World Ended: The Diary of Emma LeConte.* Edited
    by Earl Schenck Miers. New York: Oxford University Press, 1957.

Loguen, Rev. Jermain Wesley. *The Rev. J. W. Loguen, as a Slave and as a
    Freeman: A Narrative of Real Life.* New York: Negro Universities Press,
    1968.

Lyman, Darius. *Leaven for Doughfaces; or, Threescore and Ten Parables Touching
    Slavery.* Cincinnati: Bangs and Co., 1857. Reprint, Freeport, N.Y.: Books
    for Libraries Press, 1971.

Lyman, Stanford M., ed. *Selected Writings of Henry Hughes: Antebellum*

Southerner, Slavocrat, Sociologist. Jackson: University Press of Mississippi, 1985.

McKivigan, James E., ed. *The Roving Editor; or, Talks with Slaves in the Southern States, by James Redpath.* University Park: Pennsylvania State University Press, 1996.

Malet, Rev. William Wyndham. *An Errand to the South in the Summer of 1862.* London: Richard Bentley, 1863.

Martineau, Harriet. *Retrospect of Western Travel.* Vol. 1. London: Saunders and Otley, 1838.

May, Samuel J. *Some Recollections of Our Antislavery Conflict.* Boston: Fields, Osgood, & Co., 1869. Reprint, New York: Arno Press, 1968.

Meader, J. W. *The Merrimack River; its Source and its Tributaries.* Boston: B. B. Russell, 1869.

Meltzer, Milton, ed. *Voices from the Civil War: A Documentary History of the Great American Conflict.* New York: HarperTrophy, 1989.

Milbert, J[acques Gérard]. *Picturesque Itinerary of the Hudson River and the Peripheral Parts of North America.* Translated by Constance D. Sherman. N.p.: Gregg Press, 1968.

Montgomery, James. *A Practical Detail of the Cotton Manufacture of the United States of America; and the State of the Cotton Manufacture of that Country Contrasted and Compared with that of Great Britain.* In *Technology and Power in the Early American Cotton Industry: James Montgomery, the Second Edition of His "Cotton Manufacture" (1840), and the 'Justitia' Controversy about Relative Power Costs,* edited by David J. Jeremy. Philadelphia: American Philosophical Society, 1990.

Moore, John Hammond, ed. "Jared Sparks in Georgia—April 1826." *Georgia Historical Quarterly* 47 (1963): 425–35.

New York Plaindealer. "The Blessings of Slavery, Feb. 25, 1837." In *American Press Opinion,* edited by Allan Nevins, 127–29. Boston: D. C. Heath, 1928.

"Noise." *The Century Dictionary and Cyclopedia. A Work of Universal Reference in all Departments of Knowledge with a New Atlas of the World.* Vol. 5. New York: Century, 1897.

O'Brien, Michael, ed. *An Evening When Alone: Four Journals of Single Women in the South, 1827–1867.* Charlottesville: University Press of Virginia, 1993.

Olive, Johnson. *One of the Wonders of the Age; or, The Life and Times of Rev. Johnson Olive, Wake County, North Carolina.* Raleigh: Edwards, Broughton, 1886. Electronic ed., University of North Carolina, Chapel Hill, digitization project, 1998.

Olmsted, Frederick Law. *A Journey in the Seaboard Slave States, with Remarks on their Economy.* 1856. Reprint, New York: Negro Universities Press, 1968.

Pease, William H., and Jane H. Pease, eds. *The Antislavery Argument.* New York: Bobbs-Merrill, 1965.

Pennington, James W. C. *The Fugitive Blacksmith; or, Events in the History of James W. C. Pennington, Pastor of a Presbyterian Church, New York, Formerly a*

Slave in the State of Maryland, United States. London, 1849. In *Five Slave Narratives: A Compendium,* edited by William Loren Katz. New York: Arno Press, 1968.

Perdue, Charles L. Jr., Thomas E. Barden, and Robert K. Phillips, eds. *Weevils in the Wheat: Interviews with Virginia Ex-Slaves.* Charlottesville: University Press of Virginia, 1976.

Philipson, David., ed. and trans. *Reminiscences by Isaac M. Wise.* New York: Central Synagogue of New York, 1945.

Poe, Edgar Allan. *Tales.* New York: Wiley and Putnam, 1845. Electronic ed., University of North Carolina, Chapel Hill, digitization project, 1997.

Poore, Ben Perley. *Perley's Reminiscences of Sixty Years in the National Metropolis.* Vol. 1. Philadelphia: Hubbard Brothers, 1886.

Pope, John. *A Tour through the Southern and Western Territories of the United States of North-America.* Gainesville: University Presses of Florida, 1979.

The Problem of American Destiny Solved by Science and History. New York: Charles T. Evans, 1863.

Randolph, Peter. *Sketches of Slave Life; or, Illustrations of the "Peculiar Institution."* Boston: printed for the author, 1855.

Ransom, John L. *John Ransom's Andersonville Diary.* New York: Berkley Books, 1988.

Rawick, George P., ed. *The American Slave: A Composite Autobiography.* 41 vols. Westport, Conn.: Greenwood Press, 1972–79.

Rose, Willie Lee, ed. *A Documentary History of Slavery in North America.* New York: Oxford University Press, 1976.

Rosengarten, Theodore. *Tombee: Portrait of a Cotton Planter.* New York: Morrow, 1986.

Sewall, R. K. *Sketches of St. Augustine.* Gainesville: University Presses of Florida, 1976.

Sigourney, Mrs. L. H. *Whisper to a Bride.* Hartford: S. Andrus, 1851.

Sinclair, Upton. *The Jungle.* New York: Bantam Books, 1981.

Stevens, Walter Le Conte. *Sketch of Prof. John Le Conte, Sensitive Flames and Sound-Shadows. Two Articles Reprinted from the Popular Science Monthly for November, 1889.* New York: D. Appleton, 1889.

Stroyer, Jacob. *My Life in the South.* Salem, Mass., 1898. In *Five Slave Narratives: A Compendium,* edited by William Loren Katz. New York: Arno Press, 1968.

Tanner, Benj. T. *An Apology for African Methodism.* Baltimore, Md.: N.p., 1867. Electronic ed., University of North Carolina, Chapel Hill, digitization project, 2000.

Taylor, Susie King. *Reminiscences of My Life in Camp with the 33rd U.S. Colored Troops, Late 1st South Carolina Volunteers.* In *A Black Woman's Civil War Memoirs,* edited by Patricia W. Romero. Princeton, N.J.: Markus Wiener, 1997.

Teamoh, George. *God Made Man, Man Made the Slave: The Autobiography of*

George Teamoh. Edited by F. N. Boney, Richard L. Hume, and Rafia Zafar. Macon, Ga.: Mercer University Press, 1990.

Thoreau, Henry David. *A Week on the Concord and Merrimack Rivers*. Boston: Houghton, Mifflin, 1893.

Thornwell, James Henley. *National Sins. A Fast-Day Sermon: Preached in the Presbyterian Church, Columbia, S.C., Wednesday, November 21, 1860*. Columbia, S.C.: Southern Guardian Steam-Power Press, 1860.

[Timrod, Henry]. *The Poems of Henry Timrod*. Edited by Paul H. Hayne. New York: Arno Press, 1972.

Turner, T. G. *Gazetteer of the St. Joseph Valley, Michigan and Indiana, with a View of its Hydraulic and Business Capacities*. Chicago: Hazlitt & Reed, 1867.

Tuttle, Charles R. *An Illustrated History of the State of Wisconsin. Being a Complete Civil, Political, and Military History of the State*. Boston: B. B. Russell, 1875.

Walker, John. *Cahaba Prison and the Sultana Disaster*. Hamilton, Ohio: Brown and Whitaker, 1910.

[Walker, Jonathan]. *Trial and Imprisonment of Jonathan Walker, at Pensacola, Florida, for Aiding Slaves to Escape from Bondage*. Gainesville: University Presses of Florida, 1974.

Washington, Booker T. *Up from Slavery: An Autobiography*. Garden City, N.Y.: Doubleday, 1901.

Watson, John F. *Annals of Philadelphia, and Pennsylvania, in the Olden Time*. 3 vols. Philadelphia: Edwin S. Stuart, 1891.

Webb, General J. Watson. *Slavery and Its Tendencies: A Letter from General J. Watson Webb to the New York Courier and Enquirer*. Washington, D.C.: Buell & Blanchard, 1856.

Webster, T., and Mrs. Parkes. *The American Family Encyclopedia of Useful Knowledge . . . .* Edited by D. M. Reese. New York: J. C. Derby, 1856.

Whitman, Walt. *Leaves of Grass*. Edited by Sculley Bradley and Harold W. Blodgett. New York: W. W. Norton, 1973.

Wilson, John Lyde. *The Code of Honor; or, Rules for the Government of Principals and Seconds in Duelling*. Charleston, S.C.: James Phinney, 1858.

Witheral, Elizabeth. "How May an American Woman Best Show Her Patriotism?" *Ladies' Wreath* 5 (1850): 313–27.

Woloch, Nancy, ed. *Early American Women: A Documentary History, 1600–1900*. Belmont, Calif.: Wadsworth, 1992.

"Woman." *Godey's Lady's Book*, Aug. 1831, 222.

Yarbrough, George W. *Boyhood and Other Days in Georgia*. Nashville, Tenn.: M. E. Church, 1917.

Young Ladies' Seminary. *Prospectus of the Young Ladies' Seminary in Keene, New Hampshire*. N.p., 1832.

Zempel, Solveig, ed. and trans. *In Their Own Words: Letters from Norwegian Immigrants*. Minneapolis: University of Minneapolis Press, 1991.

SECONDARY SOURCES

Books and Articles

Abrahams, Roger D. *Singing the Master: The Emergence of African American Culture in the Plantation South.* New York: Pantheon, 1992.

Anderson, Benedict. *Imagined Communities: Reflections on the Origins and Spread of Nationalism.* New York: Verso, 1991.

Ashworth, John. "Free Labor, Wage Labor, and the Slave Power: Republicanism and the Republican Party in the 1850s." In *The Market Revolution in America: Social, Political, and Religious Expressions, 1800–1860,* edited by Melvyn Stokes and Stephen Conway, 128–46. Charlottesville: University Press of Virginia, 1996.

———. *Slavery, Capitalism, and Politics in the Antebellum Republic.* Vol. 1, *Commerce and Compromise, 1820–1850.* New York: Cambridge University Press, 1995.

Attali, Jacques. *Noise: The Political Economy of Music.* Translated by Brian Massumi. Minneapolis: University of Minnesota Press, 1985.

Austin, J. L. *Sense and Sensibilia.* Oxford: Oxford University Press, 1962.

Auer, J. Jeffrey, ed. *Antislavery and Disunion, 1858–1861: Studies in the Rhetoric of Compromise and Conflict.* New York: Harper and Row, 1963.

Ayers, Edward L. *Vengeance and Justice: Crime and Punishment in the Nineteenth-Century American South.* New York: Oxford University Press, 1984.

Bailey, Peter. *Popular Culture and Performance in the Victorian City.* Cambridge: Cambridge University Press, 1998.

Ball, Edward. *Slaves in the Family.* New York: Ballantine, 1999.

Barron, Robert Alex. *The Tyranny of Noise.* New York: St. Martin's Press, 1970.

Bateman, Fred, and Thomas Weiss. *A Deplorable Scarcity: The Failure of Industrialization in the Slave Economy.* Chapel Hill: University of North Carolina Press, 1981.

Baynton, Douglas C. *Forbidden Signs: American Culture and the Campaign against Sign Language.* Chicago: University of Chicago Press, 1996.

Beck, B. "The Metaphor as a Mediator between Semantic and Analogic Modes of Thought." *Current Anthropology* 19 (1978): 83–97.

Bender, John. *Imagining the Penitentiary: Fiction and the Architecture of Mind in Eighteenth-Century England.* Chicago: University of Chicago Press, 1987.

Bensel, Richard Franklin. *Yankee Leviathan: The Origins of Central State Authority in America, 1859–1877.* New York: Cambridge University Press, 1990.

Berlin, Ira. *Many Thousands Gone: The First Two Centuries of Slavery in North America.* Cambridge, Mass.: Belknap Press of Harvard University Press, 1998.

Bertelson, David. *The Lazy South.* New York: Oxford University Press, 1967.

Blackmar, Elizabeth. *Manhattan for Rent, 1785–1850*. Ithaca, N.Y.: Cornell University Press, 1989.

Blight, David W. "'Analyze the Sounds': Frederick Douglass's Invitation to Modern Historians of Slavery." In *Slave Cultures and the Cultures of Slavery*, edited by Stephan Palmié, 1–11. Knoxville: University of Tennessee Press, 1995.

Boland, Charles Michael. *Ring in the Jubilee: The Epic of America's Liberty Bell*. Riverside, Conn.: Chatham Press, 1973.

Boles, John B. *Religion in Kentucky*. Lexington: University Press of Kentucky, 1976.

Bolton, Charles C. *Poor Whites of the Antebellum South: Tenants and Laborers in Central North Carolina and Northeast Mississippi*. Durham, N.C.: Duke University Press, 1994.

Bowman, Shearer Davis. "Antebellum Planters and *Vormärz* Junkers in Comparative Perspective." *American Historical Review* 85 (Oct. 1980): 770–808.

Brenneis, Donald Lawrence. "Straight Talk and Sweet Talk: Political Discourse in an Occasionally Egalitarian Community." In *Dangerous Words: Language and Politics in the Pacific*, edited by Donald Lawrence Brenneis and Fred R. Myers, 69–84. New York: New York University Press, 1984.

Brinson, Claudia Smith. "The Cicadas' Soft Summer Song." *The State* (Columbia, S.C.), July 16, 2000.

Bruegel, Martin. "'Time That Can Be Relied Upon': The Evolution of Time Consciousness in the Mid-Hudson Valley, 1790–1860." *Journal of Social History* 28 (spring 1995): 547–64.

Carothers, J. C. "Culture, Psychiatry, and the Written Word." *Psychiatry* 22 (Nov. 1959): 308–10.

Cashman, Sean Dennis. "The Sight and Sound of Industrial America." In his *America in the Gilded Age: From the Death of Lincoln to the Rise of Theodore Roosevelt*, 1–202. New York: New York University Press, 1993.

Clark, Elizabeth B. "'The Sacred Rights of the Weak': Pain, Sympathy, and the Culture of Individual Rights in Antebellum America." *Journal of American History* 82 (Sept. 1995): 463–93.

Clark, Gregory, and S. Michael Halloran, eds. *Oratorical Culture in Nineteenth-Century America: Transformations in the Theory and Practice of Rhetoric*. Carbondale, Ill.: Southern Illinois University Press, 1993.

Classen, Constance. *Worlds of Sense: Exploring the Senses in History and across Cultures*. London: Routledge, 1993.

Clinton, Catherine, and Nina Silber, eds. *Divided Houses: Gender and the Civil War*. New York: Oxford University Press, 1992.

Cmiel, Kenneth. *Democratic Eloquence: The Fight over Popular Speech in Nineteenth-Century America*. New York: Morrow, 1990.

Coclanis, Peter A. *The Shadow of a Dream: Economic Life and Death in the South Carolina Low Country, 1670–1920*. New York: Oxford University Press, 1989.

———. "Slavery, African-American Agency, and the World We Have Lost." *Georgia Historical Quarterly* 79 (1995): 873–84.

Corbin, Alain. *The Foul and the Fragrant: Odor and the French Social Imagination.* Translated by Miriam L. Kochan. Cambridge, Mass.: Harvard University Press, 1986.

———. *Time, Desire, and Horror: Towards a History of the Senses.* Translated by Jean Bissell. Cambridge: Polity Press, 1995.

———. *Village Bells: Sound and Meaning in the Nineteenth-Century French Countryside.* Translated by Martin Thom. New York: Columbia University Press, 1998.

Counter, S. Allen. *Electromagnetic Stimulation of the Auditory System: Effects and Side-Effects.* Stockholm: Scandinavian University Press, 1993.

———. *North Pole Legacy: Black, White, and Eskimo.* Amherst: University of Massachusetts Press, 1991.

Cozzens, Peter. *This Terrible Sound: The Battle of Chickamauga.* Urbana: University of Illinois Press, 1996.

Crawford, Richard. *The American Musical Landscape.* Berkeley: University of California Press, 1993.

Cruz, Jon. *Culture on the Margins: The Black Spiritual and the Rise of American Cultural Interpretation.* Princeton, N.J.: Princeton University Press, 1999.

Curtis, Michael Kent. "The Crisis over *The Impending Crisis:* Free Speech, Slavery, and the Fourteenth Amendment." In *Slavery and the Law,* edited by Paul Finkelman, 161–205. Madison, Wisc.: Madison House, 1997.

Davis, Angela Y. *Women, Race, and Class.* New York: Vintage, 1983.

DeCredico, Mary A. *Patriotism for Profit: Georgia's Urban Entrepreneurs and the Confederate War Effort.* Chapel Hill: University of North Carolina Press, 1990.

Dew, Charles B. *Ironmaker to the Confederacy: Joseph R. Anderson and the Tredegar Iron Works.* New Haven: Yale University Press, 1966.

Dohrn-van Rossum, Gerhard. *History of the Hour: Clocks and Modern Temporal Orders.* Translated by Thomas Dunlap. Chicago: University of Chicago Press, 1996.

Farrell, Betty G. *Elite Families: Class and Power in Nineteenth-Century Boston.* Albany: State University of New York Press, 1993.

Fass, Dan. *Processing Metonymy and Metaphor.* Greenwich, Conn.: Ablex Pub. Co., 1997.

Faust, Drew Gilpin. *The Creation of Confederate Nationalism: Ideology and Identity in the Civil War South.* Baton Rouge: Louisiana State University Press, 1988.

———. "'Ours as well as that of the Men': Women and Gender in the Civil War." In *Writing the Civil War: The Quest to Understand,* edited by James M. McPherson and William J. Cooper Jr., 228–40. Columbia: University of South Carolina Press, 1998.

———. "Slavery in the American Experience." In *Before Freedom Came: African-American Life in the Antebellum South,* edited by Edward D. C.

Campbell and Kim S. Rice, 1–19. Richmond and Charlottesville: Museum
of the Confederacy and the University Press of Virginia, 1991.

Feld, Steven. *Sound and Sentiment: Birds, Weeping, Poetics, and Song in Kalui
Expression.* Philadelphia: University of Pennsylvania Press, 1982.

———. "Waterfalls of Song: An Acoustemology of Place Resounding in
Bosavi, Papua New Guinea." In *Sense of Place,* edited by Steven Feld and
Keith H. Basso, 91–136. Santa Fe, N.M.: School of American Research
Press, 1996.

Fellman, Michael. *Citizen Sherman: A Life of William Tecumseh Sherman.*
Lawrence: University Press of Kansas, 1995.

Fischer, David Hackett. *Albion's Seed: Four British Folkways in America.* New
York: Oxford University Press, 1989.

Fisher, William W., III. "Ideology and Imagery in the Law of Slavery." In
Slavery and the Law, edited by Paul Finkelman, 43–85. Madison, Wisc.:
Madison House, 1997.

Flindell, I. H. "Fundamentals of Human Response to Sound." In
Fundamentals of Noise and Vibration, edited by F. J. Fahy and J. G. Walker,
151–53. London: E & FN Spon, 1998.

Fogel, Robert W., and Stanley Engerman. *Time on the Cross: The Economics of
American Negro Slavery.* 2 vols. Boston: Little, Brown, 1976.

Foner, Eric. *Free Soil, Free Labor, Free Men: The Ideology of the Republican Party
before the Civil War.* New York: Oxford University Press, 1970.

———. *Politics and Ideology in the Age of the Civil War.* New York: Oxford
University Press, 1980.

———. *Reconstruction: America's Unfinished Revolution, 1863–1877.* New York:
Harper and Row, 1988.

Ford, Lacy K., Jr. *Origins of Southern Radicalism: The South Carolina
Upcountry, 1800–1860.* New York: Oxford University Press, 1988.

Foucault, Michel. *The Birth of the Clinic.* New York: Random House, 1975.

———. *Discipline and Punish: The Birth of the Prison.* London: Allen Lane,
1977.

Fox-Genovese, Elizabeth. *Within the Plantation Household: Black and White
Women of the Old South.* Chapel Hill: University of North Carolina Press,
1988.

Fox-Genovese, Elizabeth, and Eugene D. Genovese. "The Divine Sanction of
Social Order: Religious Foundations of the Southern Slaveholders' World
View." *Journal of the American Academy of Religion* 55 (summer 1987):
211–34.

Franklin, John Hope. *A Southern Odyssey: Travelers in the Antebellum North.*
Baton Rouge: Louisiana State University Press, 1991.

Freehling, William W. *The Reintegration of American History: Slavery and the
Civil War.* New York: Oxford University Press, 1994.

———. *The Road to Disunion.* Vol. 1, *Secessionists at Bay, 1776–1854.* New York:
Oxford University Press, 1990.

Frey, Sylvia R., and Betty Wood. *Come Shouting to Zion: African American*

Protestantism in the American South and British Caribbean to 1830. Chapel Hill: University of North Carolina Press, 1998.

Friedman, Lawrence J. *Gregarious Saints: Self and Community in American Abolitionism, 1830–1870.* Cambridge: Cambridge University Press, 1982.

Friedman, Lawrence M. *Crime and Punishment in American History.* New York: Basic Books, 1993.

Gadamer, H-G. *Truth and Method.* London: Sheed and Ward, 1975.

Gaines, Francis Pendleton. *Southern Oratory: A Study in Idealism.* University: University of Alabama Press, 1946.

Gallman, J. Matthew. *The North Fights the Civil War: The Home Front.* Chicago: Ivan R. Dee, 1994.

Genovese, Eugene D. "Marxian Interpretations of the Slave South." In *Towards a New Past: Dissenting Essays in American History,* edited by Barton J. Bernstein, 90–126. New York: Pantheon, 1968.

———. *The Political Economy of Slavery: Studies in the Economy and Society of the Slave South.* Middletown, Conn.: Wesleyan University Press, 1989.

———. *Roll, Jordan, Roll: The World the Slaves Made.* New York: Vintage, 1976.

———. *The Slaveholders' Dilemma: Freedom and Progress in Southern Conservative Thought, 1820–1860.* Columbia: University of South Carolina Press, 1992.

———. "Yeoman Farmers in a Slaveholders' Democracy." *Agricultural History* 49 (Apr. 1975): 331–42.

Gienapp, William E. *The Origins of the Republican Party, 1854–1856.* New York: Oxford University Press, 1987.

———. "The Republican Party and the Slave Power." In *New Perspectives on Race and Slavery in America: Essays in Honor of Kenneth M. Stampp,* edited by Robert H. Abzug and Stephen E. Maizlish, 51–78. Lexington: University Press of Kentucky, 1987.

Gilje, Paul A. *The Road to Mobocracy: Popular Disorder in New York City, 1765–1854.* Chapel Hill: University of North Carolina Press, 1987.

Glassie, Henry. *Patterns in the Material Folk Culture of the Eastern United States.* Philadelphia: University of Pennsylvania Press, 1968.

Gomez, Michael A. *Exchanging Our Country Marks: The Transformation of African Identities in the Colonial and Antebellum South.* Chapel Hill: University of North Carolina Press, 1998.

Goodman, Paul. *Of One Blood: Abolitionism and the Origins of Racial Equality.* Berkeley: University of California Press, 1998.

Goody, Jack, and Ian Watt. "The Consequences of Literacy." In *Literacy in Traditional Societies,* edited by Jack Goody, 27–68. Cambridge: Cambridge University Press, 1968.

Gorn, Elliot J. "'Gouge and Bite, Pull Hair and Scratch': The Social Significance of Fighting in the Southern Backcountry." *American Historical Review* 90 (Feb. 1985): 18–43.

Grammer, John M. *Pastoral and Politics in the Old South.* Baton Rouge: Louisiana State University Press, 1996.

Grant, Susan-Mary. *North over South: Northern Nationalism and American Identity in the Antebellum Era.* Lawrence: University Press of Kansas, 2000.

Green, Fletcher Melvin. *Democracy in the Old South and Other Essays.* Edited by J. Isaac Copeland. Nashville: Vanderbilt University Press, 1969.

Greenberg, Kenneth S. "The Nose, the Lie, and the Duel." In his *Honor and Slavery: Lies, Duels, Noses, Masks, Dressing as a Woman, Gifts, Strangers, Death, Humanitarianism, Slave Rebellions, the Proslavery Argument, Baseball, Hunting, and Gambling in the Old South,* 3–23. Princeton, N.J.: Princeton University Press, 1996.

Gretlund, Jan Norby. "Southern Silence?" In *Semantics of Silences in Linguistics and Literature,* edited by Gudrun M. Grabher and Ulrike Jessner, 329–37. Heidelberg: Winter, 1996.

Grier, Katherine C. *Culture and Comfort: Parlor Making and Middle-Class Identity, 1850–1930.* Washington, D.C.: Smithsonian Institution Press, 1997.

Grimsted, David. *American Mobbing, 1828–1861: Toward Civil War.* New York: Oxford University Press, 1998.

Gross, Ariela. "Pandora's Box: Slave Character on Trial in the Antebellum Deep South." In *Slavery and the Law,* edited by Paul Finkelman, 291–327. Madison, Wisc.: Madison House, 1997.

Gutman, Herbert G. "Work, Culture, and Society in Industrializing America, 1815–1919." In his *Work, Culture, and Society in Industrializing America: Essays in American Working-Class and Social History,* 3–78. Oxford: Basil Blackwell, 1976.

Gutman, Herbert G., and Richard Sutch. "Sambo Makes Good; or, Were Slaves Imbued with the Protestant Work Ethic?" In *Reckoning with Slavery: A Critical Study in the Quantitative History of American Negro Slavery,* edited by Paul A. David, Herbert G. Gutman, Richard Sutch, Peter Temin, and Gavin Wright, 55–93. New York: Oxford University Press.

Habermas, Jürgen. *The Structural Transformation of the Public Sphere: An Inquiry into a Category of Bourgeois Society.* Translated by Thomas Burger. Cambridge, Mass.: MIT Press, 1989.

Hagerman, Edward. *The American Civil War and the Origins of Modern Warfare: Ideas, Organization, and Field Command.* Bloomington: Indiana University Press, 1992.

Hahn, Steven. *The Roots of Southern Populism: Yeoman Farmers and the Transformation of the Georgia Upcountry, 1850–1890.* New York: Oxford University Press, 1983.

Hall, Edward T. *The Hidden Dimension.* New York: Anchor, 1982.

Handel, Stephen. *Listening: An Introduction to the Perception of Auditory Events.* Cambridge, Mass.: MIT Press, 1989.

Hankins, Thomas L., and Robert J. Silverman. *Instruments and the Imagination.* Princeton, N.J.: Princeton University Press, 1995.

Haskell, Thomas L. "Capitalism and the Origins of the Humanitarian Sensibility." Pt. 1. In *The Antislavery Debate: Capitalism and Abolitionism as* a *Problem in Historical Interpretation,* edited by Thomas Bender, 107–35. Berkeley: University of California Press, 1992.

Heintze, James R. *American Musical Life in Context and Practice to 1865.* New York: Garland, 1994.

Hess, Earl J. *The Union Soldier in Battle: Enduring the Ordeal of Combat.* Lawrence: University Press of Kansas, 1997.

Hibbitts, Bernard. "Making Sense of Metaphors: Visuality, Aurality, and the Reconfiguration of American Legal Discourse." 16 *Cardozo Law Review* 229 (1994). http://www.law.pitt.edu/hibbitts/meta_int.html. (15/1/00).

Hindus, Michael Stephen. *Prison and Plantation: Crime, Justice, and Authority in Massachusetts and South Carolina, 1767–1878.* Chapel Hill: University of North Carolina Press, 1980.

Hobsbawm, Eric. *The Age of Capital, 1848–1875.* New York: Scribner, 1975.

———. *On History.* New York: New Press, 1997.

Hoffert, Sylvia D. *When Hens Crow: The Woman's Rights Movement in Antebellum America.* Bloomington: Indiana University Press, 1995.

Holifield, E. Brooks. *The Gentlemen Theologians: American Theology in Southern Culture, 1795–1860.* Durham, N.C.: Duke University Press, 1978.

Howe, Daniel Walker. *The Political Culture of the American Whigs.* Chicago: University of Chicago Press, 1979.

Howes, David, ed. *The Varieties of Sensory Experience: A Reader in the Anthropology of the Senses.* Toronto: University of Toronto Press, 1991.

Hunter, Tera W. *To 'Joy My Freedom: Southern Black Women's Lives and Labors after the Civil War.* Cambridge, Mass.: Harvard University Press, 1998.

Huston, James L. "Property Rights in Slavery and the Coming of the Civil War." *Journal of Southern History* 65 (May 1999): 249–85.

Idhe, Don. *Listening and Voice: A Phenomenology of Sound.* Athens: Ohio University Press, 1976.

Ignatieff, Michael. *A Just Measure of Pain: The Penitentiary System in the Industrial Revolution, 1750–1850.* New York: Pantheon, 1978.

Isenberg, Nancy. *Sex and Citizenship in Antebellum America.* Chapel Hill: University of North Carolina Press, 1998.

Jay, Martin. *Downcast Eyes: The Denigration of Vision in Twentieth-Century French Thought.* Berkeley: University of California Press, 1993.

Jeffrey, Julie Roy. *The Great Silent Army of Abolitionism: Ordinary Women in the Antislavery Movement.* Chapel Hill: University of North Carolina Press, 1998.

Jeffrey, Thomas E. *Thomas Lanier Clingman: Fire Eater from the Carolina Mountains.* Athens: University of Georgia Press, 1998.

John, Richard R. *Spreading the News: The American Postal System from Franklin to Morse.* Cambridge, Mass.: Harvard University Press, 1993.

Johnson, Curtis D. *Islands of Holiness: Rural Religion in Upstate New York, 1790–1860.* Ithaca, N.Y.: Cornell University Press, 1989.

Johnson, Paul E. *A Shopkeeper's Millennium: Society and Revivals in Rochester, New York, 1815–1837.* New York: Hill and Wang, 1978.

Jones, Steve. *In the Blood: God, Genes, and Destiny.* London: Harper Collins, 1996.

Jordan, Winthrop D. *Tumult and Silence at Second Creek: An Inquiry into a Civil War Slave Conspiracy.* Baton Rouge: Louisiana State University Press, 1993.

Joyner, Charles. *Shared Traditions: Southern History and Folk Culture.* Urbana: University of Illinois Press, 1999.

Kahn, Douglas. "Introduction: Histories of Sound Once Removed." In *Wireless Imagination: Sound, Radio, and the Avant-Garde,* edited by Douglas Kahn and Gregory Whitehead, 1–29. Cambridge, Mass.: MIT Press, 1992.

———. *Noise, Water, Meat: A History of Sound in the Arts.* Cambridge, Mass: MIT Press, 1999.

Kamensky, Jane. *Governing the Tongue: The Politics of Speech in Early New England.* New York: Oxford University Press, 1997.

Kasson, John F. *Rudeness and Civility: Manners in Nineteenth-Century Urban America.* New York: Hill and Wang, 1990.

Kocher, A. Lawrence, and Howard Dearstyne. *Shadows in Silver: A Record of Virginia, 1850–1900, in Contemporary Photographs taken by George and Huestis Cook with Addition from the Cook Collection.* New York: Charles Scribner's Sons, 1954.

Kolchin, Peter. "Reevaluating the Antebellum Slave Community: A Comparative Perspective." *Journal of American History* 70 (1983): 579–601.

———. "Slavery and Freedom in the Civil War South." In *Writing the Civil War: The Quest to Understand,* edited by James M. McPherson and William J. Cooper Jr., 241–60. Columbia: University of South Carolina Press, 1998.

Kryter, K. D. *The Effects of Noise on Man.* New York: Academic Press, 1970.

Lakoff, G., and M. Johnson. *Metaphors We Live By.* Chicago: University of Chicago Press, 1980.

Lane, Mills. *Architecture of the Old South.* New York: Abbeville Press, 1993.

Lapansky, Phillip. "Graphic Discord: Abolitionist and Antiabolitionist Images." In *The Abolitionist Sisterhood: Women's Political Culture in Antebellum America,* edited by Jean Fagan Yellin and John C. Van Horne, 201–30. Ithaca, N.Y.: Cornell University Press, 1994.

Lehuu, Isabelle. *Carnival on the Page: Popular Print Media in Antebellum America.* Chapel Hill: University of North Carolina Press, 2000.

Lerner, Gerda. *The Grimké Sisters from South Carolina: Pioneers for Women's Rights and Abolition.* New York: Oxford University Press, 1998.

Levin, David Michael, ed. *Modernity and the Hegemony of Vision.* Berkeley: University of California Press, 1993.

Levine, Lawrence W. *Black Culture and Black Consciousness: Afro-American Folk Thought from Slavery to Freedom.* New York: Oxford University Press, 1977.

Loveland, Anne C. *Southern Evangelicals and the Social Order, 1800–1860.* Baton Rouge: Louisiana State University Press, 1980.

Lovell, John. *Black Song: The Forge and the Flame.* New York: Macmillan, 1972.

Lowe, Donald M. *History of Bourgeois Perception*. Chicago: University of
  Chicago Press, 1982.
Luraghi, Raimondo. "The Civil War and the Modernization of American
  Society: Social Structure and Industrial Revolution in the Old South
  before and during the War." *Civil War History* 18 (Sept. 1972): 230–50.
———. *The Rise and Fall of the Plantation South*. New York: Franklin Watts,
  1978.
McCoy, Drew R. *The Elusive Republic: Political Economy in Jeffersonian America*.
  Chapel Hill: University of North Carolina Press, 1980.
McCurry, Stephanie. *Masters of Small Worlds: Yeoman Households, Gender
  Relations, and the Political Culture of the Antebellum South Carolina
  Lowcountry*. New York: Oxford University Press, 1995.
———. "The Two Faces of Republicanism: Gender and Proslavery Politics in
  Antebellum South Carolina." *Journal of American History* 78 (Mar. 1992):
  1245–64.
McDonnell, Lawrence T. "Money Knows No Master: Market Relations and
  the American Slave Community." In *Developing Dixie: Modernization in a
  Traditional Society,* edited by Winfred B. Moore Jr., Joseph F. Tripp, and
  Lyon G. Tyler, 31–44. Westport, Conn.: Greenwood Press, 1988.
———. "Work, Culture, and Society in the Slave South, 1790–1861." In *Black
  and White Cultural Interaction in the Antebellum South,* edited by Ted
  Ownby, 125–48. Jackson: University Press of Mississippi, 1993.
———. " 'You Are Too Sentimental': Problems and Suggestions for a New
  Labor History." *Journal of Social History* 17 (winter 1983–84): 629–46.
McIntosh, James. *Thoreau as Romantic Naturalist: His Shifting Stance toward
  Nature*. Ithaca, N.Y.: Cornell University Press, 1974.
McMillen, Sally. "Mothers' Sacred Duty: Breast-feeding Patterns among
  Middle- and Upper-Class Women in the Antebellum South." *Journal of
  Southern History* 51 (Aug. 1985): 333–56.
McPherson, James M. "Antebellum Southern Exceptionalism: A New Look at
  an Old Question." *Civil War History* 29 (Sept. 1983): 230–44.
———. *Battle Cry of Freedom: The Civil War Era*. New York: Penguin, 1988.
Malone, Bill C. *Southern Music, American Music*. Lexington: University Press
  of Kentucky, 1979.
Marks, Stuart A. *Southern Hunting in Black and White: Nature, History, and
  Ritual in a Carolina Community*. Princeton, N.J.: Princeton University
  Press, 1991.
Marx, Karl. *The German Ideology*. In *The Marx-Engels Reader,* edited by
  Robert C. Tucker. New York: W. W. Norton, 1978.
Marx, Leo. *The Machine in the Garden: Technology and the Pastoral Ideal in
  America*. New York: Oxford University Press, 1964.
Mathews, Donald G. *Religion in the Old South*. Chicago: University of Chicago
  Press, 1977.
Meranze, Michael. *Laboratories of Virtue: Punishment, Revolution, and*

Authority in Philadelphia, 1760–1835. Chapel Hill: University of North Carolina Press, 1996.

Miller, J. D. "Effects of Noise on People." In *Hearing,* vol. 4 of *Handbook of Perception,* edited by Edward C. Carterette and Morton P. Friedman, 609–40. New York: Academic Press, 1978.

———. "General Physiological Effects of Noise." In *Hearing,* vol. 4 of *Handbook of Perception,* edited by Edward C. Carterette and Morton P. Friedman, 677–82. New York: Academic Press, 1978.

———. "General Psychological and Sociological Effects of Noise." In *Hearing,* vol. 4 of *Handbook of Perception,* edited by Edward C. Carterette and Morton P. Friedman, 641–75. New York: Academic Press, 1978.

Miller, William Lee. *Arguing about Slavery: John Quincy Adams and the Great Battle in the United States Congress.* New York: Vintage, 1998.

Mills, A. William. "Auditory Localization." In *Foundations of Modern Auditory Theory,* edited by Jerry V. Tobias, 2:301–45. New York: Academic Press, 1972.

Mills, Bruce. *Cultural Reformations: Lydia Maria Child and the Literature of Reform.* Athens: University of Georgia Press, 1994.

Myers, Fred R., and Donald Lawrence Brenneis. "Introduction: Language and Politics in the Pacific." In *Dangerous Words: Language and Politics in the Pacific,* edited by Donald Lawrence Brenneis and Fred R. Myers, 1–29. New York: New York University Press, 1984.

Nash, Roderick. *Wilderness and the American Mind.* New Haven, Conn.: Yale University Press, 1968.

Nessell, Gail E. "'The lowing of the cows and the sound of the whistle': Old Meets New on the Goodale Farm, Marlborough, Massachusetts, 1819–1858." In *The Dublin Seminar for New England Folklife Annual Proceedings,* edited by Peter Benes, 36–48. Boston: Boston University Press, 1986.

Nevins, Allan. *Ordeal of the Union.* 6 vols. Vol. 2, *A House Dividing, 1852–1857.* New York: Charles Scribner's Sons, 1947.

"North American Soundways, 1600–1800." *Uncommon Sense: A Newsletter Published by the Omohundro Institute of Early American History and Culture,* summer 1997, 35.

North, Douglass C. *The Economic Growth of the United States, 1790–1860.* 1961. Reprint, New York: W. W. Norton, 1966.

O'Brien, Michael. *Rethinking the South: Essays in Intellectual History.* Baltimore, Md.: Johns Hopkins University Press, 1988.

O'Connor, Thomas H. "Slavery in the North." In *The Meaning of Slavery in the North,* edited by David R. Roediger and Martin Henry Blatt, chap. 3. New York: Garland, 1998.

Olson, Harry Ferdinand. *Music, Physics, and Engineering.* New York: Dover, 1967.

Ong, Walter. *The Presence of the Word: Some Prolegomena for Cultural and Religious History.* New Haven: Yale University Press, 1967.

Paullin, Charles O. *Atlas of the Historical Geography of the United States.* Edited by John K. Wright. Washington, D.C., and New York: Carnegie Institution of Washington and the American Geographical Society of New York, 1932.

Persons, Stowe. *The Decline of American Gentility.* New York: Columbia University Press, 1973.

Pessen, Edward. "The Egalitarian Myth and American Social Reality: Wealth, Mobility, and Equality in the 'Era of the Common Man.'" *American Historical Review* 76 (Oct. 1971): 989–1034.

———. "How Different from Each Other Were the Antebellum North and South?" *American Historical Review* 85 (Dec. 1980): 1119–49.

Power, J. Tracy. *Lee's Miserables: Life in the Army of Northern Virginia from the Wilderness to Appomattox.* Chapel Hill: University of North Carolina Press, 1998.

Quist, John W. *Restless Visionaries: The Social Roots of Antebellum Reform in Alabama and Michigan.* Baton Rouge: Louisiana State University Press, 1998.

Raboteau, Albert J. "Slave Autonomy and Religion." *Journal of Religious Thought* 38 (fall 1981/winter 1982): 51–64.

———. *Slave Religion: The "Invisible Institution" in the Antebellum South.* New York: Oxford University Press, 1978.

Ridington, Robin. "In Doig People's Ears: Portrait of a Changing Community." *Anthropologica,* n.s., 25 (1983): 9–22.

Roark, James L. "Behind the Lines: Confederate Economy and Society." In *Writing the Civil War: The Quest to Understand,* edited by James M. McPherson and William J. Cooper Jr., 201–27. Columbia: University of South Carolina Press, 1998.

Robinson, Donald L. *Slavery in the Age of Revolution, 1765–1820.* New York: Harcourt, Brace, Jovanovich, 1971.

Rodgers, Daniel. *The Work Ethic in Industrial America, 1850–1920.* Chicago: University of Chicago Press, 1974.

Rogers, George C., Jr. *Charleston in the Age of the Pinckneys.* Columbia: University of South Carolina Press, 1989.

Rose, Tricia. *Black Noise: Rap Music and Black Culture in Contemporary America.* Hanover, N.H.: Wesleyan University Press, 1994.

Rosenbaum, Art. *Shout Because You're Free: The African American Ring Shout Tradition in Coastal Georgia.* Athens: University of Georgia Press, 1998.

Ross, Charles. "Ssh! Battle in Progress!" *Civil War Times Illustrated* 35 (Dec. 1996): 56–62.

Rothman, David J. *The Discovery of the Asylum: Social Order and Disorder in the New Republic.* Boston: Little, Brown, 1971.

Rousey, Dennis C. *Policing the Southern City: New Orleans, 1805–1889.* Baton Rouge: Louisiana State University Press, 1996.

Russell, Thomas D. "Slave Auctions on the Courthouse Steps: Court Sales of Slaves in Antebellum South Carolina." In *Slavery and the Law,* edited by Paul Finkelman, 329–64. Madison, Wisc.: Madison House, 1997.

Schafer, R. Murray, ed. *European Sound Diary.* Vancouver, B.C.: A.R.C. Publications, 1977.

———. *The Tuning of the World: Toward a Theory of Soundscape Design.* Philadelphia: University of Pennsylvania Press, 1980.

Schama, Simon. *Landscape and Memory.* New York: Vintage, 1995.

Schmidt, Leigh Eric. "From Demon Possession to Magic Show: Ventriloquism, Religion, and the Enlightenment." *Church History* 67 (June 1998): 274–304. http://www.materialreligion.org/journal/magic.html (7/6/00).

———. *Hearing Things: Religion, Illusion, and the American Enlightenment.* Cambridge, Mass.: Harvard University Press, 2000.

Schwalm, Leslie A. *A Hard Fight for We: Women's Transition from Slavery to Freedom in South Carolina.* Urbana: University of Illinois Press, 1997.

Schwartz, Hillel. "Beyond Tone and Decibel: The History of Noise." *Chronicle of Higher Education,* Jan. 9, 1998, B8.

———. "Hearing Aids: Sweet Nothings, or an Ear for an Ear." In *The Gendered Object,* edited by Pat Kirkham, chap. 4. Manchester: Manchester University Press, 1996.

Scott, James C. *Domination and the Arts of Resistance: Hidden Transcripts.* New Haven: Yale University Press, 1990.

Sears, John F. *Sacred Places: American Tourist Attractions in the Nineteenth Century.* New York: Oxford University Press, 1989.

Sellers, Charles. *The Market Revolution: Jacksonian America, 1815–1846.* New York: Oxford University Press, 1991.

Sheidley, Harlow W. *Sectional Nationalism: Massachusetts Conservative Leaders and the Transformation of America, 1815–1836.* Boston, Mass.: Northeastern University Press, 1998.

Sheriff, Carol. *The Artificial River: The Erie Canal and the Paradox of Progress, 1817–1862.* New York: Hill and Wang, 1996.

Siegel, Frederick F. *The Roots of Southern Distinctiveness: Tobacco and Society in Danville, Virginia, 1780–1865.* Chapel Hill: University of North Carolina Press, 1987.

Silber, Nina. *The Romance of Reunion: Northerners and the South, 1865–1900.* Chapel Hill: University of North Carolina Press, 1993.

Silbey, Joel H. *The Partisan Imperative: The Dynamics of American Politics before the Civil War.* New York: Oxford University Press, 1985.

Simpson, Lewis P. *The Dispossessed Garden: Pastoral and History in Southern Literature.* Athens: University of Georgia Press, 1975.

———. *Mind and the American Civil War: A Meditation on Lost Causes.* Baton Rouge: Louisiana State University Press, 1989.

Sinha, Manisha. *The Counter-Revolution of Slavery: Politics and Ideology in Antebellum South Carolina.* Chapel Hill: University of North Carolina Press, 2000.

Smilor, Raymond W. "Toward an Environmental Perspective: The Anti-Noise Campaign, 1883–1932." In *Pollution and Reform in American Cities, 1870–*

1930, edited by Martin V. Melosi, 135–51. Austin: University Press of Texas, 1980.

Smith, Bruce R. *The Acoustic World of Early Modern England.* Chicago: University of Chicago Press, 1999.

Smith, Kimberly K. *The Dominion of Voice: Riot, Reason, and Romance in Antebellum Politics.* Lawrence: University Press of Kansas, 1999.

Smith, Mark M. "Listening to the Heard Worlds of Antebellum America." *Journal of the Historical Society* 1 (spring 2000): 64–97.

———. *Mastered by the Clock: Time, Slavery, and Freedom in the American South.* Chapel Hill: University of North Carolina Press, 1997.

———. "Old South Time in Comparative Perspective." *American Historical Review* 101 (Dec. 1996): 1432–69.

———. "Time, Sound, and the Virginia Slave." In *Afro-Virginian History and Culture,* edited by John Saillant, 29–60. New York: Garland, 1999.

Solomon, L. N. "Search for Physical Correlates to Psychological Dimensions of Sound." *Journal of the Acoustical Society of America* 31 (1959): 492.

Spencer, Jon Michael. "The Rhythms of Black Folks." In *Ain't Gonna Lay My 'Ligion Down: African American Religion in the South,* edited by Alonzo Johnson and Paul Jersild, 39–51. Columbia: University of South Carolina Press, 1996.

Stachiw, Myron O. "'For the Sake of Commerce': Slavery, Antislavery, and Northern Industry." In *The Meaning of Slavery in the North,* edited by David R. Roediger and Martin Henry Blatt, chap. 2. New York: Garland, 1998.

Stanley, Amy Dru. *From Bondage to Contract: Wage Labor, Marriage, and the Market in the Age of Slave Emancipation.* New York: Cambridge University Press, 1998.

Stansell, Christine. *City of Women: Sex and Class in New York, 1789–1860.* Urbana: University of Illinois Press, 1987.

Stoller, Paul. "Sound in Songhay Cultural Experience." *American Ethnologist* 11 (Aug. 1984): 559–70.

Stuckey, Sterling. *Slave Culture: Nationalist Theory and the Foundation of Black America.* New York: Oxford University Press, 1987.

Takagi, Midori. *"Rearing Wolves to Our Own Destruction": Slavery in Richmond, Virginia 1782–1865.* Charlottesville: University Press of Virginia, 1999.

Teeters, Negley K., and John D. Shearer. *The Prison at Philadelphia, Cherry Hill: The Separate System of Penal Discipline, 1829–1913.* New York: Published for Temple University Press by Columbia University Press, 1957.

Thomas, John L. "Romantic Reform in America, 1815–1865." *American Quarterly* 17 (Oct. 1965): 656–81.

Thompson, E. P. "Rough Music." In his *Customs in Common,* 467–538. London: Merlin Press, 1991.

———. "Time, Work-Discipline, and Industrial Capitalism." *Past and Present* 38 (Dec. 1967): 56–97.

Thornton, J. Mills. *Politics and Power in a Slave Society: Alabama, 1800–1860.* Baton Rouge: Louisiana State University Press, 1978.

Tilley, Christopher. *Metaphor and Material Culture.* Oxford: Blackwell, 1999.

Trelease, Allen W. *White Terror: The Ku Klux Klan Conspiracy and Southern Reconstruction.* Baton Rouge: Louisiana State University Press, 1971.

Truax, Barry. "Soundscape Studies: An Introduction to the World Soundscape Project." *Numus West* 5 (1974): 36–39.

———, ed. *The World Soundscape Project's Handbook for Acoustic Ecology.* Vancouver, B.C.: A.R.C. Publications, 1978.

Upton, Dell. "The City as Material Culture." In *The Art and Mystery of Historical Archaeology: Essays in Honor of James Deetz,* edited by Anne Elizabeth Yentsch and Mary C. Beaudry, 51–74. Boca Raton, Fla.: CRC Press, 1992.

Varon, Elizabeth R. *We Mean to Be Counted: White Women and Politics in Antebellum Virginia.* Chapel Hill: University of North Carolina Press, 1998.

Waldstreicher, David. *In the Midst of Perpetual Fetes: The Making of American Nationalism, 1776–1820.* Chapel Hill: University of North Carolina Press, published for the Omohundro Institute of Early American History and Culture, 1997.

Warren, James Perrin. *Culture of Eloquence: Oratory and Reform in Antebellum America.* University Park: Pennsylvania State University Press, 1999.

Watson, Alan. "Seventeenth-Century Jurists, Roman Law, and the Law of Slavery." In *Slavery and the Law,* edited by Paul Finkelman, 367–78. Madison, Wisc.: Madison House, 1997.

Way, Peter. *Common Labor: Workers and the Digging of North American Canals, 1780–1860.* Baltimore, Md.: Johns Hopkins University Press, 1997.

Weinbaum, Paul O. *Mobs and Demagogues: The New York Response to Collective Violence in the Early Nineteenth Century.* Ann Arbor: University of Michigan Press, 1979.

Weiner, Annette B. "From Words to Objects to Magic: 'Hard Words' and the Boundaries of Social Interaction." In *Dangerous Words: Language and Politics in the Pacific,* edited by Donald Lawrence Brenneis and Fred R. Myers, 161–91. New York: New York University Press, 1984.

Welter, Barbara. "The Cult of True Womanhood, 1820–1860." *American Quarterly* 18 (summer 1966): 151–74.

White, Shane, and Graham White. "'At intervals I was nearly stunned by the noise he made': Listening to African American Religious Sound in the Era of Slavery." *American Nineteenth Century History* 1 (spring 2000), 34–61.

———. "Slave Hair and African American Culture in the Eighteenth and Nineteenth Centuries." *Journal of Southern History* 61 (Feb. 1995): 45–76.

———. "'Us Likes a Mixtery': Listening to African-American Slave Music." *Slavery and Abolition* 20 (Dec. 1999): 22–48.

Wilentz, Sean. *Chants Democratic: New York City and the Rise of the American Working Class, 1788–1850.* New York: Oxford University Press, 1984.

Williams, Raymond. *The Country and the City*. London: Chatto and Windus, 1973.

Wilson, Clyde N. *Carolina Cavalier: The Life and Mind of James Johnston Pettigrew*. Athens: University of Georgia Press, 1990.

———. "'Free Trade: No Debt: Separation from Banks': The Economic Platform of John C. Calhoun." In *Slavery, Secession, and Southern History*, edited by Robert Louis Paquette and Louis A. Ferleger, 81–100. Charlottesville: University Press of Virginia, 2000.

Wilson, Joan Hoff. "The Illusion of Change: Women and the American Revolution." In *The American Revolution: Explorations in the History of American Radicalism*, edited by Alfred E. Young, 383–445. Dekalb: Northern Illinois University Press, 1976.

Wood, Peter H. *Black Majority: Negroes in Colonial South Carolina from 1670 through the Stono Rebellion*. New York: W. W. Norton, 1975.

Woods, David L. *A History of Tactical Communications Techniques*. New York: Arno Press, 1974.

Wright, Gavin. *The Political Economy of the Cotton South: Households, Markets, and Wealth in the Nineteenth Century*. New York: W. W. Norton, 1978.

Wyatt-Brown, Bertram. *Southern Honor: Ethics and Behavior in the Old South*. New York: Oxford University Press, 1982.

Young, Jeffrey Robert. *Domesticating Slavery: The Master Class in Georgia and South Carolina, 1670–1837*. Chapel Hill: University of North Carolina Press, 1999.

Zboray, Ronald J. *A Fictive People: Antebellum Economic Development and the American Reading Public*. New York: Oxford University Press, 1993.

Zuczek, Richard. *State of Rebellion: Reconstruction in South Carolina*. Columbia: University of South Carolina Press, 1996.

Zuckerkandl, Victor. *Sound and Symbol: Music and the External World*. Translated by Williard R. Trask. Princeton, N.J.: Princeton University Press, 1958.

Unpublished Dissertations, Theses, Conference Papers, and Personal Correspondence

Downey, Thomas Moore. "Planting a Capitalist South: The Transformation of Western South Carolina, 1790–1860." Ph.D. diss., University of South Carolina, 2000.

Fahy, Frank. Personal communication to the author, Mar. 11, 1999.

Halttunen, Karen Lee. "Confidence Men and Painted Women: The Problem of Hypocrisy in Sentimental America, 1830–1870." Ph.D. diss., Yale University, 1979.

Haney, Gina. "In Complete Order: Social Control and Architectural Space Organization in the Charleston Back Lot." M.A. thesis, School of Architecture, University of Virginia, 1996.

Hubbs, Guy Ward. "Guarding Greensboro: A Confederate Company and the

Making of a Southern Community." Ph.D. diss., University of Alabama,
Tuscaloosa, 1999.

Johnson, Walter Livezey. "Masters and Slaves in the Market of Slavery and the
New Orleans Trade, 1804–1864." Ph.D. diss., Princeton University, 1995.

Juncker, Clara. "'Over the Water': Frances Butler Leigh's *Ten Years on a Georgia
Plantation since the War*." Paper presented at "Global Currents in Southern
History," Georgia Southern University, Oct. 21, 2000.

Kahn, Douglas. "The Big Bang of Auditory Culture." Paper in author's
possession.

Kilbride, Daniel. "Privileged Southerners, the Grand Tour, and the Roots of
Southern Regional Identity before the Civil War." Paper presented at
"Global Currents in Southern History," Georgia Southern University,
Oct. 21, 2000.

Kulik, Gary B. "The Beginnings of the Industrial Revolution in America:
Pawtucket, Rhode Island, 1672–1829." Ph.D. diss., Brown University, 1980.

McLeod, Richard A. "The Philadelphia Artisan, 1828–1850," Ph.D. diss.,
University of Missouri, 1971.

Puckett, David K. "'. . . The chains which had bound us so long were well
nigh broken': The Transition from Slavery to Freedom in Columbia, S.C.,
1850–1865." M.A. thesis, University of South Carolina, 1998.

Rath, Richard. "Sounding the Chesapeake: Indian and English Soundways in
the Settling of Jamestown." Paper presented to the Omohundro Institute of
Early American History and Culture, Mar. 16, 2000. http://way.net/rcr/
chesapeake (6/6/2000).

Schwartz, Hillel. "Hush Concerning a Right to Quiet." Paper delivered at the
Symposium on Human Rights: Changes and Challenges, Georgia Institute
of Technology, Apr. 30–May 1, 1999.

———. "The Indefensible Ear: A History." Soundscapes lecture, n.d.
http://omroep.nl/nps/radio/supplement/99/soundscapes/schwartz2.html
(9/10/00).

———. "Noise and Silence: The Soundscape and Spirituality." Paper
presented at the Inter-Religious Federation for World Peace, Seoul, Korea,
Aug. 20–27, 1995. http://nonoise.org/library/noisesil/noisesil.html
(7/18/00).

Sterne, Jonathan Edward. "The Audible Past: Modernity, Technology, and the
Cultural History of Sound." Ph.D. diss., University of Illinois,
Urbana-Champaign, 1999.

Thompson, Emily. "Mysteries of the Acoustic: Architectural Acoustics in
America, 1800–1932." Ph.D. diss., Princeton University, 1992.

Trask, Arthur Scott. "The Constitutional Republicans of Philadelphia, 1818–
1848: Hard Money, Free Trade, and State Rights." Ph.D diss., University of
South Carolina, 1998.

INDEX

Arnett, Benjamin, 247

Artisans, 113, 132, 134–36, 140, 141. *See also* Working class

Augustus, Sarah Louise, 239

Aurality, 2; and aural imagery, 2, 5, 8, 16, 126, 127, 147–49, 155, 164–65, 199, 208, 263, 265, 268–69, 272 (n. 9); and print, 2, 6, 126, 156, 164–65, 174; and emotionalism, 2, 14, 16–17; in colonial period, 9–12; and psycoacoustics, 14; and visuality, 15, 17, 28, 40, 85, 90, 206, 244, 258, 261–63, 272 (n. 8), 318 (n. 1); and listening, 21, 93–94; and religion, 59–61, 63, 70, 96, 97–98, 102–3, 113; and sense of place, 96; historical analysis of, 258–69, 273 (n. 10), 318 (n. 2)

Avirett, James Battle, 23, 31, 38, 39–40, 71

Bacon, Leonard, 161

Bagby, Arthur P., 188

Ball, Charles, 24, 72, 75, 77–78, 84–86, 91

Ball, John, Jr., 89

Bancroft, John M., 200, 203

Bancroft, Samuel, 129

Baptists, 59, 63, 64, 96

Barbour, Mary, 229

Barrett, O. S., 203

Beauregard, P. G. T., 232, 234

Beck, Alexander, 211

Beecher, Henry Ward, 145

Bell, Edwin Q., 247

Bell, Oliver, 81

Bells, 121; and religion, 10, 57–58, 86–88, 180; and urban safety, 10, 86–88, 55–56, 111–12, 214; associated with freedom, 10, 95–96, 113, 177–78, 179, 180–81; colonial, 10, 273 (n. 11), 274 (n. 16), 293 (n. 47); on antebellum southern plantations, 19, 23, 35–37, 38–39,

73–77, 82; and northern hotels, 28; and livestock, 33; as musical, 34; and travel, 45; and antebellum southern markets, 52–53, 87; slaves' use of, 87–88; in antebellum North, 95–96, 109, 110–13, 128, 134, 136–37, 138–40, 141; manufacture and sale of, 110–11; and northern factories, 113, 128, 134, 136–37, 138–40, 141; as abolitionist symbol, 160, 177–81, 184; and post-bellum South, 253, 255; English, compared to American, 273 (n. 11); and metallic composition of, 313 (n. 31)

Benezet, Anthony, 170

Benton, Harriette, 239

Benton, Thomas Hart, 151

Berry, Fannie, 240

Bingham, John A., 161

Bird, Lucy, 126

Birds, 9, 24, 25, 83, 130–31, 137, 253, 255, 258. *See also* Animals; Insects

Black, Hugh, 225

Blacksmiths, 40, 105

Blair, Frank P. Jr., 160

Bobo, William M., 3–5

Bolton, James, 36, 240

Border States. *See* Upper South

Boston Working Men's Party, 141

Bowditch, Henry Ingersoll, 96

Britain, 7, 9, 113. *See also* England

British Columbia, 104

Brooks, Preston, 190

Brown, John (slave), 73

Brown, John (abolitionist), 171, 313 (n. 37)

Brown, William W., 35–36, 77

Bruton Parish Church (Williams-burg, Va.,), 9

Bryson, Henry, 59

Boyce, James, 65–66

Buffum, Arnold, 159

Bull Run (First), 202, 203, 208, 210

Elites, northern
—in colonial period: use and understanding of bells, 10, 95–96. *See
also* Colonial America
—in antebellum period: and free
and wage labor, 12–15, 93–94; and
democracy, 12–15, 93–94, 119–20,
126, 132, 135, 141, 142–46, 157; and
slavery, 12–16, 21, 24, 148–49; daily
listening and sectional consciousness of, 12–16, 93–94, 147–49,
172–73; and humanitarianism, 15,
156, 159; and African Americans,
70, 103; and working class, 93–94,
97–100, 103, 105, 106, 109, 112–17,
134, 142–46; and industrial capitalism, 93–94, 113–14, 119, 120, 121,
124, 125, 127–30, 132–34, 141, 143,
144, 148–49; and similarities to
southern elites, 93–104, 105–6, 107,
108, 109, 110–14, 117, 134, 154–56,
185, 301 (n. 36); and gender, 97,
100–2, 104, 115; and industriousness, 98, 105–8; and temperance,
99, 114; and immigrants, 99, 132,
153, 167; and Native Americans,
103–5, 115, 116, 120, 153, 169; and
westward expansion, 107–9, 119,
128–29, 131, 161, 165, 166–68, 173,
176; and bells, 112; and urbanization, 119, 120–24, 126, 128, 130,
131–32; 148–49; as instigating
southern modernity, 164; and cultural imperialism, 173, 187; and
making slaveholders listen, 173–
74, 177–84; reach slaves' ears, 183.
See also Abolition and abolitionists;
Capitalism; Modernity; North:
antebellum; Republicans
—in postbellum period, 17; respond
to northern labor and economic
crises, 17, 218, 238, 246–58; travel
south, 251–58
Elites, southern
—in colonial period: use and under-

standing of bells, 10, 95–96. *See
also* Colonial America
—in antebellum period: and antebellum North, 5, 13–15, 16, 20, 21,
27–28, 33, 44, 48, 68, 142, 148, 186;
travel to antebellum North, 5, 27–
28; daily listening and sectional
consciousness of, 12–15, 16, 20,
33, 48, 91, 118, 147–49, 185–94;
and economic development, 13,
20–21, 23–24, 39–45, 46, 105–6,
185, 189; compared to postbellum,
24, 37–38; and dislike of northern taciturnity, 28; and rejection
of city noise, 31–32; and obedience to bells, 39; and commerce,
45–46, 53; and urbanization, 46,
47, 53, 54; and fear of class conflict, 46–49, 65–66, 69–70, 187,
193; and southern lower classes,
47–49, 53, 54, 55, 58–65, 69–70,
187, 193; and southern frontier,
49–50; and Native Americans,
50–52, 88, 90; and democracy, 56,
65–66, 119–20, 185; and household quietude, 57; and evangelical
Christianity, 57–65; respond to
northern criticisms, 186–94; and
southern society as quiet, 188,
189, 190, 192. *See also* Plantations;
Planters; Slaveholders; Slavery
—in postbellum period, 17, 28–
30; and memories of antebellum
period, 23–24, 26; compared to
antebellum, 24; and efforts to control freedpeople, 197, 243–46. *See
also* Freedpeople; Reconstruction;
Tourism)
Ellis, E. C., 173
Emancipation, 157, 173–74, 216, 239–
43. *See also* Freedom; Freedpeople
Embargo Act (1807), 110
Emerson, Ralph Waldo, 117
England, 130, 143, 150, 176, 189;
London, 45, 128. *See also* Britain

Episcopalians, 59, 63, 103
Eppes, Susan Bradford, 24, 37, 79, 192, 222, 230–31, 242, 243
Europe, 25–26
Everett, Edward, 202

Fair Oaks, 204, 205
Farmers' Register (Richmond), 34
Federalists, 53
Feld, Steven, 266
Fessenden, William Pitt, 161, 188
Finney, Charles G., 182
Fitzhugh, George, 5, 218, 223, 242, 245
Fleming, George, 76
Florida, 225; colonial, 9; and plantations, 24, 37, 192, 222; New Smyrna, 26, 52; and antebellum northerners, 26, 70, 156; as tranquil, 26, 252–55; antebellum, 29, 30, 41, 101, 151, 194; St. Augustine, 30, 50, 52–55, 87, 201, 252, 253, 254, 255; Tallahassee, 30, 50, 53, 111, 239; taming of, 30–31; and Native Americans in, 31, 49, 50; Picolata, 50; Hernando County, 51; Micanopy, 54; Pensacola, 54; Quincy, 70; Jacksonville, 163, 164, 205, 253, 254, 255, 256; Lake City, 239, 257; postbellum, 243, 252–57; Key West, 252; Gainesville, 256; Palatka, 256. *See also* Tourism
Forten, James Jr., 182
Fort Pillow, battle of (1864), 205
Foster, Stephen Symonds, 185
Franklin, Benjamin, 106
Frederick Douglass's Paper (Rochester), 119
Freedom, 11, 86, 173–74, 229, 236–37, 239–43. *See also* Emancipation; Freedpeople; Reconstruction
Freedpeople: and southern soundscape, 238–42, 243–44, 256–58; and freedom, 239–43; and qui-

etude, 241–42; and religious expression, 242–43, 252, 258; in urban areas, 243, 244–45; and intimidation of, 243–45; and northern impressions of, 257–58. *See also* Reconstruction
Frontier. *See* Westward expansion; Wilderness
Fulkes, Minnie, 81

Gag acts, 161–63, 186–87. *See also* Politics and political parties
Garrett, Angie, 82
Garrison, William Lloyd, 172, 181, 184, 185, 186
Gender: aural dimension in postbellum South, 28–29; aural dimension in antebellum South, 28–30, 53; aural dimensions in antebellum North, 97, 100–2, 104, 115, 177; during Civil War, 223–25; in postbellum North, 249, 252. *See also* Women
Georgia, 25, 26, 36, 63, 69, 73, 79, 169, 211, 226, 228, 229, 230, 231, 244; Athens, 27, 54; Macon, 57, 220; Savannah, 98, 145, 220, 221, 227; Kennesaw Mountain, 201, 202; Atlanta, 201, 206, 209, 227, 232, 233, 245; Jasper County, 239
Germany, 34
Gettysburg, battle of (1863), 198, 202, 207, 216, 309 (n. 28)
Giddings, Joshua Reed, 162
Gladdy, Mary, 229
Glass, 82, 109, 223. *See also* Housing materials
Gordon, N. B., 113, 139
Grayson, William J., 26–27, 38, 44, 53, 59, 153, 186, 233
Gregg, William, 36, 41–42, 43, 113
Greeley, Horace, 171
Green, Beriah, 183
Grimké, Angelina, 1–2, 5

medieval Europe, 45; and racial constructions, 68–71; slaveholders' views on slaves and, 68–71, 79; and elite views on, in antebellum North, 93–94, 108–9; ambient, 121; definitions of, 264

North, 2, 9–11; colonial, 9–11
—antebellum: and South and slavery, 2, 12–16, 21, 24, 120, 124–26, 136, 141, 145, 156–71; and Republican Party, 2, 13, 120, 142, 148, 172–73; and mobs, 2, 94, 142, 143; and free soil supporters, 2, 120, 146, 145–46, 148; and progress, 3, 12–15, 27, 33, 93–94, 97, 105–9, 110, 127; and nature, 3, 107, 117–18, 130–32; and poverty, 3–4, 5, 16, 158–59; hearing of, in southern ears, 5, 12–15, 16–17, 27–28, 48; and conservatism, 20, 70, 93, 155; similarities to southern sounds, 27, 70, 95–100, 103, 105, 106, 109, 112–17, 185; villages in, 27, 95; and urban environments, 27, 97–98, 99, 100, 102, 106–9, 111–13, 117–18; and hotels, 27–28, 97, 109, 112–13; and spas, 28; and servants, 28, 112–13; and industrialization, 32–33; and urbanization, 32–33, 97–98, 99, 100, 102, 106–9, 111–13, 117–18, 119, 120–24, 126, 128, 130, 131–32; and differences from South, 32–33, 120–25; and similarities to antebellum South, 93–103, 105, 106, 109, 112–17, 154–56, 185; and Sabbath, 96, 97, 113, 176; and immigrants, 99, 126, 153; and temperance, 99, 153; and markets and market revolution, 108, 109–10; and legal statutes, 109; and silence of recession in, 110, 129, 143–44, 190; and population, 120–24; and youth, 131–32; and political parties, 153, 154–55. See also Elites, northern
—postbellum, 17; and nostalgia for

antebellum North, 107, 251–52; economic and political dislocation in, 238, 246–50; and expansion of capitalism, 246–48; and regulation of urban soundscape, 249–51. See also Elites, northern: in postbellum period; Reconstruction

North American Review (Boston), 169
North Carolina, 4, 21, 25, 26, 30, 40, 43, 52, 61, 84, 142, 152, 187, 208, 239; Mecklenburg County, 30; New Bern, 229
North Star (Rochester), 150, 173
Nott, Josiah C., 69

Ogden, Samuel, 134
Ohio, 124, 161, 162, 165, 247; Cincinnati, 108, 110, 154, 213; New Lisbon, 180; Columbus, 212
Ojibway Indians, 104
Olmsted, Frederick Law, 56

Page, Samuel G., 244
Panic of 1873, 238, 246, 247, 248
Parker, Theodore, 126
Patrols, slave, 54, 72, 80, 81, 82, 83
Pawtucket Chronicle, 139
Pennington, James W. C., 72
Pennsylvania, 124, 129, 130, 255; Philadelphia, 11, 24, 27–28, 99, 103, 110, 115–16, 117, 130, 132, 133–34, 141, 143–44, 154, 164; Delaware County, 127; Chester Creek, 129; Columbia, 213–14
Pennsylvania Gazette, 9, 11–12
Perkins, John, 154
Pessen, Edward, 16
Pettigrew, James Johnston, 25, 29
Phillips, Wendell, 182, 249
Pierpont, John, 181
Plantations: and quietude, 13, 19–20, 21, 22, 23–27, 44, 57, 79, 148; and northern criticism, 14–15; and modern aspects of, 17, 20–21, 35–36; and bells and horns,

abolitionist assessment of South
and slavery, 155–57, 161–63, 166–
68, 172–73; postbellum, 245, 246;
agree with Democrats, 303 (n. 1)
Revivals. *See* Christianity, evangelical;
Religion
Revolutionary War, 9, 10, 15, 21, 100,
150, 198–99, 299 (n. 1)
Rhode Island: Pawtucket, 133, 134,
135, 138, 139, 143
Richard, John C., 164
Richards, William B., 182
Richmond Enquirer, 105–6, 192
Robinson, Calvin, 163
Rochester Ladies' Anti-Slavery
Society, 178
Rome, Italy, 25–26
Ruffin, Elizabeth, 27–28, 58
Ruffin, Thomas, 71
Russell, William Howard, 125–26

St. Michael's Church (Charleston,
S.C.), 233, 253, 254
St. Patrick's Church (New Orleans,
La.), 232
St. Phillip's Church (Charleston,
S.C.), 236
Savannah Republican, 245
Samford, William F., 153, 188
Schafer, R. Murray, 263, 267
Schama, Simon, 268
Scholfield, John, 243
Schwartz, Hillel, 264
Sea, 9, 10, 26, 31, 254
Secession, 153, 163, 193–94; bell of,
193–94
Sectionalism, 6–7, 13–15, 91, 118,
120–26, 141, 144, 147–49, 155, 258–
59; and print, 8; and aural imagery,
8, 15–17, 147–49, 159–60, 164–65.
See also Politics and political parties
Senses, 3, 4, 5–6; and historical writ-
ing on, 6, 272 (n. 6), 318 (n. 1). *See
also* Aurality; Sight and seeing;
Smell

Seward, William H., 157
Schools, 33, 110, 112
Seminole, 51, 90
Shakers, 102–3
Sheridan, Philip, 205
Sherman, William Tecumseh, 201,
228, 233, 236
Sight and seeing, 3, 6, 318 (n. 2); and
aurality, 15, 17, 28, 40, 73, 85, 90,
126, 206, 244, 261–63, 272 (n. 8),
318 (n. 1). *See also* Aurality
Sigourney, Lydia H., 101
Silence: northern criticisms of, 2,
14, 15, 19–20, 66, 67, 148, 157, 160,
161–65, 167; slaveholders' fear of,
19–20, 66, 67–68, 88–91, 147, 276
(n. 2); use of, in antebellum southern factories, 37; and economic
recession, 40–41, 110, 190; slaves'
use of, 67–68, 79–86, 229–30. *See
also* Prisons
Simmons, Mary Jane, 230
Simms, William Gilmore, 31–32
Sioux, 52, 104
Slater, Samuel, 114
Slaveholders, 12–13; and modernity,
13, 17, 41–43, 88; and conservatism, 13–15, 16, 17, 65–66, 186–94;
and challenges within southern
society, 19–20, 42, 46, 47–65, 69–
70, 81–91, 187, 193; and plantation
soundscape, 19–22, 23–27, 34, 35–
37, 38–39, 67, 71–72, 79–80, 148;
and fear of silence, 20, 66, 67–68,
88–91, 147, 184–85; and regulation
of southern soundscapes, 33, 35–
39, 41–43, 67–76, 79–80, 83; and
plantation management, 35–39;
and bells, 39; and southern boosters, 41–43; and urbanization, 47,
50, 53; and southern lower classes,
47–49, 53, 55, 57, 61–62, 69, 187,
193; and southern frontier, 49–50;
and Native Americans, 50–52, 173;
as skittish, 68, 69, 88–91, 184–85,

(nn. 13, 16); rural, 9, 19, 106–7; urban, 9, 106–7; national, 13, 15, 21, 27, 95–96, 105, 178; seasonal, 23–24; and memory, 23–25, 26; and sense of place, 23–28; objective conditions of antebellum southern, 30–33, 37. *See also* Sectionalism

South

—antebellum, 2; and agriculture, 2, 31, 33; travelers to, 2, 37, 38, 39, 156, 161, 251–58; and northerners, 2–3, 12–15, 16, 120, 146; as alien to northerners, 2–3, 201; and representations of North, 3–5, 12–13, 15–16, 23, 27, 53; and abolitionism, 4, 5, 15–16, 47–48, 154; as tranquil to elite white southerners, 4, 5, 19–22, 23–27, 31–32, 37; as industrious to elite white southerners, 5, 17, 20–21, 23–24, 35–37, 39–45; soundmarks of, 19, 20–21, 23–24, 27, 30–33, 35–37; and temperature, 21, 24, 31; urban areas in, 21, 31, 32, 37, 50, 52–57, 86–88, 154; compared to Europe, 25–26, 37; and romanticism, 26, 31–32, 34, 37; compared to North, 26–28, 32–33, 53, 120–26, 154–55, 165; and religion, 32, 57–65, 70, 166; and structural differences from North, 32–33, 120–26; and slave factory workers, 37; and recession, 40–41; and modernization, 41–45; and radical democracy, 56, 62, 65–66, 176; and political parties, 151–53, 154–55. *See also* Elites, southern; Upper South

—postbellum, 17; and planters efforts to control freedpeople, 197; as progressive, 254. *See also* Reconstruction

South Carolina, 24, 41, 58–59, 65, 75, 76–84, 86, 113, 124, 125, 151, 170, 190, 191, 193–94, 204, 223, 228, 241, 242, 244; Charleston, 11, 21, 27, 34, 41–42, 215, 222, 225, 226, 228, 233, 236, 253, 255, 257; Beaufort, 26–27, 228, 239; Barnwell, 57; Anderson District, 70; Graniteville, 113; Columbia, 193, 221, 233, 236; Fort Wagner, 203, 206, 207, 214; Morris Island, 206, 211, 228; James Island, 222; Fort Beauregard, 225; Fort Moultrie, 225; Fort Sumter, 225; Port Royal, 226, 228

South Carolina College, 243

Southern Literary Gazette (Charleston), 31, 42

Southern Literary Messenger (Richmond), 25, 27

Southern Rose (Charleston), 34

Sparks, Jared, 169–70

Spas, 28, 156, 294 (n. 67)

Stanton, Elizabeth Cady, 249

Steamboats, 20–21, 110

Stone, Lucy, 188

Stroyer, Jacob, 77

Stuart, James, 53

Sumner, Charles, 190

Sutherlin, William T., 42

Switzerland, 106

Tallahassee Patriot, 190

Tanner, Benj. T., 238

Taylor, Susie King, 228

Telegraph, 46, 189, 200, 203

Temperance, 42, 153;

Tennessee, 33, 51; Chattanooga, 210; Murfreesboro, 210; Nashville, 223

Texas, 33, 36, 126, 211, 239

Thome, James A., 185

Thompson, George Franklin, 256

Thoreau, Henry David, 117, 127

Thorpe, Margaret Newbold, 257

Tilley, Christopher, 268

Time and time-consciousness, 319 (n. 2), in antebellum South, 13, 23, 34–37, 52–53, 80, 86–88, 319 (n. 2); in Europe, 25, 34, 52; and leisure time, 55, 113–14; and God, 57–

White, George S., 114
Whittier, John Greenleaf, 101, 176
Wilkinson, David, 121
Williams, Andy, 83, 239
Williams, Charley, 77
Williams, Horatio, 83
Williams, Lewis, 239
Williams, Lou, 77
Williams, Mattie, 240
Willis, Sampson, 241
Willis, Winn, 36
Wise, Isaac M., 126
Women, 1; and political voice in
 North, 1–2, 176–77; as noisy in
 colonial period, 11; in antebellum South, 21, 24, 29–30, 38, 278
 (n. 17); and postbellum memories of antebellum South, 24; as
 southern travelers to antebellum
 North, 27–28; elite men's views of,
 in antebellum South, 29; and antebellum southern children, 29–30,
 83, 101–2; and domesticity, 29–30,
 100–101; southern, as custodians of
 quietude, 29–30, 278 (n. 16), 312
 (n. 12); and antebellum southern
 women's criticisms of men, 30; as
 plantation mistresses, 39, 70, 75,
 84; elite men's views of, in ante-

bellum North, 97, 100–2, 104, 115,
 176–77; as factory operatives, 136–
 41, 143; critique male abolitionists,
 177; and southern critique of abolitionists, 188; during Civil War,
 223–25; in postbellum North, 249,
 252. *See also* Gender
Working class: in antebellum North,
 93–94, 97–100, 103, 105, 106, 109,
 112–17; and time and bells, 106,
 112–13, 128, 134, 136–37, 138–40,
 141; and resistance, 132–40, 141–
 45, 147; supposed noiselessness of,
 134; and evangelicalism, 135, 138;
 and accommodation to modernity, 140–41, 143–44; as similar to
 southern slaves, 136, 139, 141; in
 postbellum North, 238, 246–52.
 See also Industrialism; Industry and
 industrialization
World Soundscape Project, 263

Yarbrough, William W., 26
Yeomen, 48–49, 126, 155; slave-
 holders' attitudes toward, 48–49,
 59
Yorktown, 206
Yorkville Enquirer, 223
Young, George, 83